T0189267

Ordered Algebraic Structures

Developments in Mathematics

VOLUME 7

Series Editor:

Krishnaswami Alladi, *University of Florida, U.S.A.*

Aims and Scope

Developments in Mathematics is a book series publishing

(i) Proceedings of Conferences dealing with the latest research advances,

(ii) Research Monographs, and

(iii) Contributed Volumes focussing on certain areas of special interest.

Editors of conference proceedings are urged to include a few survey papers for wider appeal. Research monographs which could be used as texts or references for graduate level courses would also be suitable for the series. Contributed volumes are those where various authors either write papers or chapters in an organized volume devoted to a topic of special/current interest or importance. A contributed volume could deal with a classical topic which is once again in the limelight owing to new developments.

Ordered Algebraic Structures

Proceedings of the Gainesville Conference
Sponsored by the University of Florida
28th February – 3rd March, 2001

Edited by

Jorge Martínez

Department of Mathematics,
University of Florida, Gainesville, Florida, U.S.A.

KLUWER ACADEMIC PUBLISHERS
DORDRECHT / BOSTON / LONDON

A C.I.P. Catalogue record for this book is available from the Library of Congress.

ISBN 978-1-4419-5225-7

Published by Kluwer Academic Publishers,
P.O. Box 17, 3300 AA Dordrecht, The Netherlands.

Sold and distributed in North, Central and South America
by Kluwer Academic Publishers,
101 Philip Drive, Norwell, MA 02061, U.S.A.

In all other countries, sold and distributed
by Kluwer Academic Publishers,
P.O. Box 322, 3300 AH Dordrecht, The Netherlands.

Printed on acid-free paper

All Rights Reserved
© 2002 Kluwer Academic Publishers
Softcover version of the original hardcover edition 2002
No part of this work may be reproduced, stored in a retrieval system, or transmitted
in any form or by any means, electronic, mechanical, photocopying, microfilming, recording
or otherwise, without written permission from the Publisher, with the exception
of any material supplied specifically for the purpose of being entered
and executed on a computer system, for exclusive use by the purchaser of the work.

For Paul Conrad:

For all the Theorems
And to Many More!

Table of Contents.

Preface.

From the 28th of February through the 3rd of March, 2001, the Department of Mathematics of the University of Florida hosted a conference on the many aspects of the field of Ordered Algebraic Structures. Officially, the title was "Conference on Lattice-Ordered Groups and f-Rings", but its subject matter evolved beyond the limitations one might associate with such a label. This volume is officially the proceedings of that conference, although, likewise, it is more accurate to view it as a complement to that event.

The conference was the fourth in what has turned into a series of similar conferences, on Ordered Algebraic Structures, held in consecutive years. The first, held at the University of Florida in Spring, 1998, was a modest and informal affair. The fifth is in the final planning stages at this writing, for March 7-9, 2002, at Vanderbilt University. And although these events remain modest and reasonably informal, their scope has broadened, as they have succeeded in attracting mathematicians from other, related fields, as well as from more distant lands.

In between last year's conference and the appearance of this volume Paul F. Conrad had his 80th birthday. He could not be convinced to travel to the conference, making it inevitable that this book should be dedicated to the occasion. Few readers will be unaware of the significance of Paul Conrad's work in the development of Ordered Algebraic Structures. We trust that he will be amused, we hope that he will be impressed, and we suspect that he will even be a little surprised (but, hopefully, not alarmed), at the sweep of the field in which he has played such a prominent role.

This volume is also, in some measure, a rendering of the "state of the art" in the field. Without analyzing or cataloguing the contents here, since they speak for themselves in any event, I wish to highlight the expository contributions of Robert Redfield on lattice-ordered fields, of Peter Jipsen and Constantine Tsinakis on residuated lattices, and the aptly subtitled "fruitful interaction" on MV-algebras by Vincenzo Marra and Daniele Mundici. In my view, these splendid accounts, each of a substantial body of mathematics, highlight at once old and new for the traditionally minded researcher in ordered algebraic structures – yes, now in lower case. Few subjects harken back to the basics about ordered algebra like ordered fields! How appealing and tempting and natural is the leap from lattice-ordered groups to residuated lattices! And how fruitful indeed, and overdue, is the interaction between ordered algebra and MV-algebras. A particular note should be added in this regard: already Andrew Glass and Vincenzo Marra, who were introduced at the conference, have collaborated on a paper, which is now submitted.

Among the research papers I should like to underscore my satisfaction at the appearance here of some of the contributions, with apologies to all. It is particularly satisfying to follow the development of two former students, Chawne Kimber and Warren McGovern, and to celebrate their continued successes. I am also pleased to be able to include the contributions of two current graduate students at Vanderbilt University, James A. Cole and Nikolaos Galatos. I was delighted to welcome to Gainesville my friend and *tocayo* Guillermo Martínez, and to then be able to publish his contribution with Alejandro Petrovich. Especially gratifying it has been to host Karim Boulabiar and to have his contribution be included in this volume. My own origins are in the so-called Third World, which always makes the inclusion of the mathematical accomplishments of colleagues coming from similar circumstances a particular pleasure.

The following presented a paper at the conference: Richard N. Ball (University of Denver), Bernhard Banaschewski (McMaster University), Karim Boulabiar (IPEST, Tunisia), Gerard Buskes (University of Mississippi), James A. Cole (Vanderbilt University), Michael R. Darnel (Indiana University South Bend), Nikolaos Galatos (Vanderbilt University), A. M. W. Glass (Cambridge University), Anthony W. Hager (Wesleyan University), Melvin Henriksen (Harvey Mudd College), W. Charles Holland (Bowling Green State University), Peter Jipsen (Vanderbilt University), Jingjing Ma (University of Houston Clear Lake), James Madden (Louisiana State University), Vincenzo Marra (University of Milan), N. Guillermo Martínez (University of Buenos Aires), Warren Wm. McGovern (Bowling Green State University), Daniele Mundici (University of Milan), Robert Raphael (Concordia University), Robert Redfield (Hamilton College), Akbar Rhemtulla (University of Alberta), Niels Schwartz (University of Passau), Constantine Tsinakis (Vanderbilt University), Piotr Wojciechowski (University of Texas El Paso).

There are many people to thank for their contributions to the success of the conference and this volume. First and foremost among them is the department's Chair, Professor Krishnaswami Alladi, for his generous support and encouragement. It had not been my intention to host a conference this year. He persuaded me to do it, and backed my efforts at every turn.

A particular note of appreciation and recognition goes to my friend and colleague Constantine Tsinakis, for his generosity and enthusiasm in supporting this series of conferences. My old friend Charles Holland came to my rescue at a somewhat embarrassing moment; hearty thanks for bailing me out.

To Kluwer Academic Publishers my gratitude for graciously agreeing to publish this volume and for making the experience, once again, free of stress and complications. To all my colleagues who either attended or submitted a contribution, or both, my heartfelt thanks for your efforts. To the many referees, and again to the authors, I appreciate your patience in putting up with my cajoling and prodding and, yes, occasional nagging. And, returning to the department, a huge vote of thanks to the people who actually (when it came to cases, as it were) made the conference possible: Brian, Gretchen, Julia, Marie, Pam, Sandy, Sharon, Sonia, Vicki, for making sure the folders for the participants were there when they needed to be; for making sure there was coffee and tea and cookies; for the handling of computer-related "issues", which would have rendered me helpless; for being gracious to make copies of manuscripts on impossibly short notice; and, last but most definitely not least, for seeing to it that the checks got to where they needed to go.

Jorge Martinez
Gainesville, Florida
February, 2002

Surveys & Expositions

The Bornological Tensor Product
of two Riesz Spaces

G. Buskes[1] and A. van Rooij

ABSTRACT.   We construct the bornological Riesz space tensor product of
two bornological Riesz spaces. This unifies the Archimedean Riesz space tensor
product and the projective tensor product, both introduced by Fremlin. We
extend the results, even in these special cases, by considering maps of bounded
variation rather than positive maps. This note is without proofs, but the proof
and complete bornology background of a similar result are discussed elsewhere in
this volume.

1   Introduction

In [3] D. H. Fremlin constructed, for any two Archimedean Riesz spaces E and F, a
Riesz tensor product

$$(E \bar{\otimes} F, \otimes),$$

where $E \bar{\otimes} F$ is an Archimedean Riesz space and $\otimes$ is a positive bilinear map
$E \times F \to E \bar{\otimes} F$. Among other things he showed that the Riesz tensor product has the
following universal property:

Theorem 1.1. (Fremlin, [3], Theorem 5.3) *For every positive bilinear map* T *of* $E \times F$
into any uniformly complete (hence Archimedean) Riesz space G *there exists a unique
positive linear* $T^{\otimes} : E \bar{\otimes} F \longrightarrow G$ *such that*

$$T(x, y) = T^{\otimes}(x \otimes y) \quad (x \in E, \, y \in F).$$

In [4] he returned to the subject and obtained the following Banach lattice version of
the above result. First, for Banach lattices E, F, G and for a bilinear map $T : E \times F \to G$
one defines

$$\| T \| := \sup\{ \| T(x, y) \| : x \in E, \, y \in F, \, \| x \| \le 1, \, \| y \| \le 1 \}.$$

T is continuous if and only if $\| T \|$ is finite. Fremlin proved:

---
[1]The first author acknowledges support from Office of Naval Research grant ONR N00014-01-1-
0322, in the Summer of 2001. Both authors acknowledge support from NATO CRG grant 940605.

J. Martínez (ed.), Ordered Algebraic Structures, 3–9.
© 2002 Kluwer Academic Publishers.

Theorem 1.2. (Fremlin, [4], Theorem 1E) *Let E, F be Banach lattices. The formula*

$$\| u \|_{|\pi|} := \inf \left\{ \sum_{k=1}^{K} \| x_k \| \| y_k \| : x_1, \ldots, x_K \in E, \; y_1, \ldots, y_K \in F, \; |u| \leq \sum |x_k| \otimes |y_k| \right\}$$

defines a Riesz norm $\| \; \|_{|\pi|}$ on $E \bar{\otimes} F$. Let T be a positive (and therefore continuous) bilinear map of $E \times F$ into a Banach lattice G and let $T^{\otimes}$ be as in Theorem 1.1. Then $T^{\otimes}$ is continuous relative to $\| \; \|_{|\pi|}$, and $\| T^{\otimes} \| = \| T \|$.

In fact, Fremlin's Theorem 1.2 above is somewhat stronger in that the condition on G is less restrictive than norm completeness. Also, to make the two theorems above look more comparable, norm completeness of E and F is not necessary in Theorem 1.2. One can generalize Theorem 1.2 to a locally convex setting. This is done, in part, by Wong in Theorem 5.3.3 in [13], though he studies the algebraic tensor product $E \otimes F$ rather than $E \bar{\otimes} F$, which results in an absence of any completeness conditions on G in Wong's Theorem. Finally, the condition of uniform completeness in Theorem 1.1 can be relaxed to allow any Riesz space G, if one further restricts the bilinear maps to be bimorphisms.

It is our purpose to not only unify these theorems, but at the same time extend them by changing the positivity condition into one of boundedness. To this end we introduce the notions of bornological Riesz spaces and of maps of bounded variation. The proofs are very laborious; this note is self contained without proofs while the background information and the proof of the main result, Theorem 2.6, appears elsewhere in this volume. In this paper we present the results only.

The basic idea is the application to Riesz spaces of the theory of bornological vector spaces as set forth in, e.g., H. Hogbe-Nlend's book [8] (see also Séminaire de Banach 1972, [9]). We note that connections between the theory of bornologies and tensor products (in the context of vector spaces in general) have been studied in both [7] and [9] and that our general definition for bornological tensor product coincides with [9]. For further information on tensor products of Banach lattices see [11] and [12]. For the standard terminology of Riesz spaces we refer to [1], [2], [10], [11] and [13].

2 Bornologies and Bounded Variation

2.1 Generalities.

The general theory of Riesz spaces, locally solid topologies on Riesz spaces, Banach lattices and positive operators has been well developed and organized in a series of books over the last decades (see [1], [2], [10], [11] and [13]). Curiously, a general theory of bornological Riesz spaces seems not to have been explored yet. (See however [5] and [6] and some passing remarks in [9] in connection with the order bornology below.) Consequently, we offer the basic definitions. Let E be an Archimedean Riesz space.

Definition & Remarks 2.1. A *Riesz bornology* on E is a collection $\mathcal{A}$ of subsets of E with the following properties.

(i) If $X \in \mathcal{A}$, $Y \subseteq X$, then $Y \in \mathcal{A}$.

(ii) If $X, Y \in \mathcal{A}$, then $X + Y \in \mathcal{A}$.

(iii) $\mathcal{A}$ covers E.

(iv) For every $X \in \mathcal{A}$ there is a Riesz disk A with $X \subseteq A \in \mathcal{A}$.

A *Riesz disk* in E is a nonempty subset A of E that is convex and solid (i.e., if $a \in A$, $x \in E$, $|x| \le |a|$, then $x \in A$.) If A is such a Riesz disk, then its linear span is $E_A := \cup_{\lambda > 0} \lambda A$. In the special case that $c \ge 0$ and $A = [-c, c]$ we write $E_A = E_c$. Always, E_A is a Riesz ideal in E. The set A determines a gauge $p_A : E_A \to [0, \infty)$ by

$$p_A(x) := \inf \{ \lambda \, : \, \lambda > 0, \, x \in \lambda A \}.$$

This p_A is a Riesz seminorm on E_A. We call the disk A *completant* if p_A is a norm on E_A and renders E_A a Banach space (hence a Banach lattice).

The Riesz disk *generated* by a nonempty set $S \subset E$ is the smallest Riesz disk containing S. It is the convex hull of $\cup_{s \in S}[-s, s]$. Put differently, it is the set

$$\left\{ x \in E \, : \, |x| \le \sum_{k=1}^{K} \lambda_k |s_k|, \; K \in \mathbb{N}, \; \lambda_1, \ldots, \lambda_K \in [0, 1], \; \sum_{k=1}^{K} \lambda_k = 1, \; s_1, \ldots, s_k \in S \right\}.$$

A *[complete] Riesz bornology* in E is a bornology having a base that consists of [completant] Riesz disks.

Our two main examples are the following.

Example 2.2. The order bounded subsets of E form a Riesz bornology, *the order bornology*. The order intervals $[-a, a]$ forming a base, the order bornology is complete as soon as E is uniformly complete. (This order bornology has been studied in, e.g., [5] and [6] by Marcel Grange.)

Example 2.3. Suppose E is a normed Riesz space. Its *von Neumann bornology* is the Riesz bornology consisting of all norm bounded sets. It is complete if E is a Banach lattice.

A *[complete] bornological Riesz space* is an Archimedean Riesz space endowed with a [complete] Riesz bornology. In practice we use the same letter to indicate a bornological Riesz space and the underlying Riesz space. To prove our main result, we need the following completeness condition on the range space G. This definition is in the same realm of ideas as *faiblement concordantes* or *fortement concordantes* as defined on pages 220 and 221 of [7]. (But the latter definitions are not good enough in our setting.) If G is uniformly complete and the bornology is the order bornology or if G is a Banach lattice and the bornology is the von Neumann bornology the extra condition is fulfilled.

Definition 2.4. A bornological Riesz space G is said to be *very complete* if its bornology $\mathcal{C}$ is generated by completant disks C with the property that

if $x \in E_C$, $x_1, x_2, \ldots \in C$ and $p_{C'}(x - x_n) \to 0$ for some $C' \in \mathcal{C}$, then $x \in C$.

The following lemma is easy.

Lemma 2.5. *Every uniformly complete Riesz space with the order bornology and every Banach lattice with its von Neumann bornology are very complete.*

2.2 The Product Bornology.

Let E, F be bornological Riesz spaces with bornologies $\mathcal{A}, \mathcal{B}$, respectively. Then the sets $A \times B$ with $A \in \mathcal{A}$, $B \in \mathcal{B}$ form a base for a natural bornology on $E \times F$.

2.3 Operators of Bounded Variation.

Let E, F, G be bornological Riesz spaces with Riesz bornologies $\mathcal{A}, \mathcal{B}, \mathcal{C}$. Let $T : E \longrightarrow G$ be linear. T is called *bounded* if $T(A) \in \mathcal{C}$ for all $A \in \mathcal{A}$. We say that T is of *bounded variation* if for every $A \in \mathcal{A}$ the set

$$\left\{ \sum_{n=1}^{N} |T(x_n)| \, : \, x_1, \ldots, x_n \in E^+, \ \sum_{n=1}^{N} x_n \in A \right\}$$

belongs to $\mathcal{C}$. Clearly, if T is of bounded variation, then it is bounded. The converse is true if T is positive.

Similarly, a bilinear map $T : E \times F \longrightarrow G$ is *bounded* if $T(A, B) \in \mathcal{C}$ for all $A \in \mathcal{A}$, $B \in \mathcal{B}$, whereas T is *of bounded variation* if for all such A and B the set

$$\left\{ \sum_{n=1}^{N} \sum_{m=1}^{M} |T(x_n, y_m)| \, : \, x_1, \ldots, x_n \in E^+, \ \sum_{n=1}^{N} x_n \in A, \ y_1, \ldots, y_m \in F^+, \ \sum_{n=1}^{N} y_m \in B \right\}$$

belongs to $\mathcal{C}$.

A bilinear map that is of bounded variation is bounded; for positive bilinear maps the converse is also valid. The product bornology above is the finest vector bornology for which the canonical map

$$E \times F \longrightarrow E \otimes F$$

is bounded.

2.4 The Bornological Riesz Tensor Product.

Let A and B be the Riesz disks in E and F, respectively. By $[A \otimes B]$ we denote the Riesz disk generated by $\{ a \otimes b : a \in A, b \in B \}$. It is easy to prove that the gauge on $(E \bar{\otimes} F)_{[A \otimes B]}$ determined by $[A \otimes B]$ is given by $p_{[A \otimes B]}(u) =$

$$(1) \quad \inf \left\{ \sum_{k=1}^{K} p_A(x_k) p_B(y_k) : x_1, \ldots, x_k \in E, \ y_1, \ldots, y_k \in F, \ |u| \le \sum_{k=1}^{K} |x_k| \otimes |y_k| \right\}.$$

This implies that, if E and F are Banach lattices and A, B are their closed unit balls, then $p_{[A \otimes B]}$ is precisely the norm $\| \ \|_{|\pi|}$ as introduced by Fremlin. From (1) and the Hahn-Banach Theorem one obtains

$$p_{[A \otimes B]}(a \otimes b) = p_A(a) p_B(b) \quad (a \in A, \ b \in B).$$

If E and F are bornological Riesz spaces, by their *bornological Riesz space tensor product* we mean the Riesz space $E \bar{\otimes} F$ endowed with the Riesz bornology that has as a base

$$\{ [A \otimes B] : A \text{ and } B \text{ are bounded Riesz disks in } E \text{ and } F, \text{ respectively} \}.$$

2.5 Results.

Theorem 2.6. *Let E, F, G be bornological Riesz spaces. Assume G to be very complete. Let T be a bilinear map $E \times F \longrightarrow G$ that is of bounded variation. Then there exists a unique linear $T^{\otimes} : E \bar{\otimes} F \longrightarrow G$ that is of bounded variation and satisfies*

$$T(x, y) = T^{\otimes}(x \otimes y) \quad (x \in E, \ y \in F).$$

Moreover, if $A \subseteq E$ and $B \subseteq F$ are bounded Riesz disks, then for every $\varepsilon > 0$ the set $T^{\otimes}([A \otimes B])$ is contained in the Riesz disk generated by $(1 + \varepsilon) T(A \times B)$. If T is positive, then so is $T^{\otimes}$.

From this, by using Lemma 2.5 (which assures the very completeness of G in each of the two cases) Fremlin's Theorems 1.1 and 1.2 can be derived as follows.

Proof. (Of 1.1) Endow E, F and G with their order bornologies. If $a \in E^+$ and $b \in F^+$, then $T([-a, a] \times [-b, b]) \subseteq [-T(a, b), T(a, b)]$, so T is bounded, hence of bounded variation. 1.1 now follows directly from our main result.   ∎

Proof. (Of 1.2) Let A, B, C be the closed unit balls of E, F, G. Give E, F, G the bornologies of all norm bounded subsets. These bornologies have bases $\{A\}, \{B\}, \{C\}$, respectively. The bornology of $E \bar{\otimes} F$ consists of all $\| \ \|_{|\pi|}$-bounded sets and has

$\{[A \otimes B]\}$ as a base. As $T(A \times B) \subseteq \|\ T\ \|\ C$, T is bounded, hence of bounded variation. Then $T^{\otimes}$ is of bounded variation and therefore norm continuous. Moreover, for all $\varepsilon > 0$, $T^{\otimes}([A \otimes B]) \subseteq (1 + \varepsilon)T(A \times B) \subseteq \|\ T\ \|\ C$, whereas $[A \otimes B]$ contains the open unit ball of $E \otimes F$. It follows that $\|\ T^{\otimes}\ \| \le \|\ T\ \|$. The reverse inequality is clear from the observation that for all $a \in E$ and all $b \in F$

$$
\begin{aligned}
\|\ T(a,b)\ \| &= \|\ T^{\otimes}(a \otimes b)\ \| \le \|\ T^{\otimes}\ \|\ \|\ a \otimes b\ \|_{|\pi|} \\
&= \|\ T^{\otimes}\ \|\ p_{[A \otimes B]}(a \otimes b) = \|\ T^{\otimes}\ \|\ p_A(a)p_B(b) \\
&= \|\ T^{\otimes}\ \|\ \|\ a\ \|\ \|\ b\ \|.
\end{aligned}
$$

■

References

[1] C. D. Aliprantis and O. Burkinshaw, *Positive Operators.* (1985) Acad. Press, New York-London.

[2] C. D. Aliprantis and O. Burkinshaw, *Locally Solid Riesz Spaces.* (1978) Acad. Press, New York-London.

[3] D. H. Fremlin, *Tensor products of Archimedean vector lattices.* Amer. J. Math. **XCIV** No 3 (1972), 777-798.

[4] D. H. Fremlin, *Tensor products of Banach lattices.* Math. Annalen **211** (1974), 87-106.

[5] M. Grange, *Bornologie de l'Ordre.* Doctoral Thesis, (1972) Université de Bordeaux.

[6] M. Grange, *Sur la bornologie de l'ordre.* Publ. du Département de Mathématiques Lyon, **10-3** (1973), 11-33.

[7] H. Hogbe-Nlend, *Complétion, Tenseurs et Nucléarité en Bornologie.* J. Math. Pures et Appl., **49** (1970), 193-288.

[8] H. Hogbe-Nlend, *Bornologies and Functional Analysis.* Math. Studies **26** (1977), North Holland, Amsterdam-New York-Oxford.

[9] C. Houzel (editor), *Séminaire Banach.* Lecture Notes in Math. **277** (1972), Springer Verlag, Berlin-Heidelberg-New York.

[10] W. A. J. Luxemburg and A. C. Zaanen, *Riesz spaces, I.* (1971) North Holland, Amsterdam-London.

[11] P. Meyer-Nieberg, *Banach Lattices.* Universitext (1991), Springer-Verlag, Berlin-Heidelberg-New York.

[12] H. H. Schaefer, *Banach Lattices and Positive Operators.* (1974) Springer-Verlag, Berlin-Heidelberg-New York.

[13] Y.-Ch. Wong, *Schwartz Spaces, Nuclear Spaces and Tensor Products.* Lecture Notes in Math. **726** (1979), Springer-Verlag, Berlin-Heidelberg-New York.

Department of Mathematics, University of Mississippi, University, MI 38677, USA
mmbuskes@hilbert.math.olemiss.edu

Catholic University, Department of Mathematics, Toernooiveld, 6525 ED Nijmegen, the Netherlands

Old and New Unsolved Problems
in Lattice-Ordered Rings that need not be f-Rings

Melvin Henriksen

A Brief History

Recall that a *lattice-ordered ring* or *ℓ-ring* $A(+, \cdot, \vee, \wedge)$ is a set together with four binary operations such that $A(+, \cdot)$ is a ring, $A(\vee, \wedge)$ is a lattice, and letting $P = \{a \in A : a \vee 0 = a\}$, we have both $P + P$ and $P \cdot P$ contained in P. For $a \in A$, we let $a^+ = a \vee 0, a^- = (-a) \vee 0$, and $|a| = a \vee (-a)$. It follows that $a = a^+ - a^-, |a| = a^+ + a^-$, and for any $a, b \in A$, $|a+b| \leq |a| + |b|$ and $|ab| \leq |a||b|$. As usual $a \leq b$ means $(b-a) \in P$. We leave it to the reader to fill in what is meant by a lattice-ordered algebra over a totally ordered field.

Examples of such rings and algebras were studied through most of the 20th century, but the first paper to consider such rings as abstract algebraic systems was [BP56] by G. Birkhoff and R. S. Pierce published in 1956. (It was actually submitted at least two years earlier but was delayed and plagued by many typographical errors.) This pioneering paper set the stage for research in this area that continues to this day. The reader of [BP56] learns that some ℓ-rings have properties that are counter-intuitive. For example there are ℓ-algebras that are two-dimensional over the real field (with its usual order) with an identity element that fails to be positive. Not a lot was done with general ℓ-rings except to convince the reader that getting a general structure theory for them is very difficult. The authors introduce a special class they call f-rings, which in the presence of the axiom of choice – indeed, just the weaker prime ideal theorem for Boolean algebras; see [FH88] – can be described as subdirect products of totally ordered rings. They are important generalizations of function rings, and the overwhelming majority of the papers on ℓ-rings are concerned with them or others that do not differ greatly from them. (E.g., ℓ-rings in which all squares are nonnegative.) An incomplete survey of what was known about f-rings up to the mid-1990's is given in [H97]. In it, research on applications to real semi-algebraic geometry is covered only superficially.

While the theory of f-rings is both rich and interesting it excludes the ring of polynomials ordered coefficientwise and the ring of $n \times n$ matrices over a totally ordered field if $n > 1$. The problems stated below are mostly rather old and are concerned with ℓ-rings that are far from being f-rings.

11

J. Martínez (ed.), Ordered Algebraic Structures, 11–17.
© 2002 *Kluwer Academic Publishers.*

A. What are the possible lattice orders on the real field $\mathbb{R}$?

The problem of whether there are any lattice orders on $\mathbb{R}$ other than the usual totally one is posed in [BP56]. In his 1976 publication, [Wi76], R. R. Wilson produced uncountably many distinct lattice orderings on $\mathbb{R}$ by making heavy use of Zorn's lemma. The nonconstructive nature of his techniques yields little insight into the nature of these many new lattice orderings, and it is unknown if he has exhibited all possible lattice orders on $\mathbb{R}$. In a follow up paper, [Wi80], Wilson shows how to construct similar pathological lattice orders on any formally real field. These orders, sometimes called Wilson orders, have been studied also by N. Schwartz in [S86] and R. Redfield in Section 5 of [R01], but neither of these authors attempt to answer question A.

The lattice orders on $\mathbb{R}$ constructed in [Wi76] have the property that for $a, b, c \in \mathbb{R}$:

if for $a \geq 0$, $a(b \vee c) = ab \vee ac$ and $a(b \wedge c) = ab \wedge ac$ then b and c are comparable.

Actually, this holds in any lattice-ordered division ring D in which $1 > 0$.

Proof. Suppose b and c are incomparable, let $\gamma = b - b \wedge c$, and let $\delta = c - b \wedge c$. Then $\gamma > 0$ and $|\gamma^{-1}| \geq \gamma^{-1}$, so $|\gamma^{-1}|\gamma \geq \gamma\gamma^{-1} = 1 > 0$. Similarly, $|\delta^{-1}|\delta \geq 1$. So

$$(|\gamma^{-1}| \vee |\delta^{-1}|)\gamma \wedge (|\gamma^{-1}| \vee |\delta^{-1}|)\delta \geq |\gamma^{-1}|\gamma \wedge |\delta^{-1}|\delta \geq 1 > 0.$$

But $\gamma \wedge \delta = 0$. So, by assumption, for all $a \geq 0$ in D, $a\gamma \wedge a\delta = a(\gamma \wedge \delta) = 0$. Then, for $a = |\gamma^{-1}| \vee |\delta^{-1}|$, we have $a\gamma \wedge a\delta > 0$, a contradiction. Hence b and c are comparable.   ∎

Some steps along the way to answering Question A would be to find out which lattice-orders on $\mathbb{R}$ are archimedean? What can be said about lattice orderings on algebraic or transcendental extensions of $\mathbb{R}$ equipped with a Wilson order? Are lattice orderings of $\mathbb{R}$ that are not total of some value, or just examples of pathology? Surely, classifying the lattice-orders on $\mathbb{R}$ is worthy of additional study.

B. Can the field of complex numbers be made into a lattice-ordered field?

This question was also posed in [BP56] after showing that this field cannot be made into a two dimensional lattice-ordered algebra over $\mathbb{R}$ endowed with its usual order. Indeed, these authors classify all of the two-dimensional lattice-ordered algebras over $\mathbb{R}$. In 1962, R. A. McHaffey published in an Iraqi journal [Mc62] a similar result for the division algebra **Qu** of real quaternions over $\mathbb{R}$. There it is shown that **Qu** of real quaternions considered as an algebra over $\mathbb{R}$ with its usual order does not admit a lattice order. Actually, it is shown that **Qu** cannot be made into a partially ordered division algebra whose positive cone has a nonempty interior. This is observed in the review [P64] by R. S. Pierce of [Mc62]. In it a typographical error is also corrected. No other proof of McHaffey's (hard to locate) result seems to appear in print, but J. Ma has obtained an alternate proofs of the result in [BP56] and [Mc62], which are available in preprint form [Ma01b]. Also, conceivably, McHaffey's conclusion need not hold if the ordering induced on $\mathbb{R}$ is not the usual one. So, Question B remains open.

A lot of the work on lattice-ordered fields and division rings uses power series techniques and apply only to those infinite dimensional over $\mathbb{R}$. For example, see [D89] and [R89]. Algebraic extensions of totally ordered fields have also been studied in [R00]. Much of this literature appears in the references in [R92].

C. Can the field $\mathbb{R}(x)$ of real rational functions be lattice-ordered while preserving the coefficientwise lattice order on the polynomial ring $\mathbb{R}[x]$?

The *coefficientwise order* on a subalgebra of the family $\mathbf{LR}[[x]]$ of all formal Laurent series $\sum_{k=-m}^{\infty} a_k x^k$, with coefficients in $\mathbb{R}$ is obtained by letting its positive cone be all such series in which $a_k \geq 0$ for all k. Then $\mathbf{LR}[[x]]$ is an ℓ-ring in which

$$\left| \sum_{k=-m}^{\infty} a_k x^k \right| = \sum_{k=-m}^{\infty} |a_k| x^k$$

and contains $\mathbb{R}[x]$ as a sub-ℓ-ring. By $\mathbb{R}(x)$ we mean, as usual,

$$\left\{ \frac{p(x)}{q(x)} \ : \ p(x), q(x) \in \mathbb{R}[x] \ \text{and} \ q(x) \neq 0 \right\},$$

added and multiplied like fractions. In [H71], I asked if the coefficientwise lattice-order on the sub-ℓ-ring $\mathbb{R}[x]$ can be extended to a lattice-order on $\mathbb{R}(x)$ and showed that a seemingly plausible method for extending this order did not work. While the description of the method given in [H71] and the reason given for its failure are basically correct, the exposition is badly garbled and especially deceptive because of a reference to an irrelevant paper. So a corrected and expanded version of part of [H71] will be given next.

Because each element of $\mathbb{R}(x)$ has an expansion into a Laurent series, there is an embedding of $\mathbb{R}(x)$ into $\mathbf{LR}[[x]]$ that induces a partial ordering onto the former. So, if $\mathbb{R}(x)$ were a sublattice of $\mathbf{LR}[[x]]$ under this induced ordering, Question C would have an affirmative answer.

Recall the identity $\cos t = \frac{1}{2}(e^{it} + e^{-it})$ and hence that

$$\sum_{k=0}^{\infty} (\cos k\alpha) x^k = \sum_{k=0}^{\infty} \frac{1}{2}((e^{i\alpha})^k + (e^{-i\alpha})^k) x^k = \frac{1}{2}\left[\sum_{k=0}^{\infty} (xe^{i\alpha})^k + \sum_{k=0}^{\infty} (xe^{-i\alpha})^k \right]$$

$$= \frac{1}{2}\left[\frac{1}{1 - xe^{i\alpha}} + \frac{1}{1 - xe^{-i\alpha}} \right] = \frac{1 - (\cos \alpha)x}{1 - (2\cos \alpha)x + x^2}$$

is in $\mathbb{R}((x))$, while its absolute value $\sum_{k=0}^{\infty} |\cos k\alpha| x^k$ is not in $\mathbb{R}((x))$ if α is not a rational multiple of π.

No argument is supplied to justify this in [H71], and the reader is confused by a reference to a paper by Benzaghou that is irrelevant for this purpose. A simple argument to justify this assertion follows.

For, if this latter series were in $\mathbb{R}((x))$, then, considered as a function of a complex variable, it would have an analytic continuation outside of the unit disk and hence would have to converge at some point of the unit circle. But since α is not a rational multiple of π, the values assumed by $|\cos k\alpha|$ are dense in the unit interval $[0,1]$. Thus if $|z| = 1$,

$$\lim_{k \to \infty} |\cos k\alpha||z|^k = \lim_{k \to \infty} |\cos k\alpha|$$

does not exist. So $\sum_{k=0}^{\infty} |\cos k\alpha|z^k$ does not converge at any point on the unit disk and hence $\sum_{k=0}^{\infty} |\cos k\alpha|x^k$ is not in $\mathbb{R}((x))$.

This does not supply a negative answer to Question C, which remains an open problem. Some sufficient conditions for the lattice orderability of rings of formal quotients are given in [Ma01a] and [R00]. A similar but different problem is posed in [CD69].

D. Lattice-orderings on algebras of matrices.

Because finite dimensional algebras can be represented as algebras of matrices, it important for us to know the possible lattice orderings of the algebra of $n \times n$ over totally ordered fields; in part because squares of nonzero matrices need not be positive if $n \geq 2$, in the usual entry-by-entry lattice-ordering of this algebra, f-ring techniques will not help much in this study. This may be the major reason for the lack of results on the structure of lattice-ordered rings and algebras of this sort. New techniques are needed for this purpose. I will not try to summarize all of the attempts along these lines, but will concentrate on new results by S. Steinberg, J. Ma, and P. Wojciechowski that have been successful and that open up new frontiers.

In 1966, E. Weinberg studied lattice-orders on rings of 2×2 matrices over the field of rational numbers [W66]. He produced infinitely many lattice orders on this ring, claimed that he had produced them all, and showed that the usual coordinatewise order is the only lattice-ordering in which the identity matrix is positive. In 2000, S. Steinberg showed in [St00] that Weinberg's list of lattice orders was incomplete, gave a complete listing, and extended Weinberg's results to the algebra of 2×2 matrices over any totally ordered field. The next year in [MW01a] J. Ma and P. Wojciechowski extended the Weinberg-Steinberg result by establishing the usual coordinatewise order is the only way to make the algebra of $n \times n$ matrices for $n \geq 2$ into a lattice ordered algebra over a totally ordered field, and in [MW01b] extended the Weinberg-Steinberg results classifying lattice-orders to the $n \times n$ case. Also, as noted above, in [Ma01b], J. Ma extends the use of these techniques to study finite dimensional lattice-ordered algebras over subfields of $\mathbb{R}$ with its usual order to get alternate proofs that neither the complex field nor the algebra of real quaternions can be lattice-ordered as a finite dimensional algebra over $\mathbb{R}$.

E. Structure spaces.

A great deal has been written on spaces of various kinds of ideals in f-rings, but very little on general ℓ-rings. Much of the former may be found by examining the references in [H97], and the subject is taken up quite generally in [HK91]. The approach taken in [HK91] is more likely to be applicable to general ℓ-rings than earlier work. A substantial step forward in the study of structure spaces has been made by J. Ma and P. Wojciechowski in [MW02] where they generalize a theorem of H. Subramanian [Su68] that applies only to commutative f-rings as follows. Suppose R and S are two ℓ-rings with strictly positive identity elements whose only nilpotent element is 0, and such that the intersection of their maximal ℓ-ideals is zero. Then an ℓ-group isomorphism of R onto S that sends the identity element of R onto the identity element of S induces a homeomorphism between their spaces of maximal ℓ-ideals in the hull-kernel topology.

This opens up the door to generalizing many theorems on structure spaces on f-rings to structure spaces on more general ℓ-rings.

F. The future.

It is my belief that the time has come for workers in lattice-ordered rings to start paying much more attention to ℓ-rings that need not be f-rings. The most exciting new developments of this sort are the recent contributions of Ma and Wojciechowski, but they have just begun to apply a cleverly used axe to the tip of a large iceberg. In addition to the large number of problems stated above, there is a need to study lattice-ordered algebras of (positive) operators on infinite dimensional vector lattices. The best way to sample what has been done is to start with the book [AB85] entitled Positive Operators.

There is much to be done.

I am indebted to the referees for some very valuable corrections and constructive criticism that serve to increase the value of this paper.

References

[AB85]   C. Aliprantis and O. Burkinshaw, *Positive Operators.* (1985) Academic Press, Inc., Orlando, FL.

[BP56]   G. Birkhoff and R. S. Pierce, *Lattice-ordered rings.* Anais. Acad. Brasil Cien. **29** (1956), 41-69.

[CD69]   P. Conrad and J. Dauns, *An embedding theorem for lattice-ordered fields.* Pacific J. Math. **30** (1969), 385-398.

[D89]   J. Dauns, *Lattice ordered division rings exist.* Ord. Alg. Struct. (Curaçao, 1988), W. C. Holland & J. Martinez, eds.; (1989) Math. Appl. **55**, Kluwer Acad. Publ., Dordrecht, 229–234.

[FH88]   D. Feldman and M. Henriksen, *f-Rings, subdirect products of totally ordered rings, and the prime ideal theorem.* Proc. Ned. Akad. Wetensh. **91** (1988), 121–126.

[H71]   M. Henriksen, *On difficulties in embedding lattice-ordered integral domains in lattice-ordered fields.* General Topology and its Relations to Modern Analysis and Algebra III; Proceedings of the Third Prague Topological Symposium (1971), Academia, Prague; (1972) Academic Press, New York, 183–185.

[H97]   M. Henriksen, *A survey of f-rings and some of their generalizations.* Ord. Alg. Struct. (Curaçao, 1995), W. C. Holland & J. Martinez, eds.; (1997) Kluwer Acad. Publ., 1-26.

[HK91]   M. Henriksen and R. Kopperman, *A general theory of structure spaces with applications to spaces of prime ideals.* Alg. Universalis **28** (1991), 349-376.

[Ma01a]   J. Ma, *The quotient rings of a class of lattice-ordered Ore domains.* Alg. Universalis, to appear.

[Ma01b]   J. Ma, *Finite-dimensional algebras that do not admit a lattice order.* Preprint.

[Mc62]   R. A. McHaffey, *A proof that the quaternions do not form a lattice-ordered algebra.* Proc. Iraqi Sci. Soc. **5** (1962), 70-71.

[MW01a]   J. Ma and P. Wojciechowski, *A proof of the Weinberg conjecture.* Proc. AMS, to appear; (preprint available).

[MW01b]   J. Ma and P. Wojciechowski, *Lattice orders on matrix algebras.* Submitted; (preprint available).

[MW02]   J. Ma and P. Wojciechowski, *Structure spaces of maximal ℓ-ideals of lattice-ordered rings.* In these Proceedings, 261-274.

[P64]   R. S. Pierce, *Review of [Mc62].* MR 27 (1964), #5706.

[R89]   R. Redfield, *Constructing lattice-ordered fields and division rings.* Bull. Austral. Math. Soc. **40** (1989), 365-369.

[R92]   R. Redfield, *Lattice-ordered fields as convolution algebras.* J. Algebra **153** (1992), 319-356.

[R00]   R. Redfield, *Unexpected lattice-ordered quotient structures.* Ordered Algebraic Structures; Nanjing (2001); Gordon and Breach, Amsterdam.

[R01]   R. Redfield, *Subfields of lattice-ordered fields that mimic maximal totally ordered subfields.* Czech. Math. J. **51 (126)** (2001), 143–161.

[S86]   N. Schwartz, *Lattice-ordered fields.* Order **3** (1986), 179-194.

[St00]   S. A. Steinberg, *On the scarcity of lattice-ordered matrix algebras, II.* Proc. AMS **128** (2000), 1605–1612.

[Su68]   H. Subramanian, *Kaplansky's theorem for f-rings.* Math. Annalen **179** (1968), 70-73.

[W66]   E. Weinberg, *On the scarcity of lattice-ordered matrix rings.* Pacific J. Math. **19** (1966), 561-571.

[Wi76]   R. R. Wilson, *Lattice orderings on the real field.* Pacific J. Math. **63** (1976), 571-577.

[Wi80]   R. R. Wilson, *Anti-f-rings.* Ordered Groups (Proc. Conf., Boise State Univ., Boise, Idaho, 1978), Lecture Notes in Pure and Appl. Math. **62** (1980), M. Dekker, New York, 47-51.

Harvey Mudd College, Claremont CA 91711, USA
Henriksen@hmc.edu

A Survey of Residuated Lattices[1]

P. Jipsen and C. Tsinakis

Dedicated to Paul Conrad
on the occasion of his 80th birthday.

ABSTRACT.   Residuation is a fundamental concept of ordered structures and categories. In this survey we consider the consequences of adding a residuated monoid operation to lattices. The resulting residuated lattices have been studied in several branches of mathematics, including the areas of lattice-ordered groups, ideal lattices of rings, linear logic and multi-valued logic. Our exposition aims to cover basic results and current developments, concentrating on the algebraic structure, the lattice of varieties, and decidability.

We end with a list of open problems that we hope will stimulate further research.

1   Introduction

A binary operation $\cdot$ on a partially ordered set $\langle P, \leq \rangle$ is said to be *residuated* if there exist binary operations $\backslash$ and $/$ on P such that for all $x, y, z \in P$,

$$x \cdot y \leq z \quad \text{iff} \quad x \leq z/y \quad \text{iff} \quad y \leq x \backslash z.$$

The operations $\backslash$ and $/$ are referred to as the *right and left residual* of $\cdot$, respectively. It follows readily from this definition that $\cdot$ is residuated if and only if it is order preserving in each argument and, for all $x, y, z \in P$, the inequality $x \cdot y \leq z$ has a largest solution for x (namely z/y) and for y (namely $x \backslash z$). In particular, the residuals are uniquely determined by $\cdot$ and $\leq$.

The system $\mathbf{P} = \langle P, \cdot, \backslash, /, \leq \rangle$ is called a *residuated partially ordered groupoid* or *residuated po-groupoid*. We are primarily interested in the situation where $\cdot$ is a monoid operation with unit element e, say, and the partial order is a lattice order. In this case

---

[1] We would like to thank Jac Cole, Nick Galatos, Tomasz Kowalski, Hiroakira Ono, James Raftery and an anonymous referee for numerous observations and suggestions that have substantially improved this survey.

J. Martínez (ed.), Ordered Algebraic Structures, 19–56.
© *2002 Kluwer Academic Publishers.*

we add the monoid unit and the lattice operations to the similarity type to get a purely algebraic structure $\mathbf{L} = \langle L, \vee, \wedge, \cdot, e, \backslash, / \rangle$ called a *residuated lattice-ordered monoid* or *residuated lattice* for short.

The class of all residuated lattices will be denoted by $\mathcal{RL}$. It is easy to see that the equivalences that define residuation can be captured by equations and thus $\mathcal{RL}$ is a finitely based variety. Our aim in this paper is to cover basic results and current developments, concentrating on the algebraic structure, the lattice of varieties and decidability results.

The defining properties that describe the class $\mathcal{RL}$ are few and easy to grasp. At the same time, the theory is also sufficiently robust to have been studied in several branches of mathematics, including the areas of lattice-ordered groups, ideal lattices of rings, linear logic and multi-valued logic. Historically speaking, our study draws from the work of W. Krull in [Kr24] and that of Morgan Ward and R. P. Dilworth, which appeared in a series of important papers [Di38], [Di39], [Wa37], [Wa38], [Wa40], [WD38] and [WD39]. Since that time, there has been substantial research regarding some specific classes of residuated structures, see for example [AF88], [Mu86], [Ha98] and [NPM99].

We conclude the introduction by summarizing the contents of the paper. Section 2 contains basic results on residuation and the universal algebraic background needed in the remainder of the paper. In Section 3 we develop the notion of a normal subalgebra and show that $\mathcal{RL}$ is an "ideal variety", in the sense that it is an equational class in which congruences correspond to "convex normal" subalgebras, in the same way that group congruences correspond to normal subgroups. Further, we provide an element-wise description of the convex normal subalgebra generated by an arbitrary subset. In Section 4, we study varieties of residuated lattices with distributive lattice reducts. As an application of the general theory developed in the preceding sections, we produce an equational basis for the important subvariety that is generated by all residuated chains. We conclude Section 4 by introducing the classes of generalized MV-algebras and generalized BL-algebras. These objects generalize MV-algebras and BL-algebras in two directions: the existence of a lower bound is not stipulated and the commutativity assumption is dropped. Thus, bounded commutative generalized MV-algebras are reducts of MV-algebras, and likewise for BL-algebras. Section 5 is concerned with the variety and subvarieties of cancellative residuated lattices, that is, those residuated lattices whose monoid reducts are cancellative. We construct examples that show that in contrast to ℓ-groups, the lattice reducts of cancellative residuated lattices need not be distributive. Of particular interest is the fact that the classes of cancellative integral generalized MV-algebras and cancellative integral generalized BL-algebras coincide, and are precisely the negative cones of ℓ-groups, hence the latter form a variety. We prove that the map that sends a subvariety of ℓ-groups to the corresponding class of negative cones is a lattice isomorphism of the subvariety lattices, and show how to translate equational bases between corresponding subvarieties. Section 6 is devoted to the study of the lattice of subvarieties of residuated lattices. We prove that there are only two cancellative commutative varieties that cover the trivial variety, namely the varieties generated by the integers and the negative integers (with zero). On the other hand, we show that there are uncountably many non-cancellative atoms of the subvariety lattice. In Section 7 we give details of a result of Ono and Komori [OK85] that shows the equational theory of $\mathcal{RL}$ is decidable. We also mention several related decidability results about subvarieties, and summarize the currently know results in a table. In the last section we list open problems that we hope will stimulate further research.

2 Basic Results

Let $\cdot$ be a residuated binary operation on a partially ordered set $\langle P, \leq \rangle$ with residuals $\backslash$ and $/$. Intuitively, the residuals serve as generalized division operations, and x/y is read as "x over y" while $y\backslash x$ is read as "y under x". In either case, x is considered the *numerator* and y is the *denominator*. When doing calculations, we tend to favor $\backslash$ since it is more closely related to applications in logic. However, any statement about residuated structures has a "mirror image" obtained by reading terms backwards (i.e. replacing $x \cdot y$ by $y \cdot x$ and interchanging x/y with $y\backslash x$). It follows directly from the definition above that a statement is equivalent to its mirror image, and we often state results in only one form.

As usual, we write xy for $x \cdot y$ and adopt the convention that, in the absence of parenthesis, $\cdot$ is performed first, followed by $\backslash$ and $/$, and finally $\vee$ and $\wedge$. We also define $x^1 = x$ and $x^{n+1} = x^n \cdot x$.

The existence of residuals has the following basic consequences.

Proposition 2.1. *Let* **P** *be a residuated po-groupoid.*

(i) *The operation $\cdot$ preserves all existing joins in each argument; i.e., if $\bigvee X$ and $\bigvee Y$ exist for $X, Y \subset P$ then $\bigvee_{x \in X, y \in Y} x \cdot y$ exists and*

$$\left(\bigvee X\right) \cdot \left(\bigvee Y\right) = \bigvee_{x \in X, y \in Y} x \cdot y.$$

(ii) *The residuals preserve all existing meets in the numerator, and convert existing joins to meets in the denominator, i.e. if $\bigvee X$ and $\bigwedge Y$ exist for $X, Y \subset P$ then for any $z \in P$, $\bigwedge_{x \in X} x\backslash z$ and $\bigwedge_{y \in Y} z\backslash y$ exist and*

$$\left(\bigvee X\right)\backslash z = \bigwedge_{x \in X} x\backslash z \quad \text{and} \quad z\backslash\left(\bigwedge Y\right) = \bigwedge_{y \in Y} z\backslash y.$$

A *residuated po-monoid* $\langle P, \cdot, e, \backslash, /, \leq \rangle$ is a residuated po-groupoid with an identity element e and an associative binary operation.

Proposition 2.2. *The following identities (and their mirror images) hold in any residuated po-monoid.*

  (i) $(x\backslash y)z \leq x\backslash yz$

 (ii) $x\backslash y \leq zx\backslash zy$

(iii) $(x\backslash y)(y\backslash z) \leq x\backslash z$

(iv) $xy\backslash z = y\backslash(x\backslash z)$

 (v) $x\backslash(y/z) = (x\backslash y)/z$

(vi) $(x\backslash e)y \leq x\backslash y$

(vii) $x(x\backslash x) = x$

(viii) $(x\backslash x)^2 = x\backslash x$

If a residuated po-groupoid **P** has a bottom element we usually denote it by 0. In this case **P** also has a top element $0\backslash 0$, denoted by 1, and, for all $x \in P$, we have

$$x0 = 0 = 0x \quad \text{and} \quad 0\backslash x = 1 = x\backslash 1.$$

An equational basis for $\mathcal{RL}$ is given, for example, by a basis for lattice identities, monoid identities, $x(x\backslash z \wedge y) \leq z$, $y \leq x\backslash(xy \vee z)$ and the mirror images of these identities. Note that the operations $\cdot, \backslash, /$ are performed before $\vee, \wedge$. The two inequalities above follow directly from the fact that $\backslash$ is a right-residual of $\cdot$. Conversely, if the inequalities above hold, then $xy \leq z$ implies $y \leq x\backslash(xy \vee z) = x\backslash z$, and $y \leq x\backslash z$ implies $xy = x(x\backslash z \wedge y) \leq z$. Hence the identities indeed capture the property of being residuated. The argument just given is a special case of the following simple but useful observation (and its dual).

Lemma 2.3. *Let r, s, t be terms in a (semi)lattice-ordered algebra, and denote the sequence of variables $y_1, \ldots, y_n$ by $\underline{y}$. Then the quasi-identity*

$$x \leq r(\underline{y}) \Rightarrow s(x, \underline{y}) \leq t(x, \underline{y})$$

is equivalent to the identity

$$s(x \wedge r(\underline{y}), \underline{y}) \leq t(x \wedge r(\underline{y}), \underline{y}).$$

Combined with the property of residuation, the lemma above allows many quasi-identities to be translated to equivalent identities. For example the class of residuated lattices with cancellative monoid reducts forms a variety (see Section 5).

An interesting special case results when the underlying monoid structure of a residuated po-monoid is in fact a group. In this case, $x\backslash y$ is term definable as $x^{-1}y$ (and likewise $x/y = xy^{-1}$), so this class coincides with the class of *po-groups*. If we add the requirement that the partial order is a lattice, then we get the variety $\mathcal{LG}$ of *lattice-ordered groups*. This shows that lattice-ordered groups (or ℓ-groups for short) are term-equivalent to a subvariety of $\mathcal{RL}$. In the language of residuated lattices, this subvariety is defined relative to $\mathcal{RL}$ by the identity $x(x\backslash e) = e$.

Other well known subvarieties of $\mathcal{RL}$ include *integral residuated lattices* ($\mathcal{IRL}$, defined by $x \leq e$), *commutative residuated lattices* ($\mathcal{CRL}$, defined by $xy = yx$), *Brouwerian algebras* ($\mathcal{BrA}$, defined by $x \wedge y = xy$), and *generalized Boolean algebras* ($\mathcal{GBA}$, defined by $x \wedge y = xy$ and $x \vee y = (x\backslash y)\backslash y$).

We briefly discuss further connections between residuated lattices and existing classes of algebras. Recall that a *reduct* of an algebra is an algebra with the same universe but with a reduced list of fundamental operations. The opposite notion of an *expansion* is given by expanding the list of fundamental operations with new operations defined on the same universe. A *subreduct* is a subalgebra of a reduct. Subreducts obtained from $\mathcal{IRL}$ by omitting the $\vee$ operation have been studied as *generalized hoops*, and adding commutativity gives *hoops*. Omitting $\vee, \wedge, \backslash$ produces partially ordered left-residuated integral monoids (polrims). Subreducts of commutative $\mathcal{IRL}$ with only $\backslash, e$ as fundamental operations are *BCK-algebras*. If we expand bounded residuated lattices by considering the bounds as constant operations, we obtain the variety of residuated 0, 1-lattices. It has subvarieties corresponding to the variety of Boolean algebras, Stone algebras, Heyting algebras, MV-algebras and intuitionistic linear logic algebras. Further expansion with unary complementation gives relation algebras and residuated Boolean monoids.

The classes mentioned above have been studied extensively in their own right. Most of them have also been studied in logical form, such as propositional logic, intuitionistic logic, multi-valued logic, BCK logic and linear logic. In this survey we consider only the algebraic viewpoint and restrict ourselves mainly to $\mathcal{RL}$ and (some of) its subvarieties.

Universal Algebraic Background. In order to describe the algebraic structure and properties of residuated lattices, we recall here some general terminology.

An n-ary operation f is compatible with a binary relation θ if

$$\text{for all } \langle a_1, b_1 \rangle, \ldots, \langle a_n, b_n \rangle \in \theta \text{ we have } f(a_1, \ldots, a_n)\,\theta\,f(b_1, \ldots, b_n).$$

An equivalence relation θ on (the underlying set of) an algebra $\mathbf{A}$ is a *congruence* if each fundamental operation (and hence each term-definable operation) of $\mathbf{A}$ is compatible with θ. The equivalence class of a is denoted by $[a]_\theta$, and the quotient algebra of $\mathbf{A}$ with respect to θ is denoted by $\mathbf{A}/\theta$. The collection of all congruences of $\mathbf{A}$ is written $\mathrm{Con}(\mathbf{A})$. It is an algebraic lattice[2] with intersection as meet.

An algebra $\mathbf{A}$ is *congruence permutable* if $\theta \circ \phi = \phi \circ \theta$ for all $\theta, \phi \in \mathrm{Con}(\mathbf{A})$. A variety is congruence permutable if each member has this property. By a result of

---

[2] An element c in a complete lattice is *compact* if for all subsets S, $c \leq \bigvee S$ implies $c \leq s_1 \vee \cdots \vee s_n$ for some $s_1, \ldots, s_n \in S$. A complete lattice is *algebraic* if every element is a join of compact elements.

Mal'cev (see [BS81]), this is equivalent to the existence of a ternary term p such that the variety satisfies the identities $p(x, y, y) = x = p(y, y, x)$. It is not difficult to show that, for residuated lattices, the term $p(x, y, z) = x/(z \backslash y \wedge e) \wedge z/(x \backslash y \wedge e)$ has this property, hence $\mathcal{RL}$ is congruence permutable.

An algebra is *congruence regular* if each congruence is determined by any one of its congruence classes (i.e. for all $\theta, \phi \in \mathrm{Con}(A)$ if $[a]_\theta = [a]_\phi$ for some $a \in A$ then $\theta = \phi$).

Any algebra that has a group reduct is both congruence permutable and congruence regular. For algebras with a constant e, a weaker version of congruence regularity is *e-regularity*: for all $\theta, \phi \in \mathrm{Con}(A)$, $[e]_\theta = [e]_\phi$ implies $\theta = \phi$. We will see below that residuated lattices are e-regular, but not regular.

3   Structure theory

Unless noted otherwise, the results in this section are due to Blount and Tsinakis [BT]. The presentation using ideal terms is new, and the proofs have been revised considerably. A general framework for the notion of ideals in universal algebras was introduced by Ursini [Ur72] (see also Gumm and Ursini in [GU84]).

A term $t(u_1, \ldots, u_m, x_1, \ldots, x_n)$ in the language of a class $\mathcal{K}$ of similar algebras with a constant e is called an *ideal term of $\mathcal{K}$ in $x_1, \ldots, x_n$* if $\mathcal{K}$ satisfies the identity $t(u_1, \ldots, u_m, e, \ldots, e) = e$. We also write the term as $t_{u_1, \ldots, u_m}(x_1, \ldots, x_n)$ to indicate the distinction between the two types of variables.

Examples of ideal terms for $\mathcal{RL}$ are $\lambda_u(x) = (u \backslash xu) \wedge e$, $\rho_u(x) = (ux/u) \wedge e$, referred to as the *left and right conjugates of x with respect to u*, as well as $\kappa_u(x, y) = (u \wedge x) \vee y$, and the fundamental operations $x \diamond y$ for $\diamond \in \{\vee, \wedge, \cdot, \backslash, /\}$.

A subset H of $\mathbf{A} \in \mathcal{K}$ is a *$\mathcal{K}$-ideal of $\mathbf{A}$* if for all ideal terms t of $\mathcal{K}$, and all $a_1, \ldots, a_m \in A$, $b_1, \ldots, b_n \in H$ we have $t^{\mathbf{A}}(a_1, \ldots, a_m, b_1, \ldots, b_n) \in H$. Note that we use the superscript $\mathbf{A}$ to distinguish the term function $t^{\mathbf{A}}$ from the (syntactic) term t that defines it.

Clearly any e-congruence class is a $\mathcal{K}$-ideal. A class $\mathcal{K}$ is called an *ideal class* if in every member of $\mathcal{K}$ every ideal is an e-congruence class. We will prove below that this is the case for residuated lattices, and that the ideals of a residuated lattice are characterized as those subalgebras that are closed under the ideal terms λ, ρ and κ.

In analogy with groups, a subset S of a residuated lattice $\mathbf{L}$ is called *normal* if $\lambda_u(x), \rho_u(x) \in S$ for all $u \in L$ and all $x \in S$. The closed interval $\{u \in L : x \leq u \leq y\}$ is denoted by $[x, y]$. As for posets, we call S *convex* if $[x, y] \subseteq S$ for all $x, y \in S$. Note that for a sub*lattice* S the property of being convex is equivalent to $\kappa_u(x, y) \in S$ for all $u \in L$ and $x, y \in S$. Thus a convex normal subalgebra is precisely a subalgebra of $\mathbf{L}$ that is closed under the $\mathcal{RL}$-ideal terms λ, ρ and κ.

Now it follows immediately that every ideal is a convex normal subalgebra, and since we observed earlier that every e-congruence class is an ideal, we have shown that every e-congruence class is a convex normal subalgebra. As we will see below, the converse requires a bit more work.

The term $d(x, y) = x\backslash y \wedge y\backslash x \wedge e$ is useful for the description of congruences. Note that in ℓ-groups, d gives the negative absolute value of $x - y$. Alternatively one could use the opposite term $d'(x, y) = x/y \wedge y/x \wedge e$.

Lemma 3.1. *Let* $\mathbf{L}$ *be a residuated lattice. For any congruence* θ *of* $\mathbf{L}$, *we have* $a\theta b$ *if and only if* $d(a, b)\theta e$.

Proof. From $a\theta b$ one infers

$$e = (a\backslash a \wedge b\backslash b \wedge e)\, \theta\, (a\backslash b \wedge b\backslash a \wedge e) = d(a, b).$$

Conversely, $d(a, b)\, \theta\, e$ implies $a\, \theta\, ad(a, b) \le a(a\backslash b) \le b$, hence $[a]_\theta \le [b]_\theta$, and similarly $[b]_\theta \le [a]_\theta$. Therefore we have $a\theta b$. ∎

In the next result, $L^- = \{x \in L : x \le e\}$ denotes the *negative part* of L.

Corollary 3.2. $\mathcal{RL}$ *is an* e-*regular variety. In fact, if* θ *and* ϕ *are congruences on a residuated lattice* $\mathbf{L}$, *then* $[e]_\theta \cap L^- = [e]_\phi \cap L^-$ *implies* $\theta = \phi$.

To see that $\mathcal{RL}$ is not regular, it suffices to consider the 3-element Brouwerian algebra $\{0 < a < e\}$ with $xy = x \wedge y$, since it has two congruences with $\{0\}$ as congruence class.

Lemma 3.3. *Suppose* M *is a convex normal submonoid of* $\mathbf{L}$. *For any* $a, b \in L$, $d(a, b) \in M$ *if and only if* $d'(a, b) \in M$.

Proof. Assume $d(a, b) \in M$. By normality $\rho_a(d(a, b)) \in M$. But

$$\rho_a(d(a, b)) \le a(a\backslash b)/a \wedge e \le b/a \wedge e \le e \in M,$$

and similarly $\rho_b(d(a, b)) \le a/b \wedge e \le e$. Since M is convex and closed under $\cdot$, we have $(a/b \wedge e)(b/a \wedge e) \in M$, and this element is below $d'(a, b) \le e$. Again by convexity, $d'(a, b) \in M$. The reverse implication is similar, with ρ replaced by λ. ∎

Lemma 3.4. *Let* H *be a convex normal subalgebra of* $\mathbf{L}$, *and define*

$$\theta_H = \{\langle a, b\rangle : d(a, b) \in H\}.$$

Then θ_H *is a congruence of* L *and* $H = [e]_{\theta_H}$.

Proof. Clearly θ_H is reflexive and symmetric. Assuming $d(a, b), d(b, c) \in H$, we have

$$d(a, b)d(b, c) \wedge d(b, c)d(a, b) \le (a\backslash b)(b\backslash c) \wedge (c\backslash b)(b\backslash a) \wedge e \le d(a, c) \le e$$

by 2.2 (iii). Since H is convex, $d(a,c) \in H$, and therefore θ_H is an equivalence relation.

Assuming $d(a,b) \in H$ and $c \in L$, it remains to show that $d(c \diamond a, c \diamond b), d(a \diamond c, b \diamond c) \in H$ for $\diamond \in \{\cdot, \wedge, \vee, \backslash, /\}$. Since H is convex and $d(x,y) \leq e \in H$, it suffices to construct elements of H that are below these two expressions.

By 2.2 (ii) $a \backslash b \leq ca \backslash cb$, hence $d(a,b) \leq d(ca, cb)$.

By 2.2 (i) and (iv) $\lambda_c(d(a,b)) \leq c \backslash (a \backslash b) c \wedge c \backslash (b \backslash a) c \wedge e \leq d(ac, bc)$.

Since $(a \wedge c) \cdot d(a,b) \leq ad(a,b) \wedge cd(a,b) \leq b \wedge c$, we have $d(a,b) \leq (a \wedge c) \backslash (b \wedge c)$, and similarly $d(a,b) \leq (b \wedge c) \backslash (a \wedge c)$. Therefore $d(a,b) \leq d(a \wedge c, b \wedge c) \leq e$. The computation for $\vee$ is the same.

By 2.2 (iii) $a \backslash b \leq (c \backslash a) \backslash (c \backslash b)$, whence $d(a,b) \leq d(c \backslash a, c \backslash b)$. Using the same lemma we get $a \backslash b \leq (a \backslash c)/(b \backslash c)$ and therefore $d(a,b) \leq d'(a \backslash c, b \backslash c)$. It follows that $d'(a \backslash c, b \backslash c)$ is in H, so by Lemma 3.3 the same holds for d. The computation for $/$ is a mirror image of the above.

Finally, $H = [e]_{\theta_H}$ holds because H is a convex subalgebra. Indeed, $a \in H$ implies $d(a,e) \in H$, and the reverse implication holds since $d(a,e) \leq a \leq d(a,e) \backslash e$.　∎

The collection of all convex, normal subalgebras of a residuated lattice $\mathbf{L}$ will be denoted by $\mathrm{CN}(\mathbf{L})$. This is easily seen to be an algebraic lattice, with meet in $\mathrm{CN}(\mathbf{L})$ given by intersections.

Theorem 3.5. [BT] *For any residuated lattice* $\mathbf{L}$, $\mathrm{CN}(\mathbf{L})$ *is isomorphic to* $\mathrm{Con}(\mathbf{L})$, *via the mutually inverse maps* $H \mapsto \theta_H$ *and* $\theta \mapsto [e]_\theta$.

Proof. By the preceding lemma, θ_H is a congruence, and we remarked earlier that $[e]_\theta$ is a convex, normal subalgebra. Since the given maps are clearly order-preserving, it suffices to show they are inverses of each other. The lemma above already proved $H = [e]_{\theta_H}$. To show that $\theta = \theta_{[e]_\theta}$, we let $H = [e]_\theta$ and observe that $[e]_{\theta_H} = H = [e]_\theta$. Since $\mathcal{RL}$ is e-regular, the result follows.　∎

The Generation of Ideals. Recall that $\mathbf{H}$ is a convex normal subalgebra of a residuated lattice $\mathbf{L}$ provided it is closed under the $\mathcal{RL}$-ideal terms κ, λ, ρ and the fundamental operations of $\mathbf{L}$. For a subset S of L, let $\mathrm{cn}(S)$ denote the intersection of all convex normal subalgebras containing S. When $S = \{s\}$, we write $\mathrm{cn}(s)$ rather than $\mathrm{cn}(\{s\})$. Clearly $\mathrm{cn}(S)$ can also be generated from S by iterating the ideal terms and fundamental operations. The next result shows that we may compute $\mathrm{cn}(S)$ by applying these terms in a particular order. Let

$$\Delta(S) = \{s \wedge e/s \wedge e : s \in S\}$$
$$\Gamma(S) = \{\lambda_{u_1} \circ \rho_{u_2} \circ \lambda_{u_3} \circ \cdots \circ \rho_{u_{2n}}(s) : n \in \omega, u_i \in L, s \in S\}$$
$$\Pi(S) = \{s_1 \cdot s_2 \cdots s_n : n \in \omega, s_i \in S\} \cup \{e\}.$$

Thus $\Gamma(S)$ is the normal closure of S, and $\Pi(S)$ is the submonoid generated by S. Note also that if $S \subseteq L^-$ then $\Delta(S) = S$.

Theorem 3.6. [BT] *The convex normal subalgebra generated by a subset S in a residuated lattice* $\mathbf{L}$ *is*

$$\operatorname{cn}(S) = \{a \in L : x \le a \le x \backslash e \text{ for some } x \in \Pi\Gamma\Delta(S)\}.$$

Proof. Let $\langle S \rangle$ denote the right hand side above. Since this is clearly a subset of $\operatorname{cn}(S)$, we need to show that $S \subseteq \langle S \rangle$ and that $\langle S \rangle$ is a convex normal subalgebra.

Suppose $s \in S$, and consider $x = s \wedge e/s \wedge e \in \Pi\Gamma\Delta(S)$. Then $x \le s$ and $s \le (e/s)\backslash e \le x\backslash e$, hence $s \in \langle S \rangle$. This shows $S \subseteq \langle S \rangle$.

Now let $a, b \in \langle S \rangle$. Then there are $x, y \in \Pi\Gamma\Delta(S)$ such that $x \le a \le x\backslash e$ and $y \le b \le y\backslash e$. Note that $\Pi\Gamma\Delta(S) \subseteq L^-$ since $\Delta(S) \subseteq L^-$ and L^- is closed under conjugation and multiplication. Hence $xy \le x, y$ and therefore $x\backslash e, y\backslash e \le xy\backslash e$ and $xy \le a, b \le xy\backslash e$. Since $xy \in \Pi\Gamma\Delta(S)$, it follows that $\langle S \rangle$ is convex and closed under $\vee$ and $\wedge$. Also, using 2.2 (i), (iv) we have

$$xyxy \le ab \le (xy\backslash e)(xy\backslash e) \le xy\backslash(xy\backslash e) = xyxy\backslash e$$

hence $\langle S \rangle$ is closed under $\cdot$. Note that even when S is empty, we have $e \in \langle S \rangle$ since e is the empty product in $\Pi(\emptyset)$.

To see that $\langle S \rangle$ is normal, we need to show that $\lambda_u(a), \rho_u(a) \in \langle S \rangle$ for any $u \in L$ and $a \in S$. We first observe, using 2.2(i), that

$$\lambda_u(p)\lambda_u(q) \le (u\backslash pu)(u\backslash qu) \wedge e \le u\backslash pu(u\backslash qu) \wedge e \le u\backslash pqu \wedge e = \lambda_u(pq).$$

Since $x \in \Pi\Gamma\Delta(S)$, there exist $z_1, \ldots, z_n \in \Gamma\Delta(S)$ such that $x = z_1 z_2 \cdots z_n$. Let $z = \lambda_u(z_1) \cdots \lambda_u(z_n)$, and note that $z \in \Pi\Gamma\Delta(S)$. By the observation above,

$$z \le \lambda_u(x) \le \lambda_u(a) \le \lambda_u(x\backslash e).$$

Moreover, $z\lambda_u(x\backslash e) \le \lambda_u(x(x\backslash e)) \le e$, hence $\lambda_u(x\backslash e) \le z\backslash e$. Therefore $\lambda_u(a) \in \langle S \rangle$. The argument for ρ_u is similar.

Finally, we have to prove that $\langle S \rangle$ is closed under $\backslash$ and $/$. By assumption, $x \le a$ and $b \le y\backslash e$, hence $a\backslash b \le x\backslash(y\backslash e) = yx\backslash e$ by 2.2 (iv). Also $a \le x\backslash e$ and $y \le b$ imply $yxa \le b$. Therefore $a\lambda_a(yx) \le a(a\backslash yxa) \le yxa \le b$, and it follows that $\lambda_a(yx) \le a\backslash b$. By the same observation above, there exists $z \in \Pi\Gamma\Delta(S)$ such that $z \le yx\lambda_a(yx) \le a\backslash b \le z\backslash e$.

For $/$, we note that $x \le a$ and $b \le y\backslash e$ imply $xyb \le a$, whence $xy \le a/b$. For the upper bound we have $a/b \le u$, where $u = (x\backslash e)/y$. We claim that $u \le x\rho_u(y)\backslash e$, then the (mirror image of the) observation above shows there exists $z \in \Pi\Gamma\Delta(S)$ such that $z \le xyx\rho_u(y) \le a/b \le z\backslash e$. Indeed, $x\rho_u(y)u \le x(uy/u)u \le xuy \le e$ by definition of u. This completes the proof. ∎

An element a of $\mathbf{L}$ is *negative* if $a \in L^-$, and a subset of L is *negative* if it is a subset of L^-.

The following corollary describes explicitly how the negative elements of a one-generated convex normal subalgebra are obtained. It is used later to find an equational basis for the variety generated by totally ordered residuated lattices. We use the notation $\gamma_{\mathbf{u}}$ where $\mathbf{u} = \langle u_1, \ldots, u_{2n} \rangle$ to denote $\lambda_{u_1} \circ \rho_{u_2} \circ \lambda_{u_3} \circ \cdots \circ \rho_{u_{2n}}$.

Corollary 3.7. *Let* $\mathbf{L}$ *be a residuated lattice and* $r, s \in L^-$. *Then* $r \in \mathrm{cn}(s)$ *if and only if for some* m, n *there exist* $\mathbf{u}_i \in L^{2n}$ $(i = 1, \ldots, m)$ *such that*

$$\gamma_{\mathbf{u}_1}(s) \cdot \gamma_{\mathbf{u}_2}(s) \cdots \gamma_{\mathbf{u}_m}(s) \leq r.$$

The final result in this section is due to N. Galatos [Ga00]. It is especially useful for finite residuated lattices, where it shows that congruences are determined by the negative central idempotent elements.

The *center* of a residuated lattice $\mathbf{L}$ is the set

$$Z(\mathbf{L}) = \{x \in L : ux = xu \text{ for all } u \in L\}.$$

This is easily seen to be a join-subsemilattice and a submonoid of $\mathbf{L}$. There is a close relationship between negative idempotent elements of $Z(\mathbf{L})$ and ideals in $\mathbf{L}$. Let $\mathrm{CI}(\mathbf{L}) = \{x \in Z(\mathbf{L})^- : x = x^2\}$ be the set of all negative central idempotent elements.

Corollary 3.8. [Ga00] $\langle \mathrm{CI}(\mathbf{L}), \vee, \cdot \rangle$ *is dually embedded in* $\mathrm{CN}(\mathbf{L})$ *via the map*

$$x \mapsto \{a \in \mathbf{L} : x \leq a \leq x \backslash e\}.$$

If $\mathbf{L}$ *is finite, this map is a dual isomorphism.*

4   Distributive Residuated Lattices

The variety of residuated lattices that satisfy the distributive law

$$x \wedge (y \vee z) = (x \wedge y) \vee (x \wedge z)$$

is denoted by $\mathcal{DRL}$. In this section we consider conditions on residuated lattices that imply the distributive law, as well as some results about subvarieties of $\mathcal{DRL}$. We begin with a generalization of a result by Berman [Be74]. A *downward join-endomorphism* on a lattice $\mathbf{L}$ is a join-preserving map $f : \mathbf{L} \to \mathbf{L}$ such that $f(x) \leq x$ for all $x \in L$. An *upward meet-endomorphism* is defined dually.

Proposition 4.1. *The following are equivalent in any lattice* $\mathbf{L}$.

(i) $\mathbf{L}$ *is distributive.*

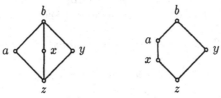

Figure 1.

(ii) *For all $a, b \in L$ with $a \leq b$ there exists a downward join-endomorphism that maps b to a.*

(iii) *The dual of (ii): For all $a, b \in L$ with $a \leq b$ there exists an upward meet-endomorphism that maps a to b.*

Proof. To see that (i) implies (ii), it suffices to check that (by distributivity) the map $x \mapsto a \wedge x$ is the required downward join-endomorphism. Conversely, suppose (ii) holds. We will show that **L** has no sublattice isomorphic to the lattices M_3 or N_5 in Figure 1, whence it follows by a well known result of Dedekind that **L** is distributive. Suppose to the contrary that either of them is a sublattice of **L**. For the elements labeled a and b, let f be the downward join-endomorphism that maps b to a. Then $f(y) \leq y$ and $a = f(b) = f(x \vee y) = f(x) \vee f(y)$, hence $f(y) \leq a$. It follows that $f(y) \leq y \wedge a = z$. Since $z \leq x$ and $f(x) \leq x$, we have $a = f(x) \vee f(y) \leq x$, which contradicts our assumptions in Figure 1.
The proof of (i)⇔(iii) is dual to the one above. ∎

Corollary 4.2. *In $\mathcal{RL}$, any of the following imply the distributive law.*

(i) $x \backslash x = e$ and $x \backslash (y \vee z) = x \backslash y \vee x \backslash z$

(ii) $x(x \backslash y \wedge e) = x \wedge y$

(iii) $x \backslash xy = y$, $xy = yx$ and $x(y \wedge z) = xy \wedge xz$

Proof. Let $\mathbf{L} \in \mathcal{RL}$ and consider $a \leq b \in L$. For (i), the map $f(x) = a(b \backslash x)$ is a downward join-endomorphism that maps b to a, and in case (ii) the map $f(x) = x(b \backslash a \wedge e)$ has the same properties. The equations in (iii) imply that $f(x) = a \backslash xb$ is an upward meet-endomorphism that maps a to b. ∎

Note that either of the conditions (i) or (ii) above imply the well known result that ℓ-groups have distributive lattice reducts.

A totally ordered residuated lattice is referred to as a *residuated chain*, and the variety generated by all residuated chains is denoted by $\mathcal{RL}^C$. Since chains are distributive lattices, this is a subvariety of $\mathcal{DRL}$. The following result provides a finite equational basis for $\mathcal{RL}^C$. A similar basis was obtained independently by C. J. van Alten [vA1] for the subvariety of integral members (defined by $x \leq e$) of $\mathcal{RL}^C$.

Theorem 4.3. [BT] $\mathcal{RL}^C$ *is the variety of all residuated lattices that satisfy*

$$e = \lambda_u((x \vee y)\backslash x) \vee \rho_v((x \vee y)\backslash y).$$

Proof. It is easy to check that the identity holds in $\mathcal{RL}^C$. Let $\mathcal{V}$ be the variety of residuated lattices defined by this identity, and let $\mathbf{L}$ be a subdirectly irreducible member of $\mathcal{V}$. To show that $\mathbf{L}$ is totally ordered, we first observe that it suffices to prove e is join-irreducible. Indeed, choosing $u = v = e$, the identity yields $e = ((x \vee y)\backslash x \wedge e) \vee ((x \vee y)\backslash y \wedge e)$, so by join-irreducibility we have either $e = ((x \vee y)\backslash x \wedge e)$ or $e = ((x \vee y)\backslash y \wedge e)$. The first case implies $y \leq x$, and the second implies $x \leq y$.

Now to prove the join-irreducibility of e, consider $a, b \in L$ such that $a \vee b = e$. We will show that $\mathrm{cn}(a) \cap \mathrm{cn}(b) = \{e\}$, and since we assumed $\mathbf{L}$ is subdirectly irreducible, it will follow that either $a = e$ or $b = e$.

Claim: If $a_1, \ldots, a_m, b_1, \ldots, b_n \in L^-$ *and* $a_i \vee b_j = e$ *for all* i, j *then*

$$a_1 a_2 \cdots a_m \vee b_1 b_2 \cdots b_n = e.$$

This is an immediate consequence of the observation that $x \vee y = x \vee z = e$ implies $x \vee yz = e$ in any ℓ-groupoid with unit (since $e = (x \vee y)(x \vee z) = x^2 \vee xz \vee yx \vee yz \leq x \vee yz \leq e$).

Claim: If $x \vee y = e$ *then* $\lambda_u(\rho_v(x)) \vee y = e$. Here we make use of the identity that was assumed to hold in $\mathbf{L}$. Given $x \vee y = e$, we have $\lambda_u(x) \vee y = \lambda_u((x \vee y)\backslash x) \vee \rho_e((x \vee y)\backslash y) = e$, and similarly $\rho_v(x) \vee y = e$. Applying the second result to the first establishes the claim.

By Corollary 3.7 the negative members of $\mathrm{cn}(a)$ are bounded below by finite products of iterated conjugates of a. By the two preceding claims, $a \vee b = e$ implies $a' \vee b' = e$ for any $a' \in \mathrm{cn}(a) \cap L^-$ and $b' \in \mathrm{cn}(b) \cap L^-$, hence $\mathrm{cn}(a) \cap \mathrm{cn}(b) = \{e\}$. ∎

Since $(x \vee y)\backslash z = x\backslash z \wedge y\backslash z$ holds in $\mathcal{RL}$, and $x \leq u\backslash xu$ whenever $ux \leq xu$ we deduce the following result.

Corollary 4.4. *An equational basis for the variety* $\mathcal{CRL}^C$, *generated by all commutative residuated chains, is given by* $xy = yx$ *and* $e = (y\backslash x \wedge e) \vee (x\backslash y \wedge e)$.

Generalized MV-Algebras and BL-Algebras. Recall that a residuated lattice is said to be *integral* if e is its top element. The variety of all integral residuated lattices is denoted by $\mathcal{IRL}$. A 0, 1-*lattice* is a bounded lattice with additional constant operations 0 and 1 denoting the bottom and top element, respectively. In a bounded residuated lattice it suffices to include 0 in the similarity type since $1 = 0\backslash 0$. For algebraic versions of logic it is natural to assume the existence of 0, since this usually denotes the logical constant *false*. In addition, it is often the case that the operation $\cdot$ is commutative, hence the residuals are related by the identity $x\backslash y = y/x$, and $x\backslash y$ is usually written $x \to y$. In logic this operation may be interpreted as a (generalized) implication.

Commutative residuated 0, 1-lattices have been studied extensively in both algebraic and logical form under various names (e.g., linear logic algebras, BCK lattices, full Lambek algebras with exchange, residuated commutative ℓ-monoids, residuated lattices). The aim of this section is to point out that several results in this area are true in the more general setting of residuated lattices, without assuming commutativity or the existence of bounds.

The algebraic version of classical propositional logic is given by *Boolean algebras*, and for intuitionistic logic it is given by *Heyting algebras*. The class $\mathcal{HA}$ of all Heyting algebras is a variety of residuated 0, 1-lattices that is axiomatized by $x \cdot y = x \wedge y$, and the class $\mathcal{BA}$ of all Boolean Algebras is axiomatized by the additional identity $(x \to 0) \to 0 = x$. This is of course a version of the classical law of double negation.

If we do not assume the existence of a constant 0, the corresponding classes of algebras are called *Brouwerian algebras* ($\mathcal{BrA}$) and *generalized Boolean algebras* ($\mathcal{GBA}$). In the latter case, the law of double negation is rewritten as $(x \to y) \to y = x \vee y$, to avoid using the constant 0. Note that the identity $xy = x \wedge y$ implies distributivity, commutativity and integrality.

A *basic logic algebra* (or BL-algebra for short) is a commutative residuated 0, 1-lattice in which the identities

$$x(x \to y) = x \wedge y \text{ and } (x \to y) \vee (y \to x) = e$$

hold. Taking $x = e$ it follows that BL-algebras are integral, and by Theorem 4.2 (ii) the first identity implies the distributive law. From the second identity and Corollary 4.4 it follows that subdirectly irreducible BL-algebras are totally ordered, but we will not assume this identity in the generalizations we consider below. For more background on BL-algebras we refer to [Ha98] and [NPM99].

The algebraic version of Lukasiewicz's multi-valued logic is given by MV-algebras. In our setting we may define MV-algebras as commutative residuated 0, 1-lattices that satisfy the identity $(x \to y) \to y = x \vee y$, though they are often defined in a slightly different but term-equivalent similarity type (see e.g. [COM00]). The class of all MV-algebras is denoted by $\mathcal{MV}$.

Standard examples of MV-algebras are Boolean algebras, and the $[0, 1]$-algebra defined on the unit interval, with $x \cdot y = \max(0, x+y-1)$ and $x \to y = \min(1, 1-x+y)$.

The latter example can be generalized to abelian ℓ-groups as follows. If $\mathbf{G} = \langle G, \wedge, \vee, \cdot, \backslash, /, e \rangle$ is an abelian ℓ-group and a a positive element, then $\Gamma(\mathbf{G}, a) = \langle [e, a], \wedge, \vee, \circ, a, \to, e \rangle$ is an MV-algebra, where

$$x \circ y = xy/a \vee e, \quad x \to y = ya/x \wedge a.$$

Chang [Ch59] proved that if $\mathbf{M}$ is a totally ordered MV-algebra then there is an abelian ℓ-group $\mathbf{G}$ and a positive element a of it such that $\mathbf{M} \cong \Gamma(\mathbf{G}, a)$. Moreover, Mundici [Mu86] generalized the result to all MV algebras and proved that Γ is an equivalence between the category of MV-algebras and the category of abelian ℓ-groups with strong unit. A good reference for MV-algebras is [COM00].

As before, we may consider the corresponding classes of generalized algebras where we assume neither commutativity nor the existence of 0. The class $\mathcal{GBL}$ of all *generalized BL-algebras* is the subvariety of $\mathcal{RL}$ defined by $y(y\backslash x \wedge e) = x \wedge y = (x/y \wedge e)y$, and the class $\mathcal{GMV}$ of all *generalized MV-algebras* is the subvariety defined by $x/(y\backslash x \wedge e) = x \vee y = (x/y \wedge e)\backslash x$. As for BL-algebras, the members of $\mathcal{GBL}$ are distributive by Theorem 4.2 (ii). Note that both of these classes include the variety of all ℓ-groups. Also, if we assume that e is the top element, the resulting integral subvarieties $\mathcal{IGBL}$ and $\mathcal{IGMV}$ can be defined by the simpler identities $y(y\backslash x) = x \wedge y = (x/y)y$ and $x/(y\backslash x) = x \vee y = (x/y)\backslash x$, respectively (and these identities imply integrality).

Lemma 4.5.

(i) *In $\mathcal{RL}$ the GBL identity $y(y\backslash x \wedge e) = x \wedge y$ is equivalent to the quasi-identity $x \le y \Rightarrow x = y(y\backslash x)$.*

(ii) *The GMV identity $x/(y\backslash x \wedge e) = x \vee y$ is equivalent to the quasi-identity $x \le y \Rightarrow y = x/(y\backslash x)$.*

Proof. The quasi-identity in (i) is equivalent to the identity $x \wedge y = y(y\backslash(x \wedge y))$. Since $\backslash$ distributes over $\wedge$ in the numerator, we may rewrite this as $x \wedge y = y(y\backslash x \wedge y\backslash y)$. Taking $x = e$ in the quasi-identity shows that all positive elements are invertible. In particular, since $e \le y\backslash y = (y\backslash y)^2$ by 2.2 (viii), we have that

$$e = (y\backslash y)((y\backslash y)\backslash e) = (y\backslash y)^2((y\backslash y)\backslash e) = y\backslash y.$$

The proof for (ii) is similar.   ∎

Residuated groupoids that satisfy the quasi-identity in (i) above and its mirror image are usually referred to as *complemented*, but since this term has a different meaning for lattices, we do not use it here.

The next result shows that $\mathcal{GMV}$ is subvariety of $\mathcal{GBL}$. In particular, it follows that GMV-algebras are distributive.

Theorem 4.6. [BCGJT] *Every GMV-algebra is a GBL-algebra.*

Proof. We make use of the quasi-equational formulation from the preceding lemma. Assume $x \leq y$ and let $z = y(y\backslash x)$. Note that $z \leq x$ and $y\backslash z \leq x\backslash z$, hence

$$
\begin{aligned}
x\backslash z &= ((y\backslash z)/(x\backslash z))\backslash(y\backslash z) \\
&= (y\backslash(z/(x\backslash z)))\backslash(y\backslash z) \quad \text{since } (u\backslash v)/w = u\backslash(v/w) \\
&= (y\backslash x)\backslash(y\backslash z) \quad \text{since } z \leq x \Rightarrow x = z/(x\backslash z) \\
&= (y(y\backslash x)\backslash z \quad \text{since } u\backslash(v\backslash w) = vu\backslash w \\
&= z\backslash z.
\end{aligned}
$$

Therefore $x = z/(x\backslash z) = z/(z\backslash z) = z$, as required. The proof of $x = (x/y)y$ is similar. ∎

5   Cancellative residuated lattices

The results in this section are from [BCGJT].

A residuated lattice is said to be *cancellative* if its monoid reduct satisfies the left and right cancellation laws: for all x, y, z, $xy = xz$ implies $y = z$, and $xz = yz$ implies $x = y$.

The class of cancellative residuated lattices is denoted by $\mathcal{C}an\mathcal{RL}$. Recall that if a residuated lattice has a bottom element 0, then $0x = 0 = x0$. Hence any nontrivial cancellative residuated lattice is infinite.

Lemma 5.1. *A residuated lattice is left cancellative if and only if it satisfies the identity* $x\backslash xy = y$.

Proof. Since $\cdot$ distributes over $\vee$, the left cancellative law is equivalent to $xz \leq xy \Rightarrow z \leq y$, and this is in turn equivalent to $z \leq x\backslash xy \Rightarrow z \leq y$. Taking $z = x\backslash xy$, we see that this implication holds if and only if $x\backslash xy \leq y$. Since the reverse inequality holds in any residuated lattice, the result follows. ∎

Corollary 5.2. $\mathcal{C}an\mathcal{RL}$ *is a variety.*

In contrast to ℓ-groups, whose lattice reducts are distributive, the next result shows that cancellative residuated lattices only satisfy those lattice identities that hold in all lattices. In [Co02] it is proved that even commutative cancellative residuated lattices are not necessarily distributive.

Given a lattice **L**, we construct a simple integral cancellative residuated lattice **L*** that has **L** as lattice subreduct.

Theorem 5.3. *Every lattice can be embedded into the lattice reduct of a simple integral member of* $\mathcal{C}an\mathcal{RL}$.

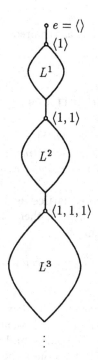

Figure 2: The structure of **L***

Proof. Suppose **L** is any lattice. We may assume that **L** has a top element 1, since any lattice can be embedded in a lattice with a top. The lattice **L*** is defined to be the ordinal sum of the cartesian power **L**^{n}, with every element of L^n above every element of L^{n+1}, for $n = 0, 1, 2, \ldots$; (see Figure 2). The operation $\cdot$ is on **L*** is simply concatenation of sequences, so the monoid reduct of **L*** is the free monoid generated by L^1. This operation is obviously cancellative, and it is residuated since each L^k has a largest element $1_k = \langle 1, 1, \ldots, 1 \rangle$. Hence the largest solution **z** of

$$\langle x_1, \ldots, x_m \rangle \cdot \mathbf{z} \leq \langle y_1, \ldots, y_n \rangle$$

is either $\mathbf{z} = \langle y_{m+1}, \ldots, y_n \rangle$ if $m \leq n$ and $x_i \leq y_i$ for $1 \leq i \leq m$, or $\mathbf{z} = 1_k$, where $k = \max(n - m + 1, 0)$. (Note that $1_0 = \langle \rangle = e$.) Thus \ exists, and the argument for / is similar.

Since **L**$^* = \mathrm{cn}(\langle 1 \rangle)$, the convex normal subalgebra generated by the coatom of **L***, it follows from Theorem 3.5 that **L*** is simple. ∎

Corollary 5.4. *The lattice of subvarieties of lattices is order-embeddable into* $\mathbf{L}(\mathcal{CanRL})$.

Proof. Given any variety $\mathcal{V}$ of lattices, define $\widehat{\mathcal{V}} = \mathbf{V}(\{\mathbf{L}^* : \mathbf{L} \in \mathcal{V}\})$, where $\mathbf{V}(\mathcal{K})$ denotes the variety generated by the class $\mathcal{K}$. Since lattice varieties are closed under the operation of ordinal sum and adding a top element, the lattice reduct of **L*** is in the same variety as **L**. Since the map $\mathcal{V} \mapsto \widehat{\mathcal{V}}$ is clearly order preserving, it suffices to prove that it is one-one. If $\mathcal{V}$ and $\mathcal{W}$ are two distinct lattice varieties, then there is a lattice identity that holds, say, in $\mathcal{V}$ but not in $\mathcal{W}$. Since the lattice reducts of $\widehat{\mathcal{V}}$ and $\widehat{\mathcal{W}}$ are members of $\mathcal{V}$ and $\mathcal{W}$, respectively, the same identity holds in $\widehat{\mathcal{V}}$, but fails in $\widehat{\mathcal{W}}$. ∎

Negative Cones of ℓ-Groups. Recall that the *negative part* of a residuated lattice **L** is $L^- = \{x \in L : x \leq e\}$. The *negative cone* of **L** is defined as $\mathbf{L}^- = \langle L^-, \vee, \wedge, \cdot, e, /^{\mathbf{L}^-}, \backslash^{\mathbf{L}^-} \rangle$, where

$$a /^{\mathbf{L}^-} b = a/b \wedge e \quad \text{and} \quad a \backslash^{\mathbf{L}^-} b = a \backslash b \wedge e.$$

It is easy to check that **L**$^-$ is again a residuated lattice. For a class $\mathcal{K}$ of residuated lattices, $\mathcal{K}^-$ denotes the class of negative cones of members of $\mathcal{K}$.

The following standard construction shows that certain cancellative monoids can be embedded in their groups of fractions (see, e.g., [Fu63]).

Lemma 5.5. *Let* **M** *be a cancellative monoid such that* $a M = M a$ *for all* $a \in M$. *Then there exists a group* **G** *and an embedding* $a \mapsto \hat{a}$ *from* **M** *to* **G** *such that every element of* **G** *is of the form* $\hat{a}\hat{b}^{-1}$ *for some* $a, b \in M$.

Theorem 5.6. *For* $\mathbf{L} \in \mathcal{RL}$ *the following statements are equivalent.*

(i) **L** *is the negative cone of an ℓ-group.*

(ii) **L** *is a cancellative integral generalized MV-algebra.*

(iii) **L** *is a cancellative integral generalized BL-algebra.*

Proof. (i)⇒(ii): for all $\mathbf{L} \in \mathcal{LG}$ and all $a, b \in L^-$, $ab/^{\mathbf{L}^-} b = abb^{-1} \wedge e = a \wedge e = a$, and $a/^{\mathbf{L}^-}(b\backslash^{\mathbf{L}^-} a) = a(b^{-1}a \wedge e)^{-1} \wedge e = (aa^{-1}b \vee a) \wedge e = a \vee b$.

(ii)⇒(iii) follows from Theorem 4.6.

(iii)⇒(i): We use Lemma 5.5 to show that any cancellative integral generalized BL-algebra is in $\mathcal{LG}^-$.

Let **L** be a cancellative integral generalized BL-algebra. For any elements $a, b \in L$ we deduce by Lemma 4.5(i) that $a(a\backslash ba) = ba$ since $b \leq e$. Hence there exists an element b_a, namely $a\backslash ba$, such that $ba = ab_a$. It follows that $La \subseteq aL$ for all elements $a \in L$, and similarly $aL \subseteq La$. Therefore the underlying monoid of L satisfies the conditions of the preceding lemma, and can be embedded into a group **G** in the prescribed manner.

We consider the following standard order on G: for all $a, b \in G$, $a \leq_G b$ if and only if $ab^{-1} \in L$. It is well known (see e.g. [Fu63], p. 13) that $\leq_G$ is a compatible partial order on G whose negative elements are precisely the elements of L. We proceed to show that $\leq_G$ is an extension of the original order $\leq$ of L. More explicitly, we prove that for all $a, b \in L$,

$$a \leq b \quad \Leftrightarrow \quad a \leq_G b \quad \Leftrightarrow \quad ab^{-1} = a/b \quad \Leftrightarrow \quad b^{-1}a = b\backslash a. \tag{$*$}$$

Let $x \leq_G y$. Thus $xy^{-1} \in L$, and $x = xy^{-1}y \leq y$ because $xy^{-1} \leq e$. Assuming $x \leq y$, $xy^{-1}y = x = x \wedge y = (x/y)y$ by the generalized basic logic identity, whence cancelling y gives $xy^{-1} = x/y$. Now if $xy^{-1} = x/y$, then $x = xy^{-1}y = (x/y)y = x \wedge y$, hence $x \leq y$. Similarly $x \leq y$ is equivalent to $y^{-1}x = y\backslash x$. Finally if $xy^{-1} = x/y$, we have that $xy^{-1} \in L$ and thus $x \leq_G y$.

The preceding conclusion allows us to drop the subscript on $\leq_G$. It remains to show that $\leq$ is a lattice order. Since any ℓ-group satisfies the identity $x \vee y = (xy^{-1} \vee e)y$, it suffices to establish the existence of all joins of the form $g \vee e$ for $g \in G$.

Let $a, b \in L$ such that $g = ab^{-1}$. We claim that $ab^{-1} \vee e = (a \vee b)b^{-1}$, where the join on the right hand side is computed in **L**. Since $a, b \leq a \vee b$, it follows that $e \leq (a \vee b)b^{-1}$ and $ab^{-1} \leq (a \vee b)b^{-1}$.

If we consider any other element of G, say cd^{-1} (where $c, d \in L$), such that both $e \leq cd^{-1}$ and $ab^{-1} \leq cd^{-1}$ hold, we have $a \leq cd^{-1}b = cd^{-1}bdd^{-1}$. Now we note that bd and d are elements of L such that $bd \leq d$. Thus $(*)$ shows that $d^{-1}bd = d\backslash bd = b_d$. Therefore $a \leq cb_dd^{-1}$, so that $ad \leq cb_d$. Similarly, working with $e \leq cd^{-1}$, we establish $bd \leq cb_d$, and hence $ad \vee bd \leq cb_d$. Since **L** is a residuated lattice, products distribute over joins, so we have $a \vee b \leq cb_dd^{-1} = cd^{-1}b$ and finally $(a \vee b)b^{-1} \leq cd^{-1}$, as desired. ∎

Corollary 5.7. $\mathcal{LG}^-$ *is a variety, defined by the identities* $xy/y = x = y\backslash yx$ *and* $(x/y)y = x \wedge y = y(y\backslash x)$. *Alternatively, the last two identities can be replaced by* $x/(y\backslash x) = x \vee y = (x/y)\backslash x$.

Corollary 5.8. *The variety* $\mathbf{V}(\mathbb{Z}^-)$ *is defined by the identities* $xy = yx$, $x = y\backslash yx$ *and* $x \wedge y = y(y\backslash x)$. *Alternatively, the last identity can be replaced by* $x \vee y = (y\backslash x)\backslash x$.

The Subvarieties of $\mathcal{LG}$ and $\mathcal{LG}^-$. We now extend the map $^- : \mathcal{LG} \to \mathcal{LG}^-$ to subclasses of $\mathcal{LG}$, and in particular to the lattice of subvarieties $\mathbf{L}(\mathcal{LG})$. We show that the image of a variety is always a variety, that every subvariety of $\mathcal{LG}^-$ is obtained in this way and that the map is order preserving. The proof is syntactical and shows how equational bases can be translated back and forth. We note that these results are related to R. McKenzie's general characterization of categorical equivalence [McK96]. For further discussion about this, we refer to [BCGJT]. Independently, C. J. van Alten [vA2] also discovered a basis for $\mathcal{LG}^-$, by proving that it is term-equivalent to the variety of cancellative generalized hoops. The correspondence between subvarieties of ℓ-groups and subvarieties of $\mathcal{LG}^-$ then follows from McKenzie's categorical equivalence.

From Subvarieties of $\mathcal{LG}^-$ to Subvarieties of $\mathcal{LG}$. In this direction, the translation is derived essentially from the definition of the negative cone. For a residuated lattice term t, we define a translated term t^- by

$$x^- = x \wedge e \qquad\qquad e^- = e$$
$$(s/t)^- = s^-/t^- \wedge e \quad (s\backslash t)^- = s^-\backslash t^- \wedge e$$
$$(st)^- = s^- t^- \qquad\quad (s \vee t)^- = s^- \vee t^- \qquad (s \wedge t)^- = s^- \wedge t^-.$$

Recall that $t^{\mathbf{L}^-}$ denotes the term-function defined by t in the algebra $\mathbf{L}^-$.

Lemma 5.9. *Let* $\mathbf{L} \in \mathcal{RL}$ *and consider any* $\mathcal{RL}$ *term* t. *For any* $a_1, \ldots, a_n \in L$,

$$t^{-\mathbf{L}}(a_1, \ldots, a_n) = t^{\mathbf{L}^-}(a_1 \wedge e, \ldots, a_n \wedge e).$$

Proof. By definition this is true for variables and the constant term e. Assume the statement holds for terms s and t. Since $(s/t)^- = s^-/t^- \wedge e$, we have

$$(s/t)^{-\mathbf{L}}(a_1, \ldots, a_n) = (s^{-\mathbf{L}}(a_1, \ldots, a_n)/^{\mathbf{L}} t^{-\mathbf{L}}(a_1, \ldots, a_n)) \wedge e$$
$$= (s^{\mathbf{L}^-}(a_1 \wedge e, \ldots, a_n \wedge e)/^{\mathbf{L}} t^{\mathbf{L}^-}(a_1 \wedge e, \ldots, a_n \wedge e)) \wedge e$$
$$= (s/t)^{\mathbf{L}^-}(a_1 \wedge e, \ldots, a_n \wedge e)$$

and similar inductive steps for $\backslash$, $\cdot$, $\vee$, $\wedge$ complete the proof. $\blacksquare$

Lemma 5.10. *For any* $\mathbf{L} \in \mathcal{RL}$, $\mathbf{L}^- \models s = t$ *if and only if* $\mathbf{L} \models s^- = t^-$.

Proof. Suppose $\mathbf{L}^- \models s = t$, and let $a_1, \ldots, a_n \in L$. By the preceding lemma, $s^{-\mathbf{L}}(a_1, \ldots, a_n) = s^{\mathbf{L}^-}(a_1 \wedge e, \ldots, a_n \wedge e) = t^{\mathbf{L}^-}(a_1 \wedge e, \ldots, a_n \wedge e) = t^{-\mathbf{L}}(a_1, \ldots, a_n)$, hence $\mathbf{L} \models s^- = t^-$. The reverse implication is similar and uses the observation that for $a_i \in L^-$, $a_i = a_i \wedge e$. ∎

Theorem 5.11. *Let $\mathcal{V}$ be a subvariety of $\mathcal{LG}^-$, defined by a set $\mathcal{E}$ of identities and let $\mathcal{W}$ be the subvariety of $\mathcal{LG}$ defined by the set of equations $\mathcal{E}^- = \{s^- = t^- : (s = t) \in \mathcal{E}\}$. Then $\mathcal{W}^- = \mathcal{V}$.*

Proof. Consider $\mathbf{M} \in \mathcal{W}^-$, which means there exists an $\mathbf{L} \in \mathcal{W}$ such that $\mathbf{M}$ is isomorphic to $\mathbf{L}^-$. Then $\mathbf{L} \models \mathcal{E}^-$, and by the previous lemma this is equivalent to $\mathbf{L}^- \models \mathcal{E}$, which in turn is equivalent to $\mathbf{L}^- \in \mathcal{V}$. Hence $\mathbf{M} \in \mathcal{V}$.

Conversely, let $\mathbf{M} \in \mathcal{V}$. Then there exists an ℓ-group $\mathbf{G}$ such that $\mathbf{M}$ is isomorphic to $\mathbf{G}^-$. ($\mathbf{G}$ is constructed as in Theorem 5.6.) Using the previous lemma again, we get that $\mathbf{G} \models \mathcal{E}^-$, hence $\mathbf{M} \in \mathcal{W}^-$. ∎

As an example, consider the variety $\mathcal{N}^-$ that is defined by the identity $x^2 y^2 \leq yx$ relative to $\mathcal{LG}^-$. The corresponding identity for the variety $\mathcal{N}$ of normal valued ℓ-groups is $(x \wedge e)^2 (y \wedge e)^2 \leq (y \wedge e)(x \wedge e)$.

From Subvarieties of $\mathcal{LG}$ to Subvarieties of $\mathcal{LG}^-$. Note that since $\cdot$ and $^{-1}$ distribute over $\vee$ and $\wedge$, any $\mathcal{LG}$ identity is equivalent to a conjunction of two identities of the form $e \leq p(g_1, \ldots, g_n)$, where p is a lattice term and $g_1, \ldots, g_n$ are group terms. Since ℓ-groups are distributive, this can be further reduced to a finite conjunction of inequalities of the form $e \leq g_1 \vee \cdots \vee g_n$.

For a term $t(x_1, \ldots, x_m)$ and a variable z distinct from $x_1, \ldots, x_m$, let

$$\bar{t}(z, x_1, \ldots, x_m) = t(z^{-1} x_1, \ldots, z^{-1} x_m).$$

Lemma 5.12. *Let $\mathbf{L}$ be an ℓ-group, and t an ℓ-group term. Then*

$$\mathbf{L} \models e \leq t(x_1, \ldots, x_m) \quad \text{iff} \quad \mathbf{L} \models x_1 \vee \cdots \vee x_m \vee z \leq e \Rightarrow e \leq \bar{t}(z, x_1, \ldots, x_m).$$

Proof. In the forward direction this is obvious. To prove the reverse implication, assume the right hand side holds and let $a_1, \ldots, a_m \in L$. Define $c = a_1^{-1} \wedge \cdots \wedge a_m^{-1} \wedge e$ and $b_i = ca_i$ for $i = 1, \ldots, m$. Then $c \leq e$ and $c \leq a_i^{-1}$, hence $b_i \leq e$. Now by assumption, $e \leq t(c^{-1} b_1, \ldots, c^{-1} b_m) = t(a_1, \ldots, a_m)$. ∎

Lemma 5.13. *Let $\mathbf{L} \in \mathcal{LG}$. For any group term g, there exist an $\mathcal{RL}$ term $\hat{g}$ such that $(g \wedge e)^{\mathbf{L}}|_{L^-} = \hat{g}^{\mathbf{L}^-}$.*

Proof. Essentially we have to rewrite group terms so that all the variables with inverses appear at the beginning of the term. This is done using conjugation: $xy^{-1} = y^{-1}(yxy^{-1}) = y^{-1}(yx/y)$. Note that L^- is closed under conjugation by arbitrary elements, since $x \leq e$ implies $yxy^{-1} \leq e$. If we also have $y \leq e$, then $yx \in L^-$ and $yx \leq y$, hence $yx/\overline{L^-} y = yx/^L y$.

To describe the translation of an arbitrary group term, we may assume that it is of the form $p_1 q_1^{-1} p_2 q_2^{-1} \cdots p_n q_n^{-1}$ where the p_i and q_i are products of variables (without inverses). By using conjugation, we write this term in the form

$$q_1^{-1} q_2^{-1} \cdots q_n^{-1} (q_n (\cdots (q_2 (q_1 p_1 / q_1) p_2 / q_2) \cdots) p_n / q_n).$$

So we can take $\hat{g} = s \backslash t$ where

$$s = q_n \cdots q_2 q_1 \quad \text{and} \quad t = q_n (\cdots (q_2 (q_1 p_1 / q_1) p_2 / q_2) \cdots) p_n / q_n.$$

■

Corollary 5.14. *Let* $g_1, \ldots, g_n$ *be group terms with variables among* $x_1, \ldots, x_m$. *For any ℓ-group* **L**,

$$\mathbf{L}^- \models \hat{g}_1 \vee \ldots \vee \hat{g}_n = e \quad \text{iff} \quad \mathbf{L} \models x_1 \vee \ldots \vee x_m \leq e \Rightarrow e \leq g_1 \vee \ldots \vee g_n.$$

For the next result, recall the discussion about identities in ℓ-groups, and the definition of $\bar{t}$ at the beginning of this subsection.

Theorem 5.15. *Let* $\mathcal{V}$ *be a subvariety of* $\mathcal{LG}$, *defined by a set* $\mathcal{E}$ *of identities, which we may assume are of the form* $e \leq g_1 \vee \ldots \vee g_n$. *Let*

$$\overline{\mathcal{E}} = \{ e = \widehat{g_1} \vee \ldots \vee \widehat{g_n} : e \leq g_1 \vee \ldots \vee g_n \text{ is in } \mathcal{E} \}.$$

Then $\overline{\mathcal{E}}$ *is an equational basis for* $\mathcal{V}^-$ *relative to* $\mathcal{LG}^-$.

Proof. By construction, any member of $\mathcal{V}^-$ satisfies all the identities in $\overline{\mathcal{E}}$. On the other hand, if $\mathbf{M} \in \mathcal{LG}^-$ is a model of the identities in $\overline{\mathcal{E}}$, then $\mathbf{M}$ is the negative cone of some $\mathbf{L} \in \mathcal{LG}$. From the reverse directions of Corollary 5.14 and Lemma 5.12 we infer that $\mathbf{L}$ satisfies the equations in $\mathcal{E}$, hence $\mathbf{M} \in \mathcal{V}^-$. ■

For example consider the variety $\mathcal{R}$ of *representable ℓ-groups* which (by definition) is generated by the class of totally ordered groups (see [AF88] for more details). An equational basis for this variety is given by $e \leq x^{-1} yx \vee y^{-1}$ (relative to $\mathcal{LG}$). Applying the translation above, we obtain $e = zx \backslash (zy/z) x \vee y \backslash z$ as as equational basis for $\mathcal{R}^-$.

Corollary 5.16. *The map* $\mathcal{V} \mapsto \mathcal{V}^-$ *from* $\mathbf{L}(\mathcal{LG})$ *to* $\mathbf{L}(\mathcal{LG}^-)$ *is a lattice isomorphism, with the property that finitely based subvarieties of* $\mathcal{LG}$ *are mapped to finitely based subvarieties of* $\mathcal{LG}^-$ *and conversely.*

Proof. Let V be any subvariety of $\mathcal{LG}$. By the above theorem, V^- is a subvariety of $\mathcal{LG}^-$, and by Theorem 5.11 every subvariety of $\mathcal{LG}^-$ is of this form. Consider $V \subseteq V' \subseteq \mathcal{LG}$, and let $\mathcal{E} \supseteq \mathcal{E}'$ be equational bases for V, V' respectively. Defining $\mathcal{E}^-$ as in the preceding theorem, we have $\mathcal{E}^- \supseteq \mathcal{E}'^-$, hence $V \subseteq V'$. Finally, the map is injective since every ℓ-group is determined by its negative cone. It follows that the map is a lattice isomorphism, and the translation $\mathcal{E} \mapsto \mathcal{E}^-$ clearly maps finite sets to finite sets.   ∎

6   Lattices of Subvarieties

We now investigate the structure of $\mathbf{L}(\mathcal{RL})$. In particular we consider which varieties are atoms in this lattices or in some of the ideals generated by particular subvarieties. Since any nontrivial ℓ-group has a subalgebra isomorphic to $\mathbb{Z}$, one obtains the well-known result that $\mathbf{V}(\mathbb{Z})$ is the only atom of $\mathbf{L}(\mathcal{RL})$. Similarly the result below is a straight forward consequence of the observation that every nontrivial integral, cancellative, residuated lattice has a subalgebra isomorphic to $\mathbb{Z}^-$.

Theorem 6.1. $\mathbf{V}(\mathbb{Z}^-)$ is the only atom in the lattice of integral subvarieties of $\mathcal{CanRL}$.

For the next result, note that the identity $x \backslash x = e$ holds in any cancellative residuated lattice.

Theorem 6.2. [BCGJT] $\mathbf{V}(\mathbb{Z})$ and $\mathbf{V}(\mathbb{Z}^-)$ are the only atoms in the lattice of commutative subvarieties of $\mathcal{CanRL}$.

Proof. Let $\mathbf{L}$ be a nontrivial cancellative, commutative residuated lattice. If $\mathbf{L}$ is integral, then $\mathbf{V}(\mathbf{L})$ contains $\mathbb{Z}^-$ by the preceding theorem. So we may assume that $\mathbf{L}$ is not integral. Let $S = \{x \in L : x \backslash e = e\}$ and consider the least congruence θ that collapses all members of S to e. By cancellativity and 2.2 (iv) $e = (x \backslash e) \backslash (x \backslash e) = x(x \backslash e) \backslash e$, hence $x(x \backslash e) \in S$ for all $x \in L$. It follows that $\mathbf{L}/\theta$ satisfies the identity $x(x \backslash e) = e$ and thus is an ℓ-group. It suffices to prove that $\mathbf{L}/\theta$ is nontrivial since any nontrivial ℓ-group contains a subalgebra isomorphic to $\mathbb{Z}$.

Note that $S \subseteq L^-$, and by commutativity, S is closed under conjugation: if $x \in S$ and $u \in L$ then

$$x = x \wedge e \le u \backslash ux \wedge e = u \backslash xu \wedge e = \lambda_u(x) \le e,$$

hence $e \le \lambda_u(x) \backslash e \le x \backslash e = e$, and it follows that $\lambda_u(x) \in S$. Furthermore, S is closed under $\cdot$ since if $x, y \in S$ then $xy \backslash e = y \backslash (x \backslash e) = y \backslash e = e$. So in the notation of Theorem 3.6 $\Pi\Gamma\Delta(S) = S$, whence $\operatorname{cn}(S) = \{u : x \le u \le x \backslash e$ for some $x \in S\}$. In particular, any negative element $u \in \operatorname{cn}(S)$ satisfies $x \le u \le e$, so $e \le u \backslash e \le x \backslash e = e$. Now choose $a \not\le e$ and consider the negative element $b = e/a \wedge e$. Then $b \notin \operatorname{cn}(S)$ since $b \backslash e \ge e \vee a > e$. By Theorem 3.5 it follows that θ does not collapse all of $\mathbf{L}$, as required.   ∎

It is not known whether there are noncommutative atoms below the variety of cancellative residuated lattices (see Problem 8.5).

Without the assumption of cancellativity, it is much easier to construct algebras that generate atoms in $\mathbf{L}(\mathcal{RL})$. An algebra is called *strictly simple* if it has no nontrivial congruences or subalgebras. It is easy to see that in a congruence distributive variety, any finite strictly simple algebra generates a variety that is an atom. In Figure 3 we list several finite strictly simple residuated lattices. In each case we give only the multiplication operation, since the residuals are determined by it. Also, in all cases $x0 = 0 = 0x$, $xe = x = ex$ and $x1 = x = 1x$ (when $x \neq e$).

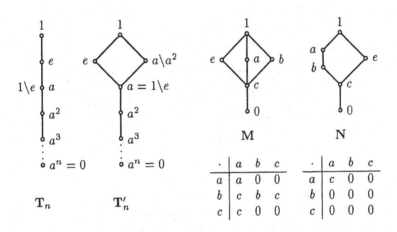

Figure 3.

We now give an example of uncountably many residuated chains that generate distinct atoms of $\mathbf{L}(\mathcal{RL})$. Let S be any subset of ω. The algebra $\mathbf{C}_S$ is based on the set $\{0, a, b, e, 1\} \cup \{c_i : i \in \omega\} \cup \{d_i : i \in \omega\}$, with the following linear order:

$$0 < a < b < c_0 < c_1 < c_2 < \cdots < \cdots < d_2 < d_1 < d_0 < e < 1$$

The operation $\cdot$ is defined by $ex = x = xe$, if $x \neq e$ then $1x = x = x1$, and if $x \notin \{e, 1\}$ then $0x = 0 = x0$, $ax = 0 = xa$, and $bx = 0 = xb$. Furthermore, for all $i, j \in \omega$, $c_i c_j = 0$, $d_i d_j = b$,

$$c_i d_j = \begin{cases} 0 & \text{if } i < j \\ a & \text{if } i = j \text{ or } (i = j+1 \text{ and } j \in S) \\ b & \text{otherwise} \end{cases} \qquad d_i c_j = \begin{cases} 0 & \text{if } i \geq j \\ b & \text{otherwise.} \end{cases}$$

This information is given in the form of an operation table in Figure 4. Depending on the chosen subset S, the elements s_i in the table are either a (if $i \in S$) or b (if $i \notin S$).

$\cdot$	1	e	d_0	d_1	d_2	d_3	$\ldots$	$\ldots$	c_3	c_2	c_1	c_0	b	a	0
1	1	1	d_0	d_1	d_2	d_3	$\ldots$	$\ldots$	c_3	c_2	c_1	c_0	b	a	0
e	1	e	d_0	d_1	d_2	d_3	$\ldots$	$\ldots$	c_3	c_2	c_1	c_0	b	a	0
d_0	d_0	d_0	b	b	b	b	$\ldots$	$\ldots$	b	b	b	0	0	0	0
d_1	d_1	d_1	b	b	b	b	$\ldots$	$\ldots$	b	b	0	0	0	0	0
d_2	d_2	d_2	b	b	b	b	$\ldots$	$\ldots$	b	0	0	0	0	0	0
d_3	d_3	d_3	b	b	b	b	$\ldots$	$\ldots$	0	0	0	0	0	0	0
$\vdots$	$\vdots$	$\vdots$	$\vdots$	$\vdots$	$\vdots$	$\vdots$			$\vdots$	$\vdots$	$\vdots$	$\vdots$	$\vdots$	$\vdots$	$\vdots$
$\vdots$	$\vdots$	$\vdots$	$\vdots$	$\vdots$	$\vdots$	$\vdots$			$\vdots$	$\vdots$	$\vdots$	$\vdots$	$\vdots$	$\vdots$	$\vdots$
c_3	c_3	c_3	b	b	s_2	a	$\ldots$	$\ldots$	0	0	0	0	0	0	0
c_2	c_2	c_2	b	s_1	a	0	$\ldots$	$\ldots$	0	0	0	0	0	0	0
c_1	c_1	c_1	s_0	a	0	0	$\ldots$	$\ldots$	0	0	0	0	0	0	0
c_0	c_0	c_0	a	0	0	0	$\ldots$	$\ldots$	0	0	0	0	0	0	0
b	b	b	0	0	0	0	$\ldots$	$\ldots$	0	0	0	0	0	0	0
a	a	a	0	0	0	0	$\ldots$	$\ldots$	0	0	0	0	0	0	0
0	0	0	0	0	0	0	$\ldots$	$\ldots$	0	0	0	0	0	0	0

Figure 4.

It is easy to check that this gives an associative operation since $xyz = 0$ whenever $e, 1 \notin \{x, y, z\}$. Now $1 = 0\backslash 0$, $d_0 = 1\backslash e$, $c_i = d_i\backslash 0$ and $d_{i+1} = c_i\backslash 0$, so the algebra is generated by 0. Distinct subsets S of ω produce nonisomorphic algebras, hence this construction gives uncountably many strictly simple algebras. However, since they are infinite, it does not yet follow that they generate distinct atoms. To complete the argument, one has to observe that any nontrivial subalgebra of an ultrapower of $\mathbf{C}_S$ contains a subalgebra isomorphic to $\mathbf{C}_S$ (since the generator 0 is definable by the universal formula $\phi(x) = \forall y(y < e \Rightarrow y^3 = x)$), and that for any distinct pair of such algebras one can find an equation that holds in one but not in the other.

Theorem 6.3. *There are uncountably many atoms in* $\mathbf{L}(\mathcal{RL})$ *that satisfy the identity* $x^3 = x^4$.

Not much is known about the global structure of $\mathbf{L}(\mathcal{RL})$. It is a dually algebraic distributive lattice, since $\mathcal{RL}$ is congruence distributive. The subvarieties of ℓ-groups and of Brouwerian algebras have been studied extensively, and for commutative integral residuated 0, 1-lattices, a recent monograph of Kowalski and Ono [KO] contains many important results.

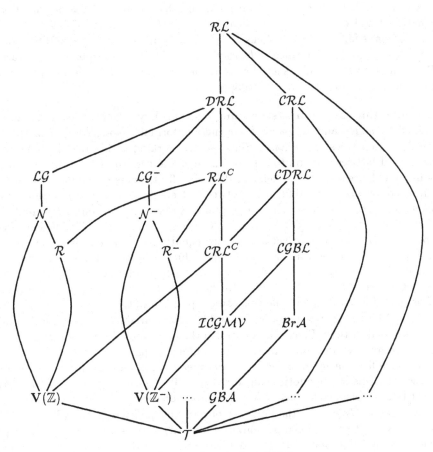

Figure 5: Some subvarieties of $\mathcal{RL}$

7 Decidability

In the first half of this section we present details of a result of Ono and Komori [OK85] which shows that the equational theory of residuated lattices is decidable and gives an effective algorithm based on a Gentzen system for the full Lambek calculus. Our approach is algebraic, and no familiarity with Gentzen systems or the Lambek calculus is assumed. In the second half we mention other results on the decidability of the equational and quasi-equational theory of various subvarieties of $\mathcal{RL}$. We note that this section is far from comprehensive, and that many related decidability results have been proved for the so-called *substructural logics* that correspond to subvarieties of $\mathcal{RL}$. The interested reader should consult the literature on relevance logic, full Lambek calculus and linear logic as a starting point.

An algebraic Gentzen system and decidability. Gentzen systems are usually defined for logics, and use pairs of sequences of formulas (so-called sequents) to specify the deduction rules of the logic. Since we are working within the algebraic theory of residuated lattices, we aim to present an algebraic version of Gentzen systems, in the hope that the reader will be persuaded by the effectiveness of the method, rather than burdened by syntactic differences between algebraic and logical notation. From an algebraic point of view, a Gentzen system is a finite set G of quasi-inequalities of the form $s_1 \leq t_1 \ \& \dots \& \ s_n \leq t_n \Rightarrow s_0 \leq t_0$, where s_i, t_i are terms. The notion of *Gentzen proof from G* is a restricted version of quasi-equational deduction that is usually simpler to work with than the standard equational deduction system of Birkhoff. In many cases it is decidable if a given inequality has a Gentzen proof, and a completeness theorem for a Gentzen system can be found to show that the concept of 'Gentzen proof from G' is in fact equivalent to 'quasi-equational proof from G'.

Let $\mathbf{T}$ be the term algebra on countably many variables $x_1, x_2, \dots$ in the language of residuated lattices. Since the equational theory of monoids is decidable (in constant time) we may restrict our attention to deciding inequalities $s \leq t$ up to associativity and multiplication by e. Effectively this means that we consider s, t as equivalence classes of terms in the quotient algebra $\mathbf{T}_M = \mathbf{T}/\equiv_M$ where $s \equiv_M s'$ if and only if the identity $s = s'$ is a consequence of the monoid identities.

We define $G_{\mathcal{RL}}$ to be the finite set of quasi-inequalities given in Table 1. Readers familiar with the traditional presentation of Gentzen rules may note that $\Rightarrow$ represents the horizontal line that is used in Gentzen rules, and that $s \leq t$ is the equivalent of a sequent (assuming $\cdot$ is considered as comma). It is straightforward to check that each quasi-inequality of $G_{\mathcal{RL}}$ holds in $\mathcal{RL}$. For example, (\left) holds since if $x \leq y$ and $uzv \leq w$ then $ux(y\backslash z)v \leq ux(x\backslash z)v \leq uzv \leq w$.

We now describe the notions of *proof-tree* and *Gentzen provable*. Recall that a *rooted tree* is a poset with a least element, called the root, and for each element, the set of all elements below it is linearly ordered. A *proof-tree* is a finite rooted tree in which each element is an inequality, and if $s_1 \leq t_1, \dots, s_n \leq t_n$ are all the covers of the

$x \leq x$	(refl)
$x \leq z \ \& \ y \leq w \Rightarrow xy \leq zw$	($\cdot$ right)
$x \leq y \ \& \ uzv \leq w \Rightarrow ux(y\backslash z)v \leq w$	($\backslash$ left)
$xy \leq z \Rightarrow y \leq x\backslash z$	($\backslash$ right)
$x \leq y \ \& \ uzv \leq w \Rightarrow u(z/y)xv \leq w$	(/ left)
$xy \leq z \Rightarrow x \leq z/y$	(/ right)
$uxv \leq w \ \& \ uyv \leq w \Rightarrow u(x \vee y)v \leq w$	($\vee$ left)
$x \leq y \Rightarrow x \leq y \vee z$	($\vee$ right)
$x \leq z \Rightarrow x \leq y \vee z$	($\vee$ right)
$uxv \leq w \Rightarrow u(x \wedge y)v \leq w$	($\wedge$ left)
$uyv \leq w \Rightarrow u(x \wedge y)v \leq w$	($\wedge$ left)
$x \leq y \ \& \ x \leq z \Rightarrow x \leq y \wedge z$	($\wedge$ right)

Table 1: The algebraic Gentzen system $G_{\mathcal{RL}}$

element $s \leq t$, then $n \in \{0, 1, 2\}$ and the quasi-inequality

$$s_1 \leq t_1 \ \& \ \ldots \ \& \ s_n \leq t_n \Rightarrow s \leq t$$

is a substitution instance of a member of $G_{\mathcal{RL}}$. Hence each element has at most 2 covers, and an inequality has no covers if and only if it is an instance of (refl). An inequality is said to be *Gentzen provable* if there exists a proof-tree with this inequality as the root.[3] The *soundness* of $G_{\mathcal{RL}}$ is the observation that any Gentzen provable inequality holds in $\mathcal{RL}$ (since $\mathcal{RL} \models G_{\mathcal{RL}}$).

We say that an inequality *matches* a quasi-inequality if it is a substitution instance of the conclusion of this quasi-inequality. Since we are considering terms equivalent modulo associativity for $\cdot$ and multiplication by e, we may re-associate freely and may match variables in a product by e. For example, the term x matches zw with z substituted by x, and w substituted by e (or vice versa). Note that for each of the members of $G_{\mathcal{RL}}$, the variables in the premises are a subset of the variables in the conclusion (this is referred to as the *subformula property* in logic). It follows that if an inequality matches a member of $G_{\mathcal{RL}}$, this determines exactly what inequalities must appear in the premises of the member. Hence the quasi-inequalities may be used as so-called rewrite rules in a search algorithm for a proof-tree of a given inequality.

For all members of $G_{\mathcal{RL}}$, the inequalities in the premises are structurally simpler than the inequalities in the conclusion. Hence the depth of a proof-tree is bounded by

---

[3] In the literature on Gentzen systems this corresponds to *cut-free provable* since the Gentzen system presented here does not mention the so-called *cut-rule* $x \leq y \ \& \ uyv \leq w \Rightarrow uxv \leq w$.

the size (defined in a suitable way) of the inequality at the root. It follows that it is decidable whether a given sequent is Gentzen provable.

The following simple examples serve to illustrate the effectiveness of this decision procedure.

$$xy \leq xy \Rightarrow xy \leq xy \vee xz \qquad \text{by ($\vee$ right) and}$$
$$xz \leq xz \Rightarrow xz \leq xy \vee xz \qquad \text{by ($\vee$ right)}$$
$$\Rightarrow x(y \vee z) \leq xy \vee xz \qquad \text{by ($\vee$ left)}$$

$$y \leq y \Rightarrow y \wedge z \leq y \qquad \text{by ($\wedge$ left)}$$
$$x \leq x \ \& \ y \wedge z \leq y \Rightarrow x(x\backslash(y \wedge z)) \leq y \qquad \text{by ($\backslash$ left)}$$
$$\Rightarrow x\backslash(y \wedge z) \leq x\backslash y \qquad \text{by ($\backslash$ right)}$$
$$z \leq z \Rightarrow y \wedge z \leq z \qquad \text{by ($\wedge$ left)}$$
$$x \leq x \ \& \ y \wedge z \leq z \Rightarrow x(x\backslash(y \wedge z)) \leq z \qquad \text{by ($\backslash$ left)}$$
$$\Rightarrow x\backslash(y \wedge z) \leq x\backslash z \qquad \text{by ($\backslash$ right)}$$
$$\Rightarrow x\backslash(y \wedge z) \leq x\backslash y \wedge x\backslash z \qquad \text{by ($\wedge$ right)}$$

An inequality such as $x \wedge (y \vee z) \leq (x \wedge y) \vee (x \wedge z)$ is not Gentzen provable since no proof tree can be found: only ($\wedge$ left) and ($\vee$ right) match this inequality, and the premises of these quasi-inequalities only match ($\vee$ left) and ($\wedge$ right). But their premises, in turn, are not instances of $x \leq x$ in all cases.

For lattice theorists it is also interesting to note that the quasi-inequalities for $\vee$ and $\wedge$ are essentially equivalent to Whitman's method for deciding if $s \leq t$ holds in all lattices.

We now prove the result of Ono and Komori [OK85] which shows that an inequality $s \leq t$ is Gentzen provable if and only if it holds in $\mathcal{RL}$. The semantical proof given here is based on a version in [OT99]. The forward implication is the soundness of the proof procedure, and follows from the observation that all the rules are valid (as quasi-inequalities) in $\mathcal{RL}$. The reverse implication is *completeness*, for which we need to prove that if $s \leq t$ is not Gentzen provable, then there is a residuated lattice in which this identity fails.

We begin with a general lemma useful for constructing residuated lattices.

Lemma 7.1. *Suppose $M = \langle M, \cdot, e^M \rangle$ is a monoid and $\mathcal{L}$ is a set of subsets of M such that*

(P_1) *$\mathcal{L}$ is closed under arbitrary intersections and*

(P_2) *for all $X \subseteq M, Y \in \mathcal{L}$ we have $X\backslash Y$ and $Y/X \in \mathcal{L}$, where*

$$X\backslash Y = \{z \in M : X\{z\} \subseteq Y\}, \quad Y/X = \{z \in M : \{z\}X \subseteq Y\} \quad and$$

$$XY = \{xy : x \in X, \ y \in Y\}.$$

Then $\mathbf{L} = (\mathcal{L}, \vee, \wedge, \cdot^{\mathbf{L}}, \backslash, /, e^{\mathbf{L}})$ is a residuated lattice, with

$$X^C = \bigcap\{Z \in \mathbf{L} : X \subseteq Z\}, \quad \text{the closure of } X, \text{ and}$$

$$X \vee Y = (X \cup Y)^C \qquad X \wedge Y = X \cap Y \qquad X \cdot^{\mathbf{L}} Y = (XY)^C \qquad e^{\mathbf{L}} = \{e^{\mathbf{M}}\}^C.$$

Proof. $\mathbf{L}$ is a lattice (in fact a complete lattice) since it is the collection of closed sets of a closure operation. By definition of $\backslash$ we have $Z \subseteq X \backslash Y$ if and only if $XZ \subseteq Y$, and for $Y \in \mathbf{L}$ this is equivalent to $X \cdot^{\mathbf{L}} Z = (XZ)^{\bar{C}} \subseteq Y$. Similarly, $Z \subseteq Y/X$ is equivalent to $Z \cdot^{\mathbf{L}} X \subseteq Y$. It remains to show that $\cdot^{\mathbf{L}}$ is associative and e is an identity. For all $X, Y \subseteq M$, $XY \subseteq (XY)^C$ implies $Y \subseteq X \backslash (XY)^C$, hence

$$XY^C \subseteq X(X \backslash (XY)^C)^C = X(X \backslash (X \cdot^{\mathbf{L}} Y)) \subseteq X \cdot^{\mathbf{L}} Y$$

where the middle equality makes use of the fact that $X \backslash (XY)^C$ is closed by (P_2). Similarly $X^C Y \subseteq X \cdot^{\mathbf{L}} Y$, hence $X^C Y^C \subseteq X^C \cdot^{\mathbf{L}} Y = (X^C Y)^C \subseteq (X \cdot^{\mathbf{L}} Y)^C = X \cdot^{\mathbf{L}} Y$. Since we also have $XY \subseteq X^C Y^C$, it follows that $(XY)^C = (X^C Y^C)^C$. Now

$$
\begin{aligned}
(X \cdot^{\mathbf{L}} Y) \cdot^{\mathbf{L}} Z &= ((XY)^C Z)^C = ((XY)^{CC} Z^C)^C = ((XY)Z)^C \\
&= (X(YZ))^C = (X^C (YZ)^C)^C = X \cdot^{\mathbf{L}} (Y \cdot^{\mathbf{L}} Z),
\end{aligned}
$$

and $e^{\mathbf{L}} \cdot^{\mathbf{L}} X = (\{1\}^C X)^C = (\{1\}^{CC} X^C)^C = (\{1\} X)^C = X^C = X$ for $X \in \mathbf{L}$. ∎

The notion of a *subterm* of a term p is defined in the standard way: p is a subterm of p, and if $s \diamond t$ is a subterm of p for $\diamond \in \{\vee, \wedge, \cdot, \backslash, /\}$ then s, t are subterms of p. Let $S(p) \subseteq \mathbf{T}_M$ be the set of subterms of p and let $M(p)$ be the submonoid of $\mathbf{T}_M$ generated by $S(p)$. For $q, q' \in M(p)$, $u \in S(p)$, define

$$[q, q', r] = \{s \in M(p) : qsq' \leq r \text{ is Gentzen provable}\}.$$

Further let

$$\mathcal{L}'(p) = \{[q, q', r] : q, q' \in M(p), \ r \in S(p)\} \quad \text{and}$$
$$\mathcal{L}(p) = \left\{\bigcap \mathcal{K} : \mathcal{K} \subseteq \mathcal{L}'(p)\right\}.$$

In the subsequent proofs we will frequently make use of the following observation:

(*) For any $X \subseteq M(p)$, $s \in S(p)$, $s \in X^C$ iff for all $q, q' \in M(p)$ and $r \in S(p)$, $X \subseteq [q, q', r]$ implies $s \in [q, q', r]$.

Lemma 7.2. $\mathbf{L}(p) = \langle \mathcal{L}(p), \vee, \wedge, \cdot, e, \backslash, /\rangle$ *is a residuated lattice, with the operations defined as in the preceding lemma.*

Proof. (P_1) holds by construction. To prove (P_2), let $X \subseteq M(p)$ and $Y \in \mathcal{L}(p)$. Now $s \in X \backslash Y$ if and only if $X\{s\} \subseteq Y$ if and only if for all $t \in X$, $ts \in Y = Y^C$. By $(*)$ this is equivalent to showing that $Y \subseteq [q, q', r]$ implies $ts \in [q, q', r]$. This last containment holds if and only if $qtsq' \leq r$ is provable if and only if $s \in [qt, q', r]$. Hence

$$s \in X \backslash Y \text{ iff } s \in \bigcap \left\{ [qt, q', r] : t \in X \text{ and } Y \subseteq [q, q', r] \right\},$$

which implies that $X \backslash Y \in \mathcal{L}(p)$, and Y/X is similar. ∎

The following result is the central part of the completeness argument. As usual, an assignment $h : \{x_1, x_2, \ldots\} \to \mathbf{L}$ extends to a homomorphism from the (quotient) term algebra $\mathbf{T}_M$ to $\mathbf{L}$, with $h(e)$ defined as $e^{\mathbf{L}}$. We now fix h to be the assignment $h(x_i) = [x_i]$, where the notation $[r]$ is shorthand for $[e, e, r]$.

Lemma 7.3. *Let* $\mathbf{L}(p)$ *and* h *be defined as above. For any subterm* t *of* p *we have* $t \in h(t) \subseteq [t]$. *In particular, if* $e \in h(t)$ *then the inequality* $e \leq t$ *is Gentzen provable.*

Proof. By induction on the structure of the subterm. If it is a variable of p, say x, then $h(x) = [x]$ by definition, and $x \in [x]$ since $x \leq x$ is Gentzen provable by (refl). Suppose s, t are subterms of p, and $s \in h(s) \subseteq [s]$, $t \in h(t) \subseteq [t]$.

$s \vee t \in h(s \vee t) \subseteq [s \vee t]$: Note that $h(s \vee t) = (h(s) \cup h(t))^C$. Let $q \in h(s) \cup h(t)$. If $q \in h(s)$, then $q \in [s]$, so $q \leq s$ is Gentzen provable. By ($\vee$ right) it follows that $q \leq s \vee t$ is Gentzen provable, hence $q \in [s \vee t]$ and therefore $h(s) \subseteq [s \vee t]$. Similarly $h(t) \subseteq [s \vee t]$, and since $[s \vee t]$ is closed, $h(s \vee t) \subseteq [s \vee t]$.

To see that $s \vee t \in h(s \vee t)$, we use observation $(*)$: Suppose $h(s) \cup h(t) \subseteq [q, q', r]$ where $q, q' \in M(p)$, $r \in S(p)$. Then $qsq' \leq r$ and $qtq' \leq r$ are Gentzen provable (since $s \in h(s)$ and $t \in h(t)$). Therefore $q(s \vee t)q' \leq r$ is Gentzen provable by ($\vee$ left) and so $s \vee t \in [q, q', r]$. By $(*)$ we conclude that $s \vee t \in (h(s) \cup h(t))^C = h(s \vee t)$.

$s \wedge t \in h(s \wedge t) \subseteq [s \wedge t]$: Let $q \in h(s \wedge t) = h(s) \cap h(t)$. Then $q \in [s] \cap [t]$, hence $q \leq s$ and $q \leq t$ are Gentzen provable. So now $q \leq s \wedge t$ is Gentzen provable by ($\wedge$ right), which shows that $q \in [s \wedge t]$.

Suppose $h(s) \subseteq [q, q', r]$. Then $qsq' \leq r$ is Gentzen provable, and by ($\wedge$ left) $q(s \wedge t)q' \leq r$ is Gentzen provable. Therefore $s \wedge t \in [q, q', r]$, and it follows from $(*)$ that $s \wedge t \in h(s)^C = h(s)$. Similarly $s \wedge t \in h(t)$, hence $s \wedge t \in h(s \wedge t)$.

$st \in h(st) \subseteq [st]$: Since $h(st) = (h(s)h(t))^C$, we have $st \in h(st)$. Now consider $r \in h(s)h(t)$. Then $r = qq'$, where $q \in h(s) \subseteq [s]$ and $q' \in h(t) \subseteq [t]$. Therefore $q \leq s$ and $q' \leq t$ are Gentzen provable, hence by ($\cdot$ right) $qq' \leq st$ is Gentzen provable, and so $r \in [st]$. It follows that $h(s)h(t) \subseteq [st]$, and since $[st]$ is closed, $h(st) \subseteq [st]$.

$s \backslash t \in h(s \backslash t) \subseteq [s \backslash t]$: Here $h(s \backslash t) = h(s) \backslash h(t) = \{q \in M(p) : h(s)\{q\} \subseteq h(t)\}$. Thus $q \in h(s \backslash t)$ implies $sq \in h(t) \subseteq [t]$, since we are assuming $s \in h(s)$. This means $sq \leq t$ is Gentzen provable, so by ($\backslash$ right) $q \in [s \backslash t]$. Therefore $h(s \backslash t) \subseteq [s \backslash t]$.

Suppose $h(t) \subseteq [q, q', r]$. Then $t \in h(t)$ implies $qtq' \leq r$ is Gentzen provable. For any $s' \in h(s) \subseteq [s]$ we have that $s' \leq s$ is Gentzen provable, so from ($\backslash$ left) we get that $s'(s\backslash t) \in [q, q', r]$. By $(*)$ it follows that $s'(s\backslash t) \in h(t)$ whenever $s' \in h(s)$, hence $h(s)\{s\backslash t\} \subseteq h(t)$. This implies $s\backslash t \in h(s)\backslash h(t) = h(s\backslash t)$.

The case for $s/t \in h(s/t) \subseteq [s/t]$ is similar.

Since we are assuming that h has been extended to a homomorphism from $\mathbf{T}_M$ to $\mathbf{L}$, we have $h(e) = e^{\mathbf{L}} = \{e\}^C$. Suppose $\{e\} \subseteq [q, q', r]$, then $qq' \leq r$ is Gentzen provable, and $e \in [q, q', r]$. Hence $(*)$ implies $e \in h(e)$. Finally, $h(e) \subseteq [e]$ holds since $\{e\} \subseteq [e]$.

The second statement is a simple consequence: if $e \in h(t)$ then $e \in [t]$ which means $e \leq t$ is Gentzen provable. ∎

Theorem 7.4. *For any $\mathcal{RL}$-term p the following statements are equivalent:*

(i) $\mathcal{RL} \models e \leq p$

(ii) $\mathbf{L}(p) \models e \leq p$

(iii) $e \leq p$ *is Gentzen provable.*

Proof. (i) implies (ii) by Lemma 7.2. Assuming (ii) holds, we have $h(e) \subseteq h(p)$. Since $e \in \{e\}^C = h(e)$, (iii) follows by Lemma 7.3. Finally, (iii) implies (i) by a standard soundness argument using the observation that $\mathcal{RL} \models G_{\mathcal{RL}}$. ∎

Since it was observed earlier that condition (iii) is decidable, and since any equation can be reduced to this form, the equational theory of $\mathcal{RL}$ is decidable. Okada and Terui [OT99] go on to prove that $\mathcal{RL}$ is generated by its finite members, and they also consider several subvarieties and expansions of $\mathcal{RL}$. For example, to decide inequalities for residuated 0, 1-lattices, one simply adds the two inequalities $x0y \leq z$ and $x \leq 1$ to $G_{\mathcal{RL}}$. In fact their results are formulated for what amounts to commutative residuated 0, 1-lattices, and the non-commutative case is only mentioned briefly at the end. To obtain a Gentzen system for $\mathcal{CRL}$, it suffices to add the so-called *exchange rule*

$$uxyv \leq w \Rightarrow uyxv \leq w \text{ corresponding to commutativity } xy = yx$$

or equivalently one replaces $\mathbf{T}_M$ by its commutative quotient algebra $\mathbf{T}_{CM}$. Other well known Gentzen quasi-inequalities are the *weakening rule*

$$uv \leq w \Rightarrow uxv \leq w \text{ corresponding to integrality } x \leq e$$

and the *contraction rule*

$$uxxv \leq w \Rightarrow uxv \leq w \text{ corresponding to } x \leq xx.$$

Adding any combination of these quasi-inequalities to $G_{\mathcal{RL}}$ gives a decision procedure for the corresponding subvariety of $\mathcal{RL}$ (as shown in [OK85], [OT99]). Note that in the presence of integrality, the contraction rule is equivalent to *idempotence* $x = xx$ and implies the identity $xy = x \wedge y$ that defines the subvariety $Br\mathcal{A}$ of Brouwerian algebras (since $xy \leq x \wedge y = (x \wedge y)(x \wedge y) \leq xy$).

The results above show that Gentzen systems are a versatile approach to proving decidability and can be adapted to cover other subvarieties of $\mathcal{RL}$.

The Finite Model Property and the Finite Embedding Property. A class $\mathcal{K}$ of algebras has the *finite model property* (FMP) if every identity that fails in some member of $\mathcal{K}$ also fails in some finite member of $\mathcal{K}$. The *strong finite model property* (SFMP) is defined in the same way, except for quasi-identities instead of identities. Since every identity is a quasi-identity, SFMP implies FMP. If we denote the finite members of $\mathcal{K}$ by $\mathcal{K}_F$, then FMP is equivalent to $\mathbf{V}(\mathcal{K}) = \mathbf{V}(\mathcal{K}_F)$, and SFMP is equivalent to $\mathbf{Q}(\mathcal{K}) = \mathbf{Q}(\mathcal{K}_F)$, where $\mathbf{Q}(\mathcal{K})$ is the quasivariety generated by $\mathcal{K}$. For a finitely based variety, FMP implies that the equational theory is decidable, and similarly for a finitely based quasivariety, SFMP implies that the quasi-equational theory is decidable. In both cases the argument is that there is an algorithm to enumerate all (quasi-)identities that are consequences of the finite basis, and there is an algorithm to enumerate all finite members of the given (quasi-)variety. Since we are assuming (S)FMP, any specific (quasi-)identity must either appear on the first list or fail in one of the algebras on the second list. Unlike a Gentzen system, the algorithm just outlined usually cannot be applied in practice, but the (S)FMP has proven valuable in establishing theoretical decidability results.

If a class $\mathcal{K}$ of structures is closed under finite products, then any universal formula is equivalent to a finite number of quasi-identities and negated equations. If, in addition, $\mathcal{K}$ contains one-element models, negated equations are not satisfied, hence $\mathcal{K}$ has a decidable universal theory if and only if it has a decidable quasi-equational theory. In particular, this argument shows that for varieties and quasivarieties of residuated lattices the decidability of the universal theory and the quasi-equational theory coincide.

Let $\mathbf{A}$ be an algebra, and B any subset of A. The partial subalgebra $\mathbf{B}$ of $\mathbf{A}$ is obtained by restricting the fundamental operations of $\mathbf{A}$ to the set B. A class $\mathcal{K}$ of algebras has the *finite embedding property* (FEP) if every finite partial subalgebra of a member of $\mathcal{K}$ can be embedded in a finite member of $\mathcal{K}$. In [Fe92] it is shown that for quasivarieties, FEP and SFMP are equivalent (see also [BvA1]).

In recent work of Blok and van Alten [BvA2], algebraic methods are used to show that $\mathcal{IRL}, \mathcal{ICRL}$, and $Br\mathcal{A}$ have the FEP, hence these varieties have decidable universal theories. Their very general results also apply to nonassociative residuated lattices and to various subreducts that are obtained when the lattice operations are omitted. Kowalski and Ono [KO] show that the variety of integral commutative residuated 0, 1-lattices is generated by its finite *simple* members, and the subvarieties defined by $x^{n+1} = x^n$ have the FEP.

Further Results. One should compare the decidability of residuated lattices with two important results about ℓ-groups.

Theorem 7.5. [HM79] *The variety of ℓ-groups has a decidable equational theory.*

Theorem 7.6. [GG83] *The (local) word problem for ℓ-groups is undecidable; i.e., there exists a finitely presented ℓ-group (with one relator) for which there is no algorithm that decides if two words are representatives of the same element.*

Corollary 7.7. *The quasi-equational theory of ℓ-groups is undecidable.*

This is also referred to as the undecidability of the global (or uniform) word problem.

Theorem 7.8.   (i) [Hi66] *The universal theory of abelian ℓ-groups is decidable.*

(ii) [We86] *The universal theory of abelian ℓ-groups is co-NP-complete.*

(iii) [Gu67] *The first-order theory of abelian ℓ-groups is hereditarily undecidable.* (See also [Bu85].)

Theorem 7.9. *The quasi-equational theory of residuated lattices is undecidable. The same holds for any subvariety that contains all powersets of finite monoids.*

Proof. Consider the class $\mathcal{K}$ of all monoid reducts of $\mathcal{RL}$. Note that a quasi-identity that uses only $\cdot$, holds in all residuated lattices if and only if it holds in $\mathcal{K}$ if and only if it holds in all subalgebras of $\mathcal{K}$.

Let $\mathcal{SG}$ be the variety of all semigroups. Any semigroup S can be embedded in some member of $\mathcal{K}$ follows. Embed S in a monoid $S_e = S \cup \{e\}$ and construct the powerset $\mathcal{P}(S_e)$, which is a complete residuated lattice with multiplication of complexes as product. The collection of singletons is closed under this operation and isomorphic to S_e.

On the other hand, $\mathcal{K} \subseteq \mathcal{SG}$, hence $\mathcal{SG}$ coincides with the class of all subalgebras of $\mathcal{K}$. But the quasi-equational theory of semigroups is undecidable, hence the same is true for residuated lattices.

For the second part of the theorem, we use the result of [GL84] that the quasi-equational theory of semigroups is recursively inseparable from the quasi-equational theory of finite semigroups.   ∎

Since $\mathcal{P}(S_e)$ is distributive, it follows that the variety of distributive residuated lattices has an undecidable quasi-equational theory. This was proved earlier by N. Galatos [Ga02] using the machinery of von Neumann n-frames. His result is somewhat stronger and also applies to many subvarieties not covered by the above argument.

Theorem 7.10. [Ga02] *The (local) word problem (and hence the quasi-equational theory) for any variety between $\mathcal{DRL}$ and $\mathcal{CDRL}$ is undecidable.*

Variety	Name	Eq. Theory	Word prob.	Univ. Th.
Residuated Lattices	$\mathcal{RL}$	FMP[OT99]		Und. 7.9
Commutative $\mathcal{RL}$	$\mathcal{CRL}$	FMP[OT99]		
Distributive $\mathcal{RL}$	$\mathcal{DRL}$		Und.[Ga02]	Und. 7.9
$\mathcal{CRL} \cap \mathcal{DRL}$	$\mathcal{CDRL}$		Und.[Ga02]	Undecidable
Modular $\mathcal{RL}$	$\mathcal{MRL}$	Und. 7.11	Undecidable	Undecidable
$\mathbf{V}$(Resid. Chains)	$\mathcal{RL}^C$			
Commutative $\mathcal{RL}^C$	$\mathcal{CRL}^C$			
Integral $\mathcal{RL}$	$\mathcal{IRL}$	FMP[OT99]	Decidable	FEP [BvA2]
Commutative $\mathcal{IRL}$	$\mathcal{CIRL}$	FMP[OT99]	Decidable	FEP [BvA1]
Cancellative $\mathcal{RL}$	$\mathcal{CanRL}$			
Comm. $\mathcal{CanRL}$	$\mathcal{CanCRL}$			
ℓ-groups	$\mathcal{LG}$	Dec.[HM79]	Und.[GG83]	Undecidable
Abelian ℓ-groups	$\mathbf{V}(\mathbb{Z})$	Decidable	Decidable	Dec.[Hi66]
Generalized BL Algs	$\mathcal{GBL}$			
Generalized MV Algs	$\mathcal{GMV}$			
Brouwerian Algs	$\mathcal{BrA}$	Decidable	Decidable	FEP[MT44]
Gen. Boolean Algs	$\mathcal{GBA}$	Decidable	Decidable	FEP

Table 2: (Un)decidability of some subvarieties of $\mathcal{RL}$

The next result shows that there are indeed varieties of residuated lattices with undecidable equational theory.

Theorem 7.11. *The variety of modular residuated lattices has an undecidable equational theory.*

Proof. Let $\mathbf{M}$ be a modular lattice and define $A = M \cup \{0, e, 1\}$ where $0 < e < x < 1$ for all $x \in M$. Then $\mathbf{A}$ is still modular (in fact in the same variety as M).

We define $\cdot$ on $\mathbf{A}$ by $x0 = 0 = 0x$, $xe = x = ex$ and $xy = 1$ if $x, y \neq 0, e$. It is easy to check that $\cdot$ is associative and residuated, hence the $\vee, \wedge$-equational theory of modular residuated lattices coincides with the equational theory of modular lattices. Since modular lattices have an undecidable equational theory ([Fr80]), the same is true for modular residuated lattices.   ∎

Table 2 summarizes what is currently known about the decidability of various subvarieties of $\mathcal{RL}$.

8 Open Problems

Problem 8.1. *Is every lattice a subreduct of* a commutative *cancellative residuated lattice (see Theorem 5.3)?*

Problem 8.2. *Are there commutative, cancellative, distributive residuated lattices that are not in* $Can\mathcal{RL}^C$? *In the noncommutative case the 2-generated free ℓ-group is an example.*

Problem 8.3. *Does* $Can\mathcal{RL}^C$ *have a decidable equational theory?*

Problem 8.4. *Is there a Weinberg-type description of free algebras in* $Can\mathcal{RL}^C$? *See e.g. Powell and Tsinakis [PT89].*

Problem 8.5. *Theorem 6.2 proves that the only two atoms in* $\mathbf{L}(\mathcal{RL})$ *that are cancellative and commutative are* $\mathcal{V}(\mathbb{Z}^-)$ *and* $\mathcal{V}(\mathbb{Z})$. *Are there any other cancellative atoms in* $\mathbf{L}(\mathcal{RL})$? *It follows from Theorem 6.1 that if this is the case then they are generated by nonintegral residuated lattices.*

Problem 8.6. *Are there uncountably many atoms in* $\mathbf{L}(\mathcal{RL})$ *that satisfy the commutative identity or the identity* $x^2 = x^3$?

Problem 8.7. *Kowalski and Ono [KO00] have shown that there are no nontrivial splitting varieties in the lattice of subvarieties of commutative integral residuated $0, 1$-lattices. What is the situation for* $\mathbf{L}(\mathcal{RL})$ *or some of the ideals determined by subvarieties of* $\mathcal{RL}$?

References

[AF88]   M. Anderson and T. Feil, *Lattice-Ordered Groups: an introduction.* D. Reidel Publishing Company, 1988.

[BCGJT] P. Bahls, J. Cole, N. Galatos, P. Jipsen, C. Tsinakis, *Cancellative residuated lattices.* preprint 2001.

[Be74]   J. Berman, *Homogeneous lattices and lattice-ordered groups.* Colloq. Math. **32** (1974), 13–24.

[Bi67]   G. Birkhoff, *Lattice Theory.* (3rd ed), Colloquium Publications **25**, Amer. Math. Soc., 1967.

[BvAl]   W. J. Blok and C. J. van Alten, *The finite embeddability property for residuated lattices, pocrims and BCK-algebras.* Preprint.

[BvA2]   W. J. Blok and C. J. van Alten, *The finite embeddability property for partially ordered biresiduated integral groupoids.* Preprint.

[BT]   K. Blount and C. Tsinakis, *The structure of Residuated Lattices.* Preprint.

[BJ72]   T. S. Blyth and M. F. Janowitz, *Residuation Theory.* (1972) Pergamon Press.

[BS81]   S. Burris and H. P. Sankappanavar, *A Course in Universal Algebra.* Springer Verlag (1981); online at http://www.thoralf.uwaterloo.ca/

[Bu85]   S. Burris, *A simple proof of the hereditary undecidability of the theory of lattice-ordered abelian groups.* Alg. Univ. **20** (1985), no. 3, 400-401.

[Ch58]   C. C. Chang, *Algebraic analysis of many valued logics.* Trans. AMS **88** (1958), 467-490.

[Ch59]   C. C. Chang, *A new proof of the completeness of the Lukasiewicz axioms.* Trans. AMS **93** (1959), 74-80.

[COM00] R. Cignoli, I. D'Ottaviano and D. Mundici, *Algebraic foundations of many-valued reasoning.* Trends in Logic—Studia Logica Library **7** (2000); Kluwer Acad. Publ., Dordrecht.

[Co02]   J. Cole, *Examples of Residuated Orders on Free Monoids.* In these Proceedings, 205-212.

[Di38]   R. P. Dilworth, *Abstract residuation over lattices.* Bull. AMS **44** (1938), 262-268.

[Di39]   R. P. Dilworth, *Non-commutative residuated lattices.* Trans. AMS **46** (1939), 426-444.

[Fe92]   I. M. A. Ferreirim, *On varieties and quasivarieties of hoops and their reducts.* Ph. D. thesis (1992); University of Illinois at Chicago.

[Fr80]   R. Freese, *Free modular lattices.* Trans. AMS **261** (1980) no. 1, 81-91.

[Fu63]   L. Fuchs, *Partially Ordered Algebraic Systems.* (1963) Pergamon Press.

[Ga00]   N. Galatos, *Selected topics on residuated lattices.* Qualifying paper (2000), Vanderbilt University.

[Ga02]   N. Galatos, *The undecidability of the word problem for distributive residuated lattices.* In these Proceedings, 231-243.

[GG83]   A. M. W. Glass and Y. Gurevich, *The word problem for lattice-ordered groups.* Trans. AMS **280** (1983) no. 1, 127-138.

[GH89]   A. M. W. Glass and W. C. Holland (editors), *Lattice-Ordered Groups*. Kluwer
          Academic Publishers, 1989, 278–307.

[GU84]   H.P. Gumm and A. Ursini, *Ideals in universal algebras*. Alg. Univ. **19** (1984)
          no. 1, 45-54.

[Gu67]   Y. Gurevich, *Hereditary undecidability of a class of lattice-ordered Abelian
          groups*. (Russian) Algebra i Logika Sem. **6** (1967) no. 1, 45-62.

[GL84]   Y. Gurevich and H. R. Lewis, *The word problem for cancellation semigroups
          with zero*. Journal of Symbolic Logic **49** (1984), 184-191.

[Ha98]   P. Hájek, *Metamathematics of Fuzzy Logic*. **4** Trends in Logic; (1998) Kluwer
          Acad. Publ., Dordrecht.

[HRT]   J. Hart, L. Rafter and C. Tsinakis, *The structure of Commutative Residuated
          lattices*. Intern'l Jour. of Alg. and Comput.; to appear.

[Hi66]   N. G. Hisamiev, *Universal theory of lattice-ordered Abelian groups*. (Russian)
          Algebra i Logika Sem. **5** (1966) no. 3, 71-76.

[HM79]   W. C. Holland and S. H. McCleary, *Solvability of the word problem in free
          lattice-ordered groups*. Houston J. Math. **5** (1979) no. 1, 99-105.

[KO]   T. Kowalski and H. Ono, *Residuated Lattices*. Preprint.

[KO00]   T. Kowalski and H. Ono, *Splittings in the variety of residuated lattices*. Alg.
          Univ. bf 44 (2000) no. 3-4, 283-298.

[Kr24]   W. Krull, *Axiomatische Begründung der algemeinen Idealtheorie*, Sitzungs-
          berichte der physikalischmedizinischen Societät zu Erlangen **56** (1924), 47-
          63.

[McK96]   R. McKenzie, *An algebraic version of categorical equivalence for varieties
          and more general algebraic categories*. In *Logic and Algebra*, A. Ursini, P.
          Aglianò, Eds.; (1996) Marcel Dekker, 211-244.

[MT44]   J. C. C. McKinsey and A. Tarski, *The algebra of topology*. Ann. of Math. (2)
          45 (1944), 141-191.

[Mu86]   D. Mundici, *Interpretation of AF C*-algebras in Lukasiewicz sentential cal-
          culus*. J. Funct. Anal. **65** (1986) no. 1, 15-63.

[NPM99]   V. Novák, I. Perfilieva, and J. Mockor, *Mathematical Principles of Fuzzy
          Logic*. **517** Kluwer Intern'l Ser. in Engin. and Comp. Sci. (1999), Kluwer
          Acad. Publ., Dordrecht.

[OK85] H. Ono and M. Komori, *Logics without the contraction rule.* Journal of Symbolic Logic **50** (1985) 169-201.

[OT99] M. Okada and K. Terui, *The finite model property for various fragments of intuitionistic linear logic.* Journal of Symbolic Logic, **64** (2) (1999), 790-802.

[PT89] W. B. Powell and C. Tsinakis, *Free products in varieties of lattice-ordered groups.* In "Lattice-Ordered Groups"; A. M. W. Glass and W. C. Holland, Eds.; (1989) Kluwer Acad. Publ., 278-307.

[Ur72] A. Ursini, *Sulle variet' di algebre con una buona teoria degli ideali.* (Italian) Boll. Un. Mat. Ital. (4) **6** (1972), 90-95.

[vA1] C. J. van Alten, *Representable biresiduated lattices.* Jour. of Alg.; to appear.

[vA2] C. J. van Alten, *The termwise equivalence of the varieties of ℓ-group cones and generalized cancellative hoops.* Preprint.

[Wa37] M. Ward, *Residuation in structures over which a multiplication is defined.* Duke Math. Jour. **3** (1937), 627-636.

[Wa38] M. Ward, *Structure Residuation.* Annals of Math., 2nd Ser. **39** (3) (1938), 558-568.

[Wa40] M. Ward, *Residuated distributive lattices.* Duke Math. Jour. **6** (1940), 641-651.

[WD38] M. Ward and R. P. Dilworth, *Residuated lattices.* Proc. Nat. Acad. of Sci. **24** (1938), 162-164.

[WD39] M. Ward and R. P. Dilworth, *Residuated lattices.* Trans. AMS **45** (1939), 335-354.

[We86] V. Weispfenning, *The complexity of the word problem for abelian ℓ-groups,* Theor. Comp. Sci. **48** (1986) no. 1, 127-132.

Department of Mathematics, Vanderbilt University, Nashville, TN 37240
peter.jipsen@vanderbilt.edu
constantine.tsinakis@vanderbilt.edu

MV-Algebras and Abelian ℓ-Groups: a Fruitful Interaction

Vincenzo Marra and Daniele Mundici [1]

Dedicated to Paul Conrad,
on the occasion of his 80th birthday:
his work has greatly influenced two mathematicians
of contiguous generations,
with origins in a specialty different from his own.

ABSTRACT.   Introduced by Chang in the late fifties, MV-algebras stand to Lukasiewicz's infinite-valued propositional logic as boolean algebras stand to the classical propositional calculus. As stated by Chang in his original paper, "MV is supposed to suggest many-valued logics ... for want of a better name". The name has stuck. After some decades of relative quiescency, MV-algebras are today intensely investigated. On the one hand, these algebras find applications in such diverse fields as error-correcting feedback codes and logic-based control theory. On the other hand, MV-algebras are interesting mathematical objects in their own right. The main aim of this paper is to show that the interaction between MV-algebras and lattice-ordered abelian groups — including the time-honored theory of magnitudes — has much to offer, not only to specialists in these two fields, but also to people interested in the fan-theoretic description of toric varieties, and in the K_0-theory of AF C^*-algebras.

[1] We are very grateful to Jorge Martinez and his collaborators for giving us the opportunity to entertain a fruitful interaction with several outstanding members of the ℓ-group community during a meeting in Gainesville.

J. Martínez (ed.), Ordered Algebraic Structures, 57–88.
© 2002 *Kluwer Academic Publishers*.

1　MV-algebras

An *MV-algebra* is a structure $\langle A, \oplus, \neg, 0 \rangle$, where $\langle A, \oplus, 0 \rangle$ is a commutative monoid with neutral element 0, and the operation $\neg$ satisfies the equations $\neg\neg x = x$, $x \oplus \neg 0 = \neg 0$ and, characteristically,

$$\neg(\neg x \oplus y) \oplus y = \neg(\neg y \oplus x) \oplus x. \tag{1}$$

(A comprehensive account on the subject may be found in [14]). The real unit interval $[0, 1]$ endowed with negation $\neg x = 1 - x$ and truncated addition $x \oplus y = \min(1, x + y)$ is an MV-algebra. The defining equations of MV-algebras, then, express *some* of the properties of this concrete model. Direct inspection, for instance, shows that equation (1) states that the maximum operation over $[0, 1]$ is commutative. The following first deep result states that, in fact, the above equations express *all* equational properties of $[0, 1]$:

Theorem 1.1 [Chang's Completeness Theorem] *MV-algebras are generated, as an equational class, by* $\langle [0, 1], \oplus, \neg, 0 \rangle$. *In symbols,* $\mathcal{MV} = \mathrm{HSP}[0, 1]$.

In logical terminology, the above result says that a formula is a theorem of Lukasiewicz's infinite-valued propositional logic if and only if it is true (i. e., evaluates constantly to 1) when interpreted in $\langle [0, 1], \oplus, \neg, 0 \rangle$. Upon adding $x \oplus x = x$ to the defining equations of MV-algebras one precisely obtains boolean algebras. Thus, Chang's theorem is a generalization of the completeness of classical propositional logic stating that $\{0, 1\}$ generates the variety of boolean algebras.

Wajsberg claimed to have a proof of Theorem 1.1 as early as 1935 ([47]). Tarski endorsed Wajsberg's claim ([45]), but the proof was never published. Later on, several proofs appeared in print. Remarkably, they span a rather wide range of techniques. The nature of the first published proof ([43]) is largely syntactical, although Rose and Rosser made decisive use of some results of Motzkin on linear inequalities. Chang's own proof ([11]) hinges on quantifier elimination for divisible ordered abelian groups. Cignoli's paper ([12]) exploits results on the variety of abelian ℓ-groups. Panti's proof ([42]) is geometrical, and rests on a technique first introduced by De Concini and Procesi in the context of toric desingularizations ([17]). The paper [13] may well be the first completely elementary proof of MV-completeness, and is a simplification of Chang's strategy.

A second significant fact is that free MV-algebras admit a transparent concrete realization. A continuous map $f : [0, 1]^n \to [0, 1]$ is a *McNaughton function* (of n variables) if and only if it is piecewise linear and each linear piece has integral coefficients. [2] We denote by $\mathcal{M}_n$ the set of McNaughton functions of n variables. The constant functions **1** and **0** belong to $\mathcal{M}_n$. Given $f, g \in \mathcal{M}_n$, define $f \oplus g = \min(f + g, \mathbf{1})$ (pointwise

---

[2] 'Linear' is meant here in the affine sense, that is, f is not necessarily homogeneous.

minimum and addition). Then it is promptly seen that $f \oplus g \in \mathcal{M}_n$. Further, define $\neg f = 1 - f$ (pointwise subtraction). Again, $\neg f \in \mathcal{M}_n$. A simple verification shows that $\langle \mathcal{M}_n, \oplus, \neg, 0 \rangle$ is an MV-algebra. Now let FORM_n be the set of terms over n variables in the language of MV-algebras. Define a function $\mathcal{E} : \mathrm{FORM}_n \to \mathcal{M}_n$ associating to any such term w the McNaughton function $\mathcal{E}(w)$ obtained by evaluating variables over the unit interval $[0, 1]$ and interpreting operation symbols as in the definition of $\langle \mathcal{M}_n, \oplus, \neg, 0 \rangle$. [3] In 1951, McNaughton proved the following result ([32]):

Theorem 1.2 [McNaughton's Representation Theorem] *For each function $f \in \mathcal{M}_n$ there exists a term $w \in \mathrm{FORM}_n$ such that $\mathcal{E}(w) = f$.*

In other words, the map $\mathcal{E}$ is surjective. It is convenient to note here that Chang's completeness theorem states the injectivity of (a certain map obtainable from) $\mathcal{E}$. More precisely, define the (Lindenbaum) equivalence relation on FORM_n as follows: $w \equiv v$ if and only if w and v may be rewritten into one another through substitution of equal terms using the above defining equations of MV-algebras. Then, by Chang's completeness theorem, the fibers of $\mathcal{E}$ are exactly the equivalence classes of this relation, and $\mathcal{E}/\equiv$ is injective. This yields the following representation theorem for finitely generated free MV-algebras: [4]

Corollary 1.3. $\langle \mathcal{M}_n, \oplus, \neg, 0 \rangle$ *is the free MV-algebra over n generators.*

The original proof of Theorem 1.2 involved a *reductio ad absurdum*. Thus, as McNaughton himself remarked in [32], it shed no light on how one might go about devising an algorithm to construct $w \in \mathrm{FORM}_n$ such that $\mathcal{E}(w) = f$, given $f \in \mathcal{M}_n$. In fact, *prima facie* it is not even clear that such an algorithm exists at all — an admittedly pedantic but spelled-out definition of a McNaughton function sounds as follows: $f \in \mathcal{M}_n$ if and only if there exists a finite set $\{l_i\}$ of linear $[0,1]$-valued n-variables functions with integral coefficients, together with a *selection function* $\chi : x \in [0,1]^n \mapsto \chi_x \in \{l_i\}$ such that for each $x \in [0,1]^n$, $f(x) = \chi_x(x)$. By definition, then, χ is an infinitary object. However, such an algorithm exists (see [36] for a proof) and the map $(\mathcal{E}/\equiv)^{-1}$ is effectively computable. [5]

As noted by Chang in [10], every MV-algebra A carries a natural lattice structure which we now define explicitly. Let us agree to write 1 as a shorthand for $\neg 0$. For $x, y \in A$, define Łukasiewicz's conjunction $x \odot y = \neg(\neg x \oplus \neg y)$. Now define $x \wedge y = (x \odot \neg y) \oplus y$ and $x \vee y = \neg(\neg x \wedge \neg y)$. It is not hard to show that $\langle A, \vee, \wedge \rangle$ is a bounded distributive lattice with 0 and 1 as top and bottom elements, respectively.

---

[3] More formally, each variable symbol X_i, $i = 1, \ldots, n$, is transformed by $\mathcal{E}$ into the ith coordinate function $\pi_i : [0,1]^n \to [0,1]$, and, by induction on subterms of w, each operation symbol in w is interpreted as pointwise application of its corresponding operation $\neg$ or $\oplus$ in $[0,1]$.

[4] Generalization to an arbitrary number of generators is straightforward.

[5] Besides effective computability, some results on the complexity of McNaughton functions are derivable from [36]. Due to lack of space, in this paper we refrain from discussing these results and their subsequent developments in the literature.

In particular, the underlying lattice orders of $\mathcal{M}_n$ and $[0,1]$ respectively coincide with pointwise order and with natural order.

We now drastically enlarge our stock of examples of MV-algebras by the following simple construction. Let G be an abelian ℓ-group with strong order unit u. For x, $y \in G$, define $\neg x = u - x$ and $x \oplus y = (x + y) \wedge u$. Then

$$\Gamma(G, u) = \langle [0, u], \oplus, \neg, 0 \rangle$$

is an MV-algebra whose lattice order coincides with the restriction order induced on $[0, u]$ from G. Further, for any morphism [6] $\eta \colon (G, u) \to (G', u')$ let us define $\Gamma(\eta)$ to be the restriction of η to $[0, u]$. Then Γ is a functor from the category $\mathcal{A}_u$ of abelian ℓ-groups with strong order unit, to the category $\mathcal{MV}$ of MV-algebras.

All MV-algebras introduced so far are images of some ℓ-group under Γ, for instance, $\Gamma(\mathbb{R}, 1) = [0, 1]$. That this is no accident is a third major result on MV-algebras:

Theorem 1.4 [Basic Categorical Equivalence] *The functor Γ is a categorical equivalence between $\mathcal{A}_u$ and $\mathcal{MV}$. In particular, to each MV-algebra M there corresponds a unique (up to isomorphism) abelian ℓ-group G with strong order unit u such that $\Gamma(G, u) = M$.*

The totally ordered case of this theorem is due to Chang ([11]) and was one of his main tools in proving the completeness theorem. The general case is due to the second author ([33], [14]).

Theorem 1.4 has the surprising consequence that, while abelian ℓ-groups with strong order unit are not even finitely axiomatizable in first-order predicate logic, [7] they are *equationally* definable up to categorical equivalence. Also notice that Γ^{-1} does *not* carry free MV-algebras into free abelian ℓ-groups — this is obvious, but it should not be overlooked. Of course, the image of a free MV-algebra under Γ^{-1} does exhibit freeness properties in the category $\mathcal{A}_u$. In loose but suggestive terms, one may say that MV-algebras are affine versions of abelian ℓ-groups, or that abelian ℓ-groups are linear homogeneous versions of MV-algebras.

Specialization of universal algebraic notions to MV-algebras is quite straightforward. An *ideal* J of an MV-algebra A is a subset containing 0, closed under $\oplus$ and downward closed (that is, $x \in J$, $y \in A$ and $y \leq x$ entail $y \in J$). Ideals, then, are precisely kernels of homomorphisms. In logical terms, ideals correspond to theories in Łukasiewicz's infinite-valued propositional calculus. Principal ideals correspond to finitely axiomatized theories. Following established terminology, let us call an MV-algebra *finitely presented* if and only if it is isomorphic to $\mathcal{M}_n/J$ for some principal

---

[6] By definition, η is a group homomorphism that also preserves the lattice structure and the strong order unit.

[7] By Gödel's completeness theorem, first-order logic is compact. Hence, in this logic one cannot faithfully express the crucial archimedean property of strong order units.

ideal J. An MV-algebra is *archimedean* if and only if the intersection of its maximal ideals is $\{0\}$.

The fourth important MV-algebraic fact that we quote in this section is the functional representation of archimedean MV-algebras.

Theorem 1.5 [Archimedean Representation] *Suppose A is an n-generated MV-algebra. Then A is archimedean if and only if it is isomorphic to the algebra of all McNaughton functions of n variables restricted to the zeroset $\mathcal{Z}_J$ of some ideal J of $\mathcal{M}_n$, where*

$$\mathcal{Z}_J = \{x \in [0,1]^n \mid \text{ for all } f \in J \, , \, f(x) = 0\}.$$

Fair attribution of this result is not an easy matter. We refer to [14] for a self-contained proof and pointers to the relevant literature. In any case, it turns out that Theorem 1.5, while of paramount importance, is just the tip of an iceberg. To plunge deeper below the surface, some terminology from rational polyhedral geometry is needed. As references on the subject we quote [25] and [52]. A finite set of rational points $\{v_0, \ldots, v_m\} \subseteq \mathbb{Q}^n$ is *affinely independent* if and only if the set of rational vectors $\{v_1 - v_0, \ldots, v_m - v_0\}$ is linearly independent over $\mathbb{R}$. A *rational simplex* is the convex hull of a finite set of affinely independent points in $\mathbb{Q}^n$, its *vertices*. A *face* of a simplex is the convex hull of a subset of its vertices. A finite set S of simplices is a *rational simplicial complex* if and only if together with a simplex it contains all its faces, and any two simplices intersect in a common face. The *support* of S is the union of all its simplices, denoted $|S|$. A *rational polyhedron* is the convex hull of a finite set of points in $\mathbb{Q}^n$. Any (rational) polyhedron can be triangulated, i.e., it is the support of some (rational) simplicial complex. More generally, a *rational polyhedral set* is the support of a rational simplicial complex.

It can be shown that J is principal if and only if $\mathcal{Z}_J$ is a rational polyhedral set in $[0,1]^n$. While not exactly trivial, this fact has nothing to do with MV-algebras in particular, being an easy consequence of a standard result in polyhedral geometry: a subset of $\mathbb{R}^n$ is a (rational) polyhedron if and only if it is bounded and it is the intersection of finitely many affine (rational) half-spaces. On the contrary, the next statement is genuinely MV-algebraic — in fact, it is the last piece of the classical theory of MV-algebras that we illustrate here ([49]).

Theorem 1.6 [Wójcicki's Theorem] *Every finitely presented MV-algebra is archimedean.*

Corollary 1.7. *An n-generated MV-algebra is finitely presented if and only if it is isomorphic to the algebra of McNaughton functions of n variables restricted to some rational polyhedral set in $[0,1]^n$.*

2   Abelian ℓ-groups

In the light of Theorem 1.4, it comes as no surprise that significant portions of the literature on MV-algebras and abelian ℓ-groups [8] have essentially the same mathematical content. Except for the Γ functor, this is certainly true for the body of MV-algebraic knowledge expounded in the previous section. We synoptically indicate the corresponding ℓ-group-theoretical results.

Notation. By an ℓ-*function* we mean a piecewise linear homogeneous continuous map $f : \mathbb{R}^n \to \mathbb{R}$ such that each of its pieces has integral coefficients. We write $\mathcal{F}_n$ to denote the set of all ℓ-functions of n variables. When equipped with pointwise addition and order, $\mathcal{F}_n$ is an abelian ℓ-group.

(i) Chang's Completeness Theorem ⊣⊢ The variety of abelian ℓ-groups is generated by $\mathbb{Z}$. [9] (This is due to Weinberg, [48]).

(ii) McNaughton's Representation Theorem ⊣⊢ Every ℓ-function of n variables belongs to the free abelian ℓ-group over n generators. (This is due to Beynon, [4]).

(iii) Corollary 1.3 ⊣⊢ $\mathcal{F}_n$ is the free abelian ℓ-group over n generators.

(iv) Archimedean Representation ⊣⊢ An n-generated ℓ-group is archimedean if and only if it is isomorphic to the ℓ-group of all ℓ-functions of n variables restricted to the zeroset $\mathcal{Z}_J$ of some ℓ-ideal J of $\mathcal{F}_n$. (We are not aware of a specific reference for abelian ℓ-groups, but strongly related work is due to Yosida ([51]) in the case of vector lattices with strong order unit).

(v) Wójcicki's Theorem ⊣⊢ Every finitely presented abelian ℓ-group is archimedean. (This is due to Baker, [3]).

Entr'acte, part I. Just as some material from polyhedral geometry was needed for the statement of Corollary 1.7, here we need some notions from the theory of fans to state its analogue for ℓ-groups. We refer to [25] for details. [10] Let $\{\vec{v}_1, \ldots, \vec{v}_m\} \subseteq \mathbb{Z}^n$ be a finite set of integral vectors, and assume they are linearly independent over $\mathbb{R}$. The *rational simplicial cone* σ generated by $\{\vec{v}_1, \ldots, \vec{v}_m\}$ is their positive linear hull over $\mathbb{R}$, i. e.,

$$\sigma = \langle \vec{v}_1, \ldots, \vec{v}_m \rangle = \{p_1 \vec{v}_1 + \cdots + p_m \vec{v}_m \mid 0 \le p_1, \ldots, p_m \in \mathbb{R}\}.$$

---

[8] We adopt the traditional abbreviation for abelian lattice-ordered groups.

[9] Or equivalently, the variety is generated by $\mathbb{R}$, the additive groups of real numbers with natural order.

[10] The monograph [25] gives a comprehensive account of the celebrated vocabulary translating the language of fans into that of toric varieties, and vice versa.

We often omit the adjective 'rational' — all simplicial cones appearing in this paper are of this type. A *face* of a simplicial cone $\sigma = \langle \vec{v}_1, \ldots, \vec{v}_m \rangle$ is the simplicial cone spanned by a subset of $\{\vec{v}_1, \ldots, \vec{v}_m\}$. The *dimension* of a simplicial cone σ in $\mathbb{R}^n$ is the dimension (as a real vector space) of the linear subspace of $\mathbb{R}^n$ spanned by σ. A *rational simplicial fan*, in short, a *fan*, is a finite set Σ of simplicial cones such that every face of every cone in Σ belongs to Σ, and any two cones in Σ intersect in a common face. The *support* of Σ, denoted $|\Sigma|$, is the union of all its cones. Continuing our synopsis we have

6. Corollary 1.7 ⊣⊢ An n-generated abelian ℓ-group is finitely presented if and only if it is isomorphic to the ℓ-group of ℓ-functions of n variables restricted to the support of some fan in $\mathbb{R}^n$. (Baker-Beynon)

Notwithstanding these close matchings, the reader should not be misled into thinking that MV-algebras and abelian ℓ-groups share one and the same theory: as a matter of fact, the lack of a strong order unit has some rather far-reaching consequences. For instance, it is not always possible to represent archimedean ℓ-groups as spaces of real-valued functions ([26], Chapter 5). We now give a less well-known and perhaps deeper example.

In the following arrow-theoretic definitions, homomorphisms are morphisms in the category of abelian ℓ-groups; hence, kernels are ℓ-ideals. An abelian ℓ-group P is *projective* if and only if for every homomorphism $h : P \to G_1$ and for every surjective homomorphism $s : G_2 \to G_1$, there exists a homomorphism $k : P \to G_2$ such that $s \circ k = h$. Similarly, upon restriction to the appropriate categories, one can speak of projective MV-algebras or, equivalently, projective abelian ℓ-groups with strong order unit. However, while it is a result of Beynon that a finitely generated abelian ℓ-group is projective if and only if it is finitely presented ([5]), we shall now describe a finitely presented non-projective MV-algebra.

To this purpose, let us first recall that if $f : A \to B$ is a morphism in an arbitrary category, then a morphism $g : B \to A$ in the same category is called a *retraction* of f if and only if $g \circ f$ is the identity (on A). In this case, f is called a *section* of g.

Consider now the free MV-algebra over two generators. By Corollary 1.3, this coincides with $\mathcal{M}_2$. Let J be the ideal of all functions of $\mathcal{M}_2$ vanishing over the frontier of the unit square $[0,1]^2$, i.e., the closed polygonal line T of vertices $(0,0)$, $(0,1)$, $(1,1)$, $(1,0)$, $(0,0)$. Then, by Corollary 1.7, the MV-algebra P defined by

$$P = \mathcal{M}_2/J$$

is finitely presented. We claim that P is not projective. For suppose it is (*absurdum hypothesis*). It follows that the canonical quotient homomorphism $q \colon \mathcal{M}_2 \to P$ is a retraction of some (necessarily injective) homomorphism $k : P \to \mathcal{M}_2$. Let **MaxSpec** be the contravariant functor from MV-algebras to their maximal spectra with the hull-kernel topology. Then **MaxSpec** $(\mathcal{M}_2)$ is homeomorphic to $[0,1]^2$ with the natural topology, and **MaxSpec** (P) is homeomorphic to the polygonal line T, a closed subspace of $[0,1]^2$ which is not simply connected. (Recall that a topological space E is *simply connected* if and only if whenever $C \subseteq E$ is a homeomorphic image of the circle, then C is continuously contractible to a point of E, without leaving E). Let **MaxSpec** $(k) = r$, and **MaxSpec** $(q) = i$. Then, by contravariance, $i : T \to [0,1]^2$ is the inclusion map, and $r : [0,1]^2 \to T$ is a retraction of i. In topological terminology, T is a deformation retract of $[0,1]^2$. Since $[0,1]^2$ is simply connected while T is not, a standard homotopy argument shows that no such r can exist. Namely, let $\pi_1(X)$ denote the fundamental group of the pathwise connected topological space X. Then $\pi_1([0,1]^2) = \{0\}$ is the trivial group, and $\pi_1(T) = \mathbb{Z}$. Since π_1 is covariant, it follows that if $i : T \to [0,1]^2$ were to admit a retraction, then $\pi_1(i) : \mathbb{Z} \to \{0\}$ would have to be injective, a contradiction. Hence, P is not projective, and our claim is settled.

Of course, in lieu of this functorial argument one could also directly use Brouwer's Fixed Point Theorem to prove that the retraction r cannot exist. A topological space X has the fixed point property if and only if every homeomorphism $h : X \to X$ fixes a point, i. e., there is $x \in X$ such that $h(x) = x$. It is not hard to show that retractions preserve the fixed point property of topological spaces. But $[0,1]^2$ has the fixed point property by Brouwer's classical result, while T does not: rotation of T through an angle of $\pi/2$ fixes none of its points.

Whether one uses fundamental groups or fixed points is largely a matter of taste — the point is that, in sharp contrast to the case of abelian ℓ-groups, a study of projectivity in MV-algebras immediately entails delicate topological questions concerning homotopy invariants. We may add that while we were able to consider general homeomorphisms in showing that projectivity did *not* hold in a special case, to prove positive statements one has to restrict attention to the appropriate morphisms in the category of maximal spectra. These are *McNaughton homeomorphisms*, i. e., maps

$$h : \mathbf{MaxSpec}\,(A) \longrightarrow \mathbf{MaxSpec}\,(B),$$

whose scalar components are McNaughton functions. Likewise, the appropriate retractions are *McNaughton retractions*. To the best of our knowledge, no satisfactory characterization of projective MV-algebras or projective abelian ℓ-groups with strong order unit is known.

3   Commensurability and Starring

Throughout this section, let G be an abelian ℓ-group, and

$$G^* = G^+ - \{0\}.$$

If $S \subseteq G$, we denote by $[[S]]$, $[S]$ and $\langle S \rangle$ the submonoid, the subgroup and the ℓ-subgroup of G generated by S, respectively. Given $a_1, a_2, f \in G^*$, let us write $f \mid a_1$ if and only if there exists $n \in \mathbb{N} = \{1, 2, 3, \ldots\}$ such that $a_1 = nf$ if and only if $a_1 \in [[f]]$. The set G^* is partially ordered by $|$. If $m_1 a_1 = m_2 a_2$ for some $m_1, m_2 \in \mathbb{N}$, then there exists a unique greatest common divisor $d \in G^*$ with respect to the divisibility relation $|$, that is, a greatest lower bound of $\{a_1, a_2\}$ in the partially ordered set $\langle G^*, | \rangle$. Of course, one can easily prove this statement applying Euclid's technique of successive subtractions. Thereby, one obtains as an extra bonus an explicit construction of d showing that $d \in [[\{a_1, a_2\}]]$. We remark in passing that this 'explicit construction' is still very far from effective computability — just think of the special case $G = \mathbb{R}$. The following definition is a generalization of classical commensurability to abelian ℓ-groups.

Definition 3.1. *Let $P \subseteq G^*$ be finite. Given another finite subset $Q \subseteq G^*$, let us write $Q \leq P$ if and only if $P \subseteq [[Q]]$; we say that Q refines P. Further, let us say that P is commensurable in G if and only if there exists a finite set $U \subseteq G^*$ such that $U \leq P$ and $[[U]]$ is free (as a monoid) over U (equivalently, U is linearly independent in the $\mathbb{Z}$-module G; equivalently, $[U]$ is a free abelian group with U as a basis). We call U a set of commensurators for P. If every finite $P \subseteq G^*$ is commensurable, G is called ultrasimplicial. (Reasons for adopting this terminology are given after Definition 5.8).*

The question arises, which abelian ℓ-groups are ultrasimplicial. The answer is that they all are ([31]), and we now attempt to trace a history of this result.

The case $G = \mathbb{Z}$ is due to Euclid ([24], Book VII, Proposition 2, and Book X, Proposition 3). In this case, of course, everything is effectively computable. The totally ordered case was proved by Elliott in [22] using completely elementary means. The proof again employs successive subtractions, and is constructive. Using geometric techniques, the second author proved in [35] that free abelian ℓ-groups are ultrasimplicial. In [29], Theorem 3(ii), Handelman claims that the ultrasimplicial property is preserved under quotients. Together with the above mentioned result of [35], this would immediately yield that every abelian ℓ-group is ultrasimplicial. Unfortunately, Handelman's proof has a gap — in our terminology, he implicitly assumes that whenever P is linearly dependent and commensurable, it is possible to find a set of commensurators U such that $|U| < |P|$ (where $|S|$ denotes the cardinality of a set S). This is true in the totally ordered case, but false in general (see [31], Example on p. 874). The second author in a joint paper with Panti constructively proved that every 3-generated abelian ℓ-group is ultrasimplicial ([39]). Results for other classes of ℓ-groups are also obtained in [38]. Reverting to an algebraic elementary approach, the first author proved the general result in [31], independently of any previous work.

The key tool for the construction of commensurators in abelian ℓ-groups will be introduced in Definition 3.2 below.

Notation. If V is a finite set, we denote by $\mathbf{P}(V)$ the set of subsets of V. An *abstract simplicial complex* on V is a set $\mathcal{S}_V \subseteq \mathbf{P}(V)$ such that $A \in \mathcal{S}_V$ and $B \subseteq A$ entail

$B \in \mathcal{S}_V$. A simplex $S \in \mathcal{S}_V$ is said to be *k-dimensional*, or a *k-simplex*, whenever $|S| = k + 1$.

Definition 3.2 [Alexander's Starring for ℓ-groups] *For any subset* $P = \{p_1, \ldots, p_m\}$ *of* G^* *let the abstract simplicial complex* $\mathcal{S}_P \subseteq \mathbf{P}(P)$ *be defined by stipulating that for all* $S \in \mathbf{P}(P)$,

$$S \in \mathcal{S}_P \quad \text{iff} \quad \bigwedge S > 0.$$

Let $S = \{s_1, \ldots, s_k\}$ *be a simplex of* $\mathcal{S}_P$. *Define* $S^* = \{s_1^*, \ldots, s_k^*, \bigwedge S\} - \{0\}$, *where* $s_i^* = s_i - \bigwedge S$ *for each* $i \in \{1, \ldots, k\}$. *Further, let* $P^* = (P - S) \cup S^*$. *We say that the abstract simplicial complex* $\mathcal{S}_{P^*}$ *is obtained from* $\mathcal{S}_P$ *via stellar subdivision (in G along S). By abuse of language, we apply the same terminology to P and P^*. If Q is obtained from P by a finite (possibly zero) number of stellar subdivisions, we write* $Q \preceq P$. *We write* $Q \preceq_2 P$ *to denote that* $Q \preceq P$ *and all stellar subdivisions are along 1-simplices, in which case we speak of* binary starring.

Remark. (Same notation as in the preceding definition). If $Q \preceq P$, it is easily seen that $Q \subseteq G^*$, $Q \leq P$, and $Q \subseteq \langle P \rangle$. The latter is the closest analogue we can exhibit of the crucial, albeit trivial, fact that $d \in [\{a_1, a_2\}]$ if $d = \gcd(a_1, a_2)$ $(d, a_1, a_2 \in \mathbb{Z})$. However, we have no analogue at all of the fact that, when $G = \mathbb{Z}$, *greatest* (hence, canonical) commensurators exist. It is of course possible that no natural analogue exists.

Stellar subdivisions were introduced by James W. Alexander in [1] for geometric simplices, and in a purely combinatorial version in [2]. How Definition 3.2 encompasses subdivisions of geometric simplices will be clarified in the sequel. Its inclusion here is justified by our next statement.

Theorem 3.3. *Let* $P \subseteq G^*$ *be finite. There exists* $U \subseteq G^*$ *such that* $U \preceq_2 P$ *and* $[[U]]$ *is free over* U *(as a monoid). Thus, in particular, every abelian ℓ-group is ultrasimplicial.*

This is the result proved in [31]. Notice that binary starring in general ℓ-groups plays the role of subtraction in totally ordered groups.

4 MV-partitions

Before it can be accepted as a good generalization of a time-honored concept, Definition 3.1 must prove its value. Hence, paraphrasing George Birkhoff's notorious question to his son Garrett ([6]), we ask ourselves: "What can we prove with generalized commensurability that we can't prove without?". We offer two examples, one in this section and the other in the next. While the theory at hand is admittedly in its infancy, we regard these results, which are consequences of Theorem 3.3, as a promising beginning.

BOOLEAN PARTITIONS, REVISITED.   As mentioned in Section 1, the variety of boolean algebras is the subvariety of MV-algebras satisfying $x \oplus x = x$. For boolean algebras, the lattice join coincides with truncated sum. Let us remark that $\Gamma(H, u)$ is a finite boolean algebra if and only if H is the cardinal sum of finitely many copies of $\mathbb{Z}$ and $u = (1, \ldots, 1)$.

Let B be a boolean algebra, and denote by 1 and 0 its top and bottom elements, respectively. A (boolean) *partition* of B is a finite set $\Pi = \{b_1, \ldots, b_n\}$ of elements of B, the *blocks* of Π, such that

BP1. $b_i \wedge b_j = 0$ for all $i, j \in \{1, \ldots, n\}$, $i \neq j$,

BP2. $\bigvee \Pi = 1$,

BP3. $b_i \neq 0$ for all $i \in \{1, \ldots, n\}$.

This is just a reformulation of ordinary set-theoretical partitions, of course. Notice, however, that we restrict ourselves to finite partitions throughout. We now collect some very easy remarks, with the intent of providing motivation for what follows.

While in many cases condition BP3 is neglected, we call the reader's attention to the fact that if partitions were to allow a zero block, then they would not satisfy the following independence property:

Observation 4.1. *Let Π be a partition of B, and let S be the subalgebra generated by Π. Then every $b \in S$ is uniquely expressible as a join of distinct blocks of Π, up to the order of blocks.*

If in the above statement one includes 0 amongst the blocks of Π, it is still true, of course, that S is atomic, but it is no longer true that the chosen generators have the stated independence property.

Another easily proved but fundamental fact is that the set of partitions of any boolean algebra naturally carries a very rich order structure:

Proposition 4.2. *Let $\Pi(B)$ denote the set of partitions of a finite boolean algebra B. Given Δ and Σ in $\Pi(B)$, define $\Delta \leq \Sigma$ if and only if every block of Σ is a join of blocks of Δ (equivalently, every block of Δ is less or equal than some block of Σ in the natural order of B). Then $\leq$ is a (not necessarily distributive) lattice order on $\Pi(B)$.*

RIESZ PARTITIONS.   We tackle the question whether MV-algebras admit a natural theory of partitions.

Notation. Throughout the rest of this section, M shall denote an MV-algebra with top element $u = 1$ and bottom element 0. Further, we let H be the unique abelian ℓ-group with strong order unit u such that $\Gamma(H, u) = M$.

A first attempt at generalizing boolean partitions to MV-algebras is likely to produce something close to the following notion.

Definition 4.3. *A finite subset of elements* $\Pi = \{b_1, \ldots, b_n\} \subseteq M$ *is a Riesz partition of* M *if and only if there exists natural numbers* $m_1, \ldots, m_n \in \mathbb{N}$, *the multiplicities of the blocks, such that*

RP1. $m_1 b_1 + \cdots + m_n b_n = u$, *where the addition is that of* H,

RP2. $b_i \neq 0$.

We denote by $\Pi_R(M)$ *the set of Riesz partitions of* M. *Given* Δ *and* Σ *in* $\Pi_R(M)$, *we say that* Δ *refines* Σ, *written* $\Delta \leq \Sigma$, *if and only if* Σ *is in the submonoid of* H *generated by* Δ.

Condition RP2 is BP3 *verbatim*. Condition RP1 is an attempt to generalize BP1 and BP2 — indeed, if M is a boolean algebra, both conditions follow from RP1. This would *not* be the case if one replaced the group sum $+$ with the truncated sum $\oplus$. Multiplicities are necessarily equal to 1 in the boolean case, which is of course a manifestation of the idempotency of the binary operations. The notion of refinement we have introduced coincides in the boolean case with the order of the lattice of partitions defined in Proposition 4.2. In a nutshell, the main idea behind the above definition is that set-theoretical union of blocks should be generalized through group-theoretical addition. Once this is granted, there is only one sensible way of defining refinement.

We mention in passing that RP1 is expressible in the language of MV-algebras, without resorting to the Γ functor (see [33, Section 3]). However, we find it most expedient to freely mix the language of MV-algebras and that of ℓ-groups, and shall continue to do so in the following without further notice.

The analogy between boolean and Riesz partitions is rather weak. Specifically, no uniqueness result generalizing Observation 4.1 seems attainable. Furthermore, multiplicities are not uniquely determined. Perhaps worst of all, refinement is *not* a partial order, but merely a pre-order. For instance, if in $\mathbb{Q}$ we select 1 as a strong order unit, then $1/2$ and $1/4$ are blocks of a Riesz partition with multiplicities 1 and 2, respectively. On the other hand, $1/4$ is the single block of a partition with multiplicity 4. However, while these partitions refine one another, they are different. Hence, refinement is not antisymmetric.

Nonetheless, there is something positive one can say about Riesz partitions:

Proposition 4.4. *Given* $\Sigma_1, \Sigma_2 \in \Pi_R(M)$, *there exists* $\Delta \in \Pi_R(M)$ *such that* $\Delta \leq \Sigma_1$ *and* $\Delta \leq \Sigma_2$. *In words, any two Riesz partitions admit a joint refinement.*

The proof of this statement is very simple, but we insist on it because it finally explains why we call Riesz partitions 'Riesz'. Let a_i and b_j be the blocks of Σ_1 and Σ_2, respectively. Let p_i and q_j be some sets of multiplicities for a_i and b_j, respectively. Then

$$p_1 a_1 + \cdots + p_m a_m = q_1 b_1 + \cdots + q_n b_n = u$$

holds. Hence, the existence of a joint refinement amounts to the Riesz Decomposition Property of abelian ℓ-groups. (Recall that a partially ordered abelian group G satisfies the Riesz Decomposition Property if and only if for every $a, b, c, d \in G^+$ such that $a + b = c + d$, there exist $e_1, e_2, e_3, e_4 \in G^+$ such that $a = e_1 + e_2$, $b = e_3 + e_4$, $c = e_1 + e_3$, $d = e_2 + e_4$. This easily generalizes to finite sets of elements. The Riesz Interpolation Property, which we shall have occasion to mention in Section 5, is a formally distinct but equivalent statement.)

Joint refinement is of paramount importance in many classical applications of boolean partitions. We merely mention the case of conditional probability measures as axiomatized, for instance, in [9]. Thereby, we take the stance that any proposed generalization of the boolean theory should at least admit joint refinements.

MV-PARTITIONS. Notwithstanding Proposition 4.4, Riesz partitions do not seem to us a satisfactory many-valued generalization of the boolean notion. This is not to say that their theory is uninteresting. Quite on the contrary, it is possible to obtain some strikingly strong results — the reader should compare Proposition 4.4 with Lemma 5.7 in the next section. But the point we want to make here is that their theory has nothing to do with *lattice* order, its natural setting being that of dimension groups (see Section 5). In such generality, of course, the notion of block is very far from its intuitive (i. e., boolean) counterpart. In an attempt to overcome this difficulty, the second author proposed the following definition (see, e.g., [37]).

Definition 4.5. *A Riesz partition* $\Pi = \{b_1, \ldots, b_n\} \subseteq M$ *is an MV-partition if and only if its blocks are linearly independent (when viewed as elements of the $\mathbb{Z}$-module H). We denote by* $\Pi_{MV}(M)$ *the set of MV-partitions of M.*

This requirement is quite strong. In particular:

- Multiplicities of blocks are uniquely determined.

- Every element in the submonoid generated by an MV-partition has a unique expression, up to the order of summands, as a positive sum of distinct blocks (cfr. Observation 4.1).

- Refinement is a partial order on $\Pi_{MV}(M)$ (cfr. Proposition 4.1).

In fact, it may even be thought that linear independence is too strong a requirement, leaving us in such a shortage of partitions as to prevent the development of a satisfactory theory. This is not the case:

Theorem 4.6. *The poset* $\Pi_{MV}(M)$ *is lower directed. In other words, any two MV-partitions admit a joint refinement.*

Unlike its Riesz analogue (Proposition 4.4), this is not easy to prove. Let a_i and b_j be the blocks of $\Sigma_1, \Sigma_2 \in \Pi_{MV}(M)$, respectively. Let p_i and q_j be the unique multiplicities of a_i and b_j, respectively. Then, as in the Riesz case, $\sum p_i a_i = \sum q_j b_j = u$ holds. Unfortunately, the Riesz joint refinement of Σ_1 and Σ_2 is generally not in $\Pi_{MV}(M)$.

However, consider the set $P = \Sigma_1 \cup \Sigma_2$. As a finite subset of H^*, P is commensurable by Theorem 3.3. Let $U \subseteq H^*$ be a set of commensurators for P. Then U is an MV-partition which is a joint refinement of Σ_1 and Σ_2, as is easily verified. Thus, generalized commensurability is essential in proving a crucial property of generalized partitions.

SCHAUDER BASES IN FINITELY PRESENTED MV-ALGEBRAS.   Encouraged by Theorem 4.6, we further elaborate on the notion of MV-partition. We show that in free (more generally, in finitely presented) MV-algebras one can isolate a significant class of MV-partitions known as Schauder bases. The latter were introduced by the second author in [35], and are now an important item in the MV-algebraist's toolbox.

Let S be a rational simplicial complex in $[0,1]^n$. If $|S| = [0,1]^n$, we call S *complete*. Let v be a vertex of S. Then $v = (r_1/s_1, \ldots, r_n/s_n)$ for uniquely determined positive integers r_i, s_i such that $s_i \neq 0$ and r_i and s_i are relatively prime. The least common multiple of the set $\{s_i\}$ is said to be the *denominator* of v, written $\mathrm{den}(v)$. The *star* $\mathrm{st}(v)$ of v is the set of simplices in S of which v is a face. The *Schauder hat* at v in S is the unique continuous piecewise-linear function $h_v : [0,1]^n \to [0,1]$ which attains rational value $1/\mathrm{den}(v)$ at v, vanishes outside $\mathrm{st}(v)$, and is linear on each simplex of S. In general, h_v has linear pieces with non-integral coefficients. This is unfortunate, because it implies $h_v \notin \mathcal{M}_n$. In attempting to remedy this situation, one discovers a connection with the following important notion. Let S be an n-dimensional simplex in S with vertices $v_0, \ldots, v_n$. Let us again write $v_j = (r_1^j/s_1^j, \ldots, r_n^j/s_n^j)$, r_i^j and s_i^j relatively prime positive integers, $s_i^j \neq 0$. Passing to homogeneous coordinates, we obtain integral vectors $\vec{v}_j = (r_1^j, \ldots, r_n^j, \mathrm{den}(v_j)) \in \mathbb{Z}^{n+1}$. We then say that S is *unimodular* if and only if $\{\vec{v}_0, \ldots, \vec{v}_n\}$ is a basis of the free abelian group $\mathbb{Z}^{n+1}$. A complete rational simplicial complex S is *unimodular* if and only if all its n-simplices are unimodular. In this case, we call S a *unimodular partition* of $[0,1]^n$.

Definition 4.7. *Let S be a unimodular partition of $[0,1]^n$. We call the set of Schauder hats at the vertices of S a Schauder basis of $\mathcal{M}_n$, and we denote it by $\mathbf{H}_S$.*

Proposition 4.8. *Let S be a unimodular partition of $[0,1]^n$.*

(i) *For every vertex $v \in S$, $h_v \in \mathcal{M}_n$. Hence, $\mathbf{H}_S \subseteq \mathcal{M}_n$.*

(ii) *Every Schauder basis of $\mathcal{M}_n$ is an MV-partition of $\mathcal{M}_n$.*

(iii) *Let $f \in \mathcal{M}_n$. Then $f \in [[\mathbf{H}_S]]$ if and only if f is linear over each simplex of S.*

Unimodularity is of the utmost importance in diophantine approximation and integral combinatorial optimization — see for instance [44], and the historical remarks on Farey sequences and related references in [35]. Schauder hats are known as virtual support functions in the theory of polyhedra ([25], [52]).

Refinements of partitions.   Let us recall some standard notions of refinement for abstract simplicial complexes.

Definition 4.9. *Given two abstract simplicial complexes $\mathcal{S}_V$, $\mathcal{T}_V \subseteq \mathbf{P}(V)$, we say that $\mathcal{S}_V$ refines $\mathcal{T}_V$, written $\mathcal{S}_V \leq \mathcal{T}_V$, if and only if every simplex of $\mathcal{T}_V$ is a union of simplices of $\mathcal{S}_V$. Let $S \in \mathcal{S}_V$. The star of S is the set $\mathrm{st}(S) = \{F \in \mathcal{S}_V \mid S \subseteq F\}$, and the closed star of S is the set $\mathrm{clst}(S) = \{F \in \mathcal{S}_V \mid F \subseteq F' \in \mathrm{st}(S)\}$. The stellar subdivision along S of $\mathcal{S}_V$ is the abstract simplicial complex $\mathcal{S}_{V^*}$, where $V^* = V \cup \{w\}$ and $w \notin V$, such that*

$$\mathcal{S}_{V^*} = (\mathcal{S}_V - \mathrm{st}(S)) \cup w \cdot (\mathrm{clst}(S) - \mathrm{st}(S)),$$

where $w \cdot (\mathrm{clst}(S) - \mathrm{st}(S)) = \{T' \mid T' = T \cup \{w\}, T \in (\mathrm{clst}(S) - \mathrm{st}(S))\}$.

These notions all apply to rational simplicial complexes with the same support, and yield the classical geometric concept of Alexander's starring. Thus, while formally involved, the above definition has intuitive geometric content in the case of rational simplicial complexes.

It is important to notice that stellar subdivisions do *not* preserve unimodularity of rational simplicial complexes. For instance, consider the unimodular 1-simplex $S = \{0, 1\}$ in $\mathbb{R}$. The new vertex $w = 1/3$ determines a stellar subdivision along S that does not yield a unimodular complex — indeed, the 1-simplex $\{1/3, 1\}$ is not unimodular, for the set $\{\vec{v}_0 = (1, 3), \vec{v}_1 = (1, 1)\}$ is not a basis of $\mathbb{Z}^2$. We are thus led to a further specialization.

Definition 4.10. *Let $\mathcal{S}$ be a unimodular partition of $[0, 1]^n$. Let $S = (v_0, \ldots, v_m)$ be an m-simplex of $\mathcal{S}$. Passing to homogeneous coordinates, let $\{\vec{v}_0, \ldots, \vec{v}_m\}$ be the corresponding integral vectors in $\mathbb{Z}^{m+1}$, where we assume $\mathrm{den}(v_i)$ is the last coordinate of $\vec{v}_i$, for all $i \in \{0, \ldots, m\}$. The barycentre of S is the unique point $w \in [0, 1]^n$ whose homogeneous coordinates are given by $\vec{w} = \sum_{i=0}^{m} \vec{v}_i$. If the simplicial complex $\mathcal{T}$ is obtained from $\mathcal{S}$ via a finite (possibly zero) number of stellar barycentric subdivisions (i. e. new vertices are barycentres), we write $\mathcal{T} \preceq \mathcal{S}$. If, in addition, all subdivisions are along 1-simplices, we write $\mathcal{T} \preceq_2 \mathcal{S}$ and speak of binary starring. (The barycentre of a rational 1-simplex is known as the Farey mediant of its vertices).*

It turns out that stellar barycentric subdivisions preserve unimodularity. It follows that if $\mathcal{T} \preceq \mathcal{S}$ and $\mathcal{S}$ is a unimodular partition of $[0, 1]^n$, then so is $\mathcal{T}$.

Recall from Definitions 3.1 and 3.2 that we adopt the same terminology of the preceding definitions in the case of finite sets of strictly positive elements in abelian ℓ-groups. Moreover, we have the notion of refinement of MV-partitions, which one could extend to cover stellar subdivisions of partitions following Definition 3.2 in the obvious way.

With all these notions of refinement, a book-keeping proposition is in order:

Proposition 4.11. *Let* $\mathcal{U}_1$ *and* $\mathcal{U}_2$ *be unimodular partitions of* $[0,1]^n$. *The following are equivalent.*

(i) $\mathbf{H}_{\mathcal{U}_1} \leq \mathbf{H}_{\mathcal{U}_2}$ *in the sense of Definition 3.1.*

(ii) $\mathcal{U}_1 \leq \mathcal{U}_2$ *in the sense of Definition 4.9.*

(iii) $\mathbf{H}_{\mathcal{U}_1} \leq \mathbf{H}_{\mathcal{U}_2}$ *in the sense of Definition 4.3.*

Further, the following are equivalent.

4. $\mathbf{H}_{\mathcal{U}_1} \preceq \mathbf{H}_{\mathcal{U}_2}$ *in the sense of Definition 3.2.*

5. $\mathcal{U}_1 \preceq \mathcal{U}_2$ *in the sense of Definition 4.10.*

Moreover, the following are equivalent.

6. $\mathbf{H}_{\mathcal{U}_1} \preceq_2 \mathbf{H}_{\mathcal{U}_2}$ *in the sense of Definition 3.2.*

7. $\mathcal{U}_1 \preceq_2 \mathcal{U}_2$ *in the sense of Definition 4.10.*

Hence, in particular, our notation is consistent.

As is easily seen, not all MV-partitions of $\mathcal{M}_n$ are Schauder bases. We would very much like to characterize in a purely algebraic manner Schauder bases amongst MV-partitions of $\mathcal{M}_n$, but at present we are unable to do so. However, we can get a glimpse of the true algebraic significance of Schauder hats returning for a moment to boolean algebras.

Observation 4.12. *A partition* Π *of a boolean algebra* B *generates* B *if and only if* B *is finite (hence atomic), and the blocks of* Π *are precisely the atoms of* B.

By analogy, one may thus take blocks of generating MV-partitions as generalized atoms. In particular, Schauder bases turn out to be generating sets for $\mathcal{M}_n$, which, of course, is not the case for every MV-partition. Loosely speaking, Schauder hats should then play the role of atoms, or points, in boolean algebras.

Let us indicate how one can prove that Schauder bases are generating. The crucial fact is the following:

Theorem 4.13 [The De Concini-Procesi Lemma] *Let* $\mathcal{U}_1$ *and* $\mathcal{U}_2$ *be unimodular partitions of* $[0,1]^n$. *There exists a third unimodular partition* $\mathcal{R}$ *such that*

(i) $\mathcal{R} \preceq_2 \mathcal{U}_1$, *i. e.,* $\mathcal{R}$ *is obtained from* $\mathcal{U}_1$ *via a finite sequence of binary starrings.*

(ii) $\mathcal{R} \leq \mathcal{U}_2$, *i. e.,* $\mathcal{R}$ *refines* $\mathcal{U}_2$.

Now let $f \in \mathcal{M}_n$ be a McNaughton function, and let $\mathcal{U}$ be a unimodular partition of $[0,1]^n$. A standard argument in piecewise-linear topology shows that there exists a rational, complete simplicial complex $\mathcal{S}$ such that f is linear over each simplex of $\mathcal{S}$. It is known that every rational simplicial complex $\mathcal{S}$ has a unimodular refinement $\mathcal{S}^u$ ([35] or [25] for both statements). By Theorem 4.13, there exists a unimodular partition $\mathcal{R}$ such that $\mathcal{R} \preceq_2 \mathcal{U}$ and $\mathcal{R} \leq \mathcal{S}^u$. By Proposition 4.8, this implies $f \in [[\mathbf{H}_{\mathcal{R}}]]$. But $\mathcal{R} \preceq_2 \mathcal{U}$ implies $\langle \mathbf{H}_{\mathcal{R}} \rangle = \langle \mathbf{H}_{\mathcal{U}} \rangle$, so that we obtain $f \in \langle \mathbf{H}_{\mathcal{U}} \rangle$. Hence, every Schauder basis of $\mathcal{M}_n$ generates $\mathcal{M}_n$. Notice that an iteration of the same argument for a finite set of elements instead of a single f proves that $\mathbf{\Gamma}^{-1}(\mathcal{M}_n)$ is ultrasimplicial, Schauder hats sufficing as commensurators.

As mentioned in Section 1, Theorem 4.13 lies at the heart of Panti's geometric proof of Chang's Completeness Theorem ([42]). The interested reader can find a self-contained proof in Panti's paper. The De Concini-Procesi Lemma, as shown in their original paper ([17]), is a key tool for the elimination of points of indeterminacy in toric varieties. Remarkably enough, it is also a basic ingredient in our present analysis of non-boolean partitions and their refinements.

The purely algebraic study of commensurability led us to an investigation of partitions, which in turn brought us to the geometry of finitely presented objects. Armed with the notion of a Schauder basis as a particularly appealing type of MV-partition, we have shown that every finite subset of a free MV-algebra may be commensurated by Schauder hats. Thus we have finally come full circle. We begin another tour in the next section from an entirely different starting point.

5   AF C^*-algebras and dimension groups

Every quantum physical system with finitely many degrees of freedom admits exactly one irreducible Hilbert space representation. This is no longer true of systems with infinitely many degrees of freedom, whence the latter are beyond the scope of the Hilbert space formalism. Through a rich and interesting history which we shall not even attempt to sketch, [11] the appropriate mathematical abstraction of arbitrary quantum physical systems emerged with the theory of C^*-algebras. While nowadays the literature on C^*-algebras is immense, our main interest lies in a small subclass of structures, known as approximately finite-dimensional (AF) C^*-algebras. This is so because, as we shall see in this section, AF C^*-algebras are closely related to certain partially ordered abelian groups.

In an attempt to avoid tedious bibliographical remarks every other line, we quote some blanket references. For the general theory of C^*-algebras, see [18]. For the K-theory of operator algebras, see [7]. For the relationships between AF C^*-algebras and partially ordered abelian groups see [27] and [19].

---

[11] the interested reader is referred to, e.g., [23]

$\mathbf{K_0}$ OF RINGS. Let M be a commutative monoid. It is both well-known and easy to prove that if M is cancellative, there exists an embedding $\iota : M \to \mathbf{K}(M)$ into an abelian group $\mathbf{K}$ having the following universal property. If $f : M \to A$ is an homomorphism into an abelian group A, there exists a unique homomorphism $g : \mathbf{K}(M) \to A$ such that $f = g \circ \iota$. The proof just mimicks the construction of the group of integers from the natural numbers. The abelian group $\mathbf{K}(M)$ is known as the *Grothendieck group* of M. Its importance springs from the construction we presently introduce. Let R be a ring, which we always assume to have a unit element 1. Let $\mathcal{C}_R$ be the class of finitely generated projective (right) R-modules. Given $P \in \mathcal{C}_R$, let $[P]$ denote the isomorphism class of P. Let $\mathcal{P}_R = \{[P] \mid P \in \mathcal{C}_R\}$. Define a binary operation on $\mathcal{P}_R$ by stipulating

$$[P] + [Q] = [P \oplus Q].$$

It is a simple exercise to check that the definition does not depend on the choice of representatives in $[P]$ and $[Q]$. Further, the operation makes $\mathcal{P}_R$ into a monoid M. In many important cases, one must use a coarser equivalence than isomorphism in order to make M cancellative. Two modules $P, Q \in \mathcal{C}_R$ are *stably isomorphic* if and only if there exist finitely generated free modules F_P and F_Q such that $P \oplus F_P$ is isomorphic to $Q \oplus F_Q$. Let $\mathcal{S}_R = \{[[P]] \mid P \in \mathcal{C}_R\}$, where $[[P]]$ denotes the stable isomorphism class of P. Defining addition as in the previous case, one obtains a cancellative monoid whose Grothendieck group is denoted $\mathbf{K}_0(R)$. Grothendieck's construction is a powerful unifying tool — a list of important problems of modern mathematics can be rephrased in terms of $\mathbf{K}_0$ of some ring.

C^*-ALGEBRAS. We shall consider $\mathbf{K}_0$ of (a small class of) C^*-algebras, the latter being involutive rings endowed with additional structure. All algebras considered henceforth shall be defined over the field of complex numbers $\mathbb{C}$. A ring A is a *(complex) algebra* if it is also a vector space over $\mathbb{C}$ — with the ring and vector space additions coinciding — and for every $\lambda \in \mathbb{C}$, $x, y \in A$, $\lambda(xy) = (\lambda x)y = x(\lambda y)$ holds. An algebra A is a *-*algebra* if and only if it is equipped with a map $* : A \to A$ which is involutive (i. e., $x^{**} = x$), conjugate linear (i. e., it is a group homomorphism on $\langle A, + \rangle$ such that $(\lambda x)^* = \bar{\lambda}x^*$, $\lambda \in \mathbb{C}$), and reverse multiplicative (i. e., $(xy)^* = y^*x^*$). A *Banach algebra* is an algebra A endowed with a norm $\| \cdot \|$ making it a Banach space — that is, A is complete with respect to $\| \cdot \|$ — and such that $\|xy\| \leq \|x\|\|y\|$. One also makes the (harmless) assumption that $\| \cdot \|$ is normalized, $\|1\| = 1$. A *Banach *-algebra* is a Banach algebra which is simultaneously a *-algebra. Finally, by a *Hilbert space* we mean a complex vector space equipped with an inner product, complete with respect to the induced norm. [12]

Definition 5.1. *A C^*-algebra is a Banach *-algebra A such that $\|x^*x\| = \|x\|^2$.*

---

[12]Thus, we do not assume separability.

Examples. 1.   Let $B(H)$ denote the algebra of all bounded linear operators on a Hilbert space H, where product coincides with composition. For each $T \in B(H)$ one defines the norm of T by recalling that $\|T(x)\| < c\|x\|$ for all $x \in H$ and some fixed real number c independent of x. The greatest lower bound of all real numbers c satisfying the latter condition is the norm of T, again denoted $\|T\|$. Further, T admits an adjoint operator, the unique linear operator T^* such that $\langle y, T(x) \rangle = \langle T^*(y), x \rangle$ for all $x, y \in H$, where $\langle \cdot, \cdot \rangle$ is the inner product of H. With these operations, $B(H)$ is a C^*-algebra. Also notice that a *-subalgebra of $B(H)$ which is norm-closed is itself a C^*-algebra.

2.   Let $\mathbb{M}_n$ denote the algebra of $n \times n$ matrices with complex entries. The natural involution on $\mathbb{M}_n$, namely, the conjugate transpose of a given matrix, is a *-map. In order to make $\mathbb{M}_n$ into a C^*-algebra, we still need a compatible norm. Identifying $\mathbb{M}_n$ with $B(\mathbb{C}^n)$, we can use the norm of $B(\mathbb{C}^n)$ as defined above in the general case. We thus obtain a C^*-algebra.

3.   We can extend the latter example by noticing that any finite direct product $A_1 \times \cdots \times A_k$ of C^*-algebras is again a C^*-algebra under componentwise operations and supremum norm. We thereby obtain the class of C^*-algebras of the form $\mathbb{M}_{n_1} \times \cdots \times \mathbb{M}_{n_k}$, known as *matricial algebras*.

4.   Let $C(X, \mathbb{C})$ denote the set of complex-valued continuous functions on a compact Hausdorff topological space X. Clearly, this is a complex algebra under pointwise operations. A natural *-map is given by setting $f^* = \bar{f}$. A natural compatible norm is obtained by taking suprema. Equipped with these operations $C(X, \mathbb{C})$ is a commutative C^*-algebra.

The following basic result states that the above list is exhaustive for various relevant classes of C^*-algebras. Here, 'isomorphic' means 'isomorphic as C^*-algebras':

Theorem 5.2.

(i) *Every C^*-algebra is isomorphic to a norm closed *-subalgebra of $B(H)$ for some Hilbert space H.*

(ii) *Every simple (i.e., ideal-free) finite-dimensional C^*-algebra is isomorphic to some $\mathbb{M}_n$.*

(iii) *Every finite-dimensional C^*-algebra is isomorphic to a matricial algebra.*

(iv) *Every commutative C^*-algebra is isomorphic to $C(X, \mathbb{C})$ for some compact Hausdorff space X, uniquely determined up to homeomorphism.*

AF C^*-algebras and their $\mathbf{K}_0$-groups.   The above (essentially Gel'fand-Naimark) Theorem accounts for the extraordinary development of the subject in the last fifty years. Along the way, a certain class of C^*-algebras has emerged as an important benchmark for many theories and conjectures.

Definition 5.3 [AF C^*-algebras] *An approximately finite-dimensional (for short, AF) C^*-algebra A is the norm-closure of an ascending sequence $R_1 \subseteq R_2 \subseteq \ldots$ of finite-dimensional C^*-algebras, where R_i is a $*$-subalgebra of R_{i+1} (with the same unit, according to our standing assumption that all rings are unital).*

Thus, an AF C^*-algebra A contains a norm-dense $*$-subalgebra $R = \cup R_i$, in the notation of the preceding definition. By part 3 in the above theorem, each R_i is a matricial algebra. Complex $*$-algebras of the form $R = \cup R_i$ are called *ultramatricial.*

We are now going to illustrate why AF C^*-algebras are a tractable class. Let A be an AF C^*-algebra. In case A is finite-dimensional, it is matricial and hence isomorphic to $\mathbb{M}_{n_1} \times \cdots \times \mathbb{M}_{n_k}$. Thus, A carries a finite set of natural numerical invariants, the tuple $(n_1, \ldots, n_k)$. Moreover, there is no need to search any further, for it is clear that such 'dimension tuples' are complete invariants. However, there is another way of getting at such tuples which is definitely fancier, but will be the key tool to deal with more general cases. Since a C^*-algebra A is, in particular, a ring, $\mathbf{K}_0(A)$ is defined. Now, when $A \cong \mathbb{M}_{n_1} \times \cdots \times \mathbb{M}_{n_k}$, it can be shown by direct computation that $\mathbf{K}_0(A) = \mathbb{Z}^k$, the free abelian group over k generators. Of course, this is not enough to pin A down up to isomorphism. Applying $\mathbf{K}_0$, we lose track of the tuple $(n_1, \ldots, n_k)$. We can remedy this situation as follows. If R is a ring, $\mathbf{K}_0(R)$ contains the distinguished subset $\mathcal{S}_R$ of stable isomorphism classes of finitely generated projective R-modules. For $p, q \in \mathbf{K}_0(R)$, define $p \leq q$ if and only if $q - p \in \mathcal{S}_R$. Then $\mathbf{K}_0(R)$ is pre-ordered by $\leq$. Further, the stable isomorphism class of R itself, denoted $[[R]]$, is a strong order unit in $\mathbf{K}_0(R)$. We thus obtain the *(pre)-ordered Grothendieck group* of R, denoted $\langle \mathbf{K}_0(R), [[R]] \rangle$. Henceforth, we shall use this modified definition of $\mathbf{K}_0$. Returning to the case of finite-dimensional C^*-algebras, and resuming our previous notation, it can be shown that $\langle \mathbf{K}_0(A), [[A]] \rangle$ is in fact *partially* ordered, and it is isomorphic (as a partially ordered group with strong order unit) to $\langle \mathbb{Z}^k, (n_1, \ldots, n_k) \rangle$, where the positive cone of $\mathbb{Z}^k$ is the set of tuples with positive entries. In other words, $\mathbb{Z}^k$ denotes the cardinal product of k copies of the integers. It is then clear that the (ordered, unital) $\mathbf{K}_0$-group of finite-dimensional C^*-algebras is a complete invariant — two such algebras are isomorphic (as C^*-algebras) if and only if their associated invariants are isomorphic (as partially ordered abelian groups with strong order unit).

More generally, if A is *approximately* finite-dimensional, it can still be classified up to isomorphism by a similar invariant. Indeed, suppose first that $R = \cup R_i$ is an ultramatricial $*$-algebra (hence not necessarily a C^*-algebra). Applying $\mathbf{K}_0$ to each R_i we obtain ℓ-groups $\mathbb{Z}^{n_i}$ with order units u_i. The crucial point is now that $\mathbf{K}_0$ is a *functor* from rings to abelian groups, so that each inclusion $\iota : R_i \to R_{i+1}$ is mapped onto a group homomorphism $\mathbf{K}_0(\iota) : \mathbb{Z}^{n_i} \to \mathbb{Z}^{n_{i+1}}$. (*Caution:* $\mathbf{K}_0$ does *not* carry injective maps into injective maps). Moreover, $\mathbf{K}_0(\iota)$ is unit-, as well as order-preserving. (*Caution:* $\mathbf{K}_0(\iota)$ is *not* an ℓ-homomorphism — it merely preserves order).

Thus, $\mathbf{K}_0$ is in fact a functor from rings to partially ordered abelian groups with strong order unit. Hence, to every ultramatricial $*$-algebra $R = \cup R_i$ we can associate a direct system of the form

$$\varphi_i : \langle \mathbb{Z}^{n_i}, u_i \rangle \to \langle \mathbb{Z}^{n_{i+1}}, u_{i+1} \rangle,$$

where the *transition maps* φ_i are order-preserving, unit-preserving group homomorphisms. Let $G = \lim \mathbb{Z}^{n_i}$ be the limit abelian group. Elements of G are sequences of tuples of integers (an n_i-tuple at position i), modulo the equivalence relation which identifies eventually coinciding sequences. Declare an element of G positive if and only if it can be represented by an eventually pointwise positive sequence (whereby *all* the members of its equivalence class are eventually pointwise positive). This makes G into a partially ordered group with strong order unit, the latter being represented by the equivalence class of the sequence $\{u_i\}$. In the light of its origin, it is natural to call G a *countable unital dimension group*.

We would now like to associate a countable unital dimension group to an AF C^*-algebra, not just to an ultramatricial $*$-algebra. This turns out to be possible because $\mathbf{K}_0$ *is invariant under completion*. More precisely, suppose A is an AF C^*-algebra, and let R be a norm-dense ultramatricial $*$-subalgebra. Then $\mathbf{K}_0(A)$, as a partially ordered abelian group with strong order unit, is isomorphic to $\mathbf{K}_0(R)$.

The reason why it is worthwhile bearing with this rather complex construction is the following celebrated result ([8], [21]):

Theorem 5.4 [Elliott's Classification Theorem] *Let R_1, R_2 be ultramatricial $*$-algebras. Further, assume A_1, A_2 are AF C^*-algebras such that R_i is a norm-dense $*$-subalgebra of A_i, $i = 1, 2$. The following conditions are equivalent.*

(i) *$R_1 \cong R_2$ as rings.*

(ii) *$R_1 \cong R_2$ as $*$-algebras.*

(iii) *$\langle \mathbf{K}_0(R_1), [[R_1]] \rangle \cong \langle \mathbf{K}_0(R_2), [[R_2]] \rangle$ as partially ordered groups with strong order unit.*

(iv) *$A_1 \cong A_2$ as C^*-algebras.*

(v) *$\langle \mathbf{K}_0(A_1), [[A_1]] \rangle \cong \langle \mathbf{K}_0(A_2), [[A_2]] \rangle$ as partially ordered groups with strong order unit.*

It follows that, up to isomorphism, AF C^*-algebras are classified by countable unital dimension groups. We give a few examples of an AF C^*-algebra A, and its corresponding countable unital dimension group, which we denote D_A.

For each integer $k = 1, 2, 3, \ldots$, we use the term "factor of type I_k" to denote any C^*-algebra isomorphic to the algebra $\mathbb{M}_k$ of all $k \times k$ complex matrices (Example 2 above). As we have already seen, when A is a factor of type I_k then D_A is the naturally ordered additive group of integers, with the element k as the strong unit.

As explained in [30, Sections 11 and 12], (also see [19, p.13]) the CAR algebra is the (countably) infinite C^*-tensor product of factors, all of type I_2. The CAR algebra is a convenient mathematization of the ideal Fermi gas. Its Hilbert space representations correspond to the representations of the Canonical Anticommutation Relations. If A denotes the CAR algebra, then $D_A = \langle \mathbb{Z}[1/2], 1 \rangle$, i.e., the additive naturally ordered group of dyadic rationals with 1 as the strong unit.

Similarly, Glimm's universal UHF algebra (see [30, 12.1.7]) is the infinite tensor product of countably many copies of each factor I_k, $k = 1, 2, 3, \ldots$. When A is Glimm's universal UHF algebra, $D_A = \langle \mathbb{Q}, 1 \rangle$.

If A is commutative, then we can identify A with the C^*-algebra $C(X, \mathbb{C})$ of all continuous complex-valued functions defined over a uniquely determined separable totally disconnected compact Hausdorff space X. In this case, D_A is the abelian ℓ-group $C(X, \mathbb{Z})$ of all continuous integer-valued functions over X (where $\mathbb{Z}$ has the discrete topology), with the constant function 1 as the strong order unit. Many more interesting examples can be found in the literature.

DIMENSION GROUPS. It is now natural to investigate dimension groups in their own right, without any cardinality restrictions and without considering strong order units. To this purpose we prepare the following

Definition 5.5 [Dimension groups] *A simplicial group is a cardinal product of finitely many copies of* $\mathbb{Z}$. *(In other words, a simplicial group is a finitely generated free* $\mathbb{Z}$-*module S with positive cone the set of those elements which have positive coordinates with respect to a free basis of S). A dimension group is a direct limit of simplicial groups with order-preserving transition maps.*

Note that we are allowing general direct limits over directed index sets, whereas up to this point we had used direct sequences only.

After Elliott's classification of AF C^*-algebras, intensive research aiming at an intrinsic characterization of dimension groups was undertaken. Elementary preservation theorems for direct limits ensure that every dimension group is directed, unperforated ($nx \geq 0$ and $n \in \mathbb{N}$ imply $x \geq 0$) and has the Riesz Interpolation Property. It turns out that, conversely, these properties are enough to characterize dimension groups:

Theorem 5.6 [Effros-Handelman-Shen] *A directed, unperforated partially ordered abelian group with the Riesz Interpolation Property is a dimension group.*

This is a major result which has triggered much interesting research on dimension groups and AF C^*-algebras. The original proof is in [20], but a completely self-contained version can be found in [28].

Remark. It is fair to add at this point that, while the value and beauty of the Bratteli-Effros-Elliott-Handelman-Shen theory is beyond doubt, the problem of classifying dimension groups up to isomorphism turns out to be just as hard as that of classifying AF C^*-algebras — dimension groups are extremely rich mathematical structures.

As an immediate consequence of Theorem 5.6, one has that every abelian ℓ-group is a dimension group. The converse inclusion does not hold. Indeed, focusing attention on the way a dimension group can be approximated by simplicial groups, one is naturally led to consider an interesting intermediate class of ordered groups, namely the class of ultrasimplicial groups, where the order-preserving transition maps appearing in Definition 5.5 are one-one. Of the following two inclusions

$$\ell - \text{groups} \subseteq \text{ultrasimplicial groups} \subseteq \text{dimension groups} \qquad (2)$$

the second is trivial. The first inclusion will be established by Theorem 5.10 below. We shall also see that both inclusions are proper.

A fourth interesting class of ordered groups arises from the observation that, while simplicial groups are lattice-ordered, the order-preserving transition maps appearing in Definition 5.5 are not lattice homomorphisms — they merely send positive elements into positive elements. Thus one may naturally impose the stronger condition that transition maps are (one-one) ℓ-homomorphisms, and investigate the resulting class of abelian ℓ-groups. Lack of space prevents us from pursuing this line of research here.

In the course of proving Theorem 5.6, Effros, Handelman and Shen derived the following lemma:

Lemma 5.7 [Strong Riesz Decomposition] *Let D be a dimension group, and let a_i, $b_j \in D^* = D^+ - \{0\}$ be strictly positive elements, finite in number. Assume there exist natural numbers m_i, $n_j \in \mathbb{N}$ such that*

$$\sum m_i a_i = \sum n_j b_j . \qquad (3)$$

Then there exist a finite set $U = \{u_1, \ldots, u_k\} \subseteq D^*$ and positive integers p_{is} and q_{jt} such that the following conditions are satisfied.

(i) $a_i = p_{i1} u_1 + \cdots + p_{ik} u_k$ *and* $b_j = q_{j1} u_1 + \cdots + q_{jk} u_k$.

(ii) $\sum_i m_i p_{is} = \sum_j n_j q_{jt}$.

Remark 1. Condition 1 simply says that a_i and b_j are contained in the submonoid generated by U, but it singles out specific sets of coefficients p_{is} and q_{jt} that witness this fact. These, of course, are not unique, but they satisfy the property stated in Condition 2. In words, Condition 2 says that rewriting each side of equation (3) as a linear polynomial in the variables U using the span given by Condition 1 yields a termwise identity. One may then say that U *annihilates* equation (3).

Remark 2. The reader should compare the above lemma with Proposition 4.4. It is clear that Lemma 5.7 is a very strong form of joint refinement, definitely a non-trivial statement even when applied to ℓ-groups. However, notice that Lemma 5.7 is weaker than Theorem 3.3, since U is generally linearly dependent. Hence, Theorem 3.3 can be considered as a strengthening of Lemma 5.7 in the special case of ℓ-groups. Lemma 5.7 is a crucial tool for the proof of Theorem 5.6, because it grants the existence of a *directed* family of positive cones of simplicial groups whose direct limit gives the positive cone of the dimension group at stake. For details, see [28].

Ultrasimplicial groups. In [22], Elliott proposed the following terminology.

Definition 5.8. *A dimension group is* ultrasimplicial *if and only if it is a direct limit of simplicial groups with order-preserving* injective *transition maps.*

This is also the origin of our own terminology at the end of Definition 3.1. Notice that the latter applies *verbatim* to dimension groups. Hence, we now have two definitions of what it means for a dimension group to be ultrasimplicial. Following established mathematical tradition, we show they are equivalent:

Proposition 5.9. *Let D be a dimension group. D is the direct limit of simplicial groups with order-preserving injective transition maps if and only if every finite subset of D^* is commensurable.*

Indeed, if D is such a direct limit, then every finite subset of its positive cone is commensurable, because this trivially holds in simplicial groups, and injective maps preserve linear independence. To prove the converse we generalize the argument given in [29] for the countable case. Let us write

$$\mathcal{U} = \{[U] \mid U \subseteq D^* \, , \, U \text{ linearly independent and finite}\}.$$

Order these groups simplicially by choosing as positive cone the monoid $[[U]]$. (Notice that this is *not* the order inherited from D by restriction). A moment's reflection shows that $\mathcal{U}$ is partially ordered by refinement. Since we are assuming that D is ultrasimplicial in the sense of Definition 3.1, $\mathcal{U}$ is lower directed: to obtain a lower bound of $[U_1]$ and $[U_2]$, let us commensurate $U_1 \cup U_2$ by, say, C. Then both $[U_1]$ and $[U_2]$ can be embedded in $[C]$ by injective order-preserving group homomorphisms. Injectivity is a consequence of the linear independence of C, while preservation of order is due to the fact that C positively spans both U_1 and U_2. It is now easy to show that the limit of the direct system $\mathcal{U}$ precisely coincides with D. The proposition is proved.

Ultrasimplicial dimension groups have a more transparent structure than general dimension groups, because direct systems with injective maps are merely unions over a directed index set. The question arises, which dimension groups are ultrasimplicial. To the best of our knowledge, no satisfactory characterization is available. However, as we mentioned in Section 3, every abelian ℓ-group is ultrasimplicial. In the light of the above proposition, this can be rephrased either locally (Definition 3.1) or globally (Definition 5.8). Thus:

Theorem 5.10. *Every abelian ℓ-group G is the union of the directed system of its simplicial subgroups, in symbols,*

$$G = \cup_\iota \mathbb{Z}^{n_\iota}.$$

The positive cone G^+ is the union of the positive cones $(\mathbb{Z}^{n_\iota})^+$.

Remark. This is a second non-trivial application of the notion of generalized commensurability. In the case of ℓ-groups, the Effros-Handelman-Shen Theorem holds with injective maps. This statement has a counterpart for AF C^*-algebras whose $\mathbf{K}_0$-group is lattice-ordered.

Not much more is known beyond the lattice case, but we add a few remarks.

In [22], Elliott proves that lexicographic products of (ultrasimplicial) dimension groups are (ultrasimplicial) dimension groups. Let us show that finite lexicographic products of ℓ-groups are ultrasimplicial. We use a result of Handelman which is interesting in its own right. Let D be a dimension group. By definition, an *o-ideal* I of D is a directed subgroup of D which is also *convex*, in the sense that $x, z \in I$ and $x \leq y \leq z$ imply $y \in I$. A group homomorphism $f: D \to E$ is a homomorphism of dimension groups if and only if it is order-preserving. Let us partially order homomorphic images by declaring an element positive if and only if some element in its inverse image is positive. Then every homomorphic image of D is a dimension group, and o-ideals are precisely kernels of homomorphisms. Now, Handelman proves ([29]):

Proposition 5.11. *Suppose I is an o-ideal of the dimension group D, and let D/I be the quotient of D by I. If D/I and I are ultrasimplicial, then so is D.*

Let $P = H \vec{\otimes} K$ denote the lexicographic product of the abelian ℓ-groups H and K. It can be shown that P is a dimension group. Of course, P need not be lattice-ordered. It is an exercise to check that K is an o-ideal of P, and that $P/K = H$. By Theorem 3.3, both H and K are ultrasimplicial. Hence, by the above proposition, P is ultrasimplicial, and we have

Corollary 5.12. *Finite lexicographic products of abelian ℓ-groups are ultrasimplicial.*

In taking such lexicographic products we are not straying too far from lattice order — these structures are a proper subclass of Conrad's pseudo ℓ-groups ([15], [16]). More generally, pseudo ℓ-groups are dimension groups. This was proved by Teller ([46]), and it is not easy to show. It could be interesting to investigate pseudo ℓ-groups and their associated AF C^*-algebras in the light of the functional-analytical developments illustrated in this paper. We would like to know, for instance, whether all such groups are ultrasimplicial. It could also be worthwhile to identify the corresponding AF C^*-algebras.

Dually to Proposition 5.11, it is only natural to ask whether the ultrasimplicial property is preserved by homomorphic images. As we already mentioned, Handelman's paper [29] includes an unwarranted claim to this effect. It is tantalizing indeed that this seemingly innocuous question is, as far as we know, still open.

By the above corollary, there exist ultrasimplicial dimension groups that are not ℓ-groups. On the other hand, Elliott exhibited in [22] a dimension group which is *not* ultrasimplicial, whence we conclude that both inclusions in (2) are proper. Nevertheless, it seems surprisingly difficult to exhibit a natural class of non-ultrasimplicial dimension groups. Many questions remain open.

SCHAUDER HATS FOR ABELIAN ℓ-GROUPS. In Section 3, we introduced the notion of Schauder hats in free MV-algebras. We now sketch a parallel development for abelian ℓ-groups.

Entr'acte, part II. By definition, a fan in $\mathbb{R}^n$ is *complete* if and only if its support is $\mathbb{R}^n$. An integral vector is *primitive* if and only if its coordinates are relatively prime. Let $\sigma \subseteq \mathbb{R}^n$ be a simplicial cone. Then σ can always be written as $\sigma = \langle \vec{v}_1, \ldots, \vec{v}_m \rangle$ for uniquely determined primitive linearly independent integral vectors $\vec{v}_i$. We call such vectors the *vertices* of σ. We say that a simplicial n-dimensional cone σ in $\mathbb{R}^n$ is *unimodular* if and only if the integral matrix whose columns are the vertices of σ has determinant ± 1. A complete fan is *unimodular* if and only if all its n-dimensional cones are. Let $\vec{v}$ be a vertex of a complete unimodular fan Σ in $\mathbb{R}^n$. The *Schauder hat* at $\vec{v}$ in Σ is the unique continuous piecewise-linear homogeneous function $h_{\vec{v}} \colon \mathbb{R}^n \to \mathbb{R}$ such that

(i) $h_{\vec{v}}(\vec{v}) = 1$,

(ii) $h_{\vec{v}}(\vec{u}) = 0$ for every vertex $\vec{u} \neq \vec{v}$ of (any cone in) Σ,

(iii) $h_{\vec{v}}$ is linear homogeneous on each cone of Σ.

Again, as a consequence of the unimodularity of Σ, a Schauder hat is guaranteed to have linear pieces with integral coefficients. Thus, $h_{\vec{v}}$ is an element of $\mathcal{F}_n$, the free abelian ℓ-group over n generators. [13] We denote by $\mathbf{H}_\Sigma$ the set of all Schauder hats at the vertices of the complete unimodular fan Σ, and call it a *Schauder basis* of $\mathcal{F}_n$. (This is always a finite set, because of our definition of fan).

Now one can define various notions of refinement for fans. A fan carries a natural structure of abstract simplicial complex given by inclusion of faces. Hence, Definition 4.9 applies. Furthermore, using the obvious homogeneous analogue of the barycentre of a simplex, one can define barycentric stellar subdivisions of cones. We leave a detailed development to the interested reader. For the sake of completeness, however, we define stellar subdivisions along 2-dimensional cones through homogeneous Farey mediants. *Aliter*, we explicitly define *binary starring* for fans. Let $\sigma = \langle \vec{v}_1, \ldots, \vec{v}_m \rangle$ be an m-dimensional cone of a complete unimodular fan Σ in $\mathbb{R}^n$.

---

[13] Linear combinations of the $h_{\vec{v}}$'s with integral coefficients precisely yield the group of Σ-linear support functions, in the sense of [41, p.66].

We assume the $\vec{v}_i$'s are the vertices of σ. Then $\tau = \langle \vec{v}_i, \vec{v}_j \rangle$, for $1 \leq i < j \leq m$, is a 2-dimensional face of σ. We call $\vec{w} = \vec{v}_i + \vec{v}_j$ the *(homogeneous) Farey mediant* of τ. Let $\sigma_i = \langle \vec{w}, \vec{v}_1, \ldots, \vec{v}_{i-1}, \vec{v}_{i+1}, \ldots, \vec{v}_m \rangle$, $\sigma_j = \langle \vec{w}, \vec{v}_1, \ldots, \vec{v}_{j-1}, \vec{v}_{j+1}, \ldots, \vec{v}_m \rangle$. Define Σ^* as the fan obtained from Σ by replacing each cone $\rho \in \Sigma$ of which τ is a face by the two cones σ_i, σ_j, along with all their faces. Then we say that Σ^* is obtained from Σ by *binary starring*. In a similar fashion one can define general stellar barycentric subdivisions. Again, it is important to notice that stellar barycentric subdivisions (in particular, binary starrings) preserve unimodularity of fans. All our previous notions of refinement make sense in the linear context, save partition refinement, for which a strong order unit is needed. Hence, we shall freely use our previous notations.

We could now derive the linear analogue of the De Concini-Procesi Lemma, thus showing that Schauder bases are generating. But this is just a straightforward translation of Theorem 4.13. What we would like to do instead is to give an indication of how one might go beyond the finitely presented case. This is work in progress, and proofs will appear elsewhere.

For the sake of visualization, let us consider a finitely generated archimedean ℓ-group $G = \mathcal{F}_n / J$, J an ℓ-ideal of $\mathcal{F}_n$. Then (Theorem 1.5) G is isomorphic to the ℓ-group of restrictions of functions in $\mathcal{F}_n$ to the zeroset $\mathcal{Z}_J$. We consider the restriction of a Schauder basis of $\mathcal{F}_n$ to $\mathcal{Z}_J$. Trivially, this is a generating set for G, but it need not be linearly independent in G. One obvious reason for this is that some hats restrict to the zero function, but even once these are discarded, non-trivial dependencies may remain. However, one might hope that there is a binary starring refinement of the given basis of $\mathcal{F}_n$ such that its restriction to $\mathcal{Z}_J$ is linearly independent, once zeros are discarded. Hence, even in the non-archimedean case, we introduce the following terminology:

Definition 5.13. *Let J be an ℓ-ideal of an abelian ℓ-group G. A finite set of strictly positive elements P is said to J-split if and only if one can partition P into two blocks, say $P = Z \cup L$, $Z \cap L = \emptyset$, such that $Z \subseteq J$ and L/J is linearly independent in G/J.*

We can prove:

Theorem 5.14 [Generalized De Concini-Procesi Lemma] *Suppose $\mathbf{H}_{\Sigma_1}$ and $\mathbf{H}_{\Sigma_2}$ are two Schauder bases of $\mathcal{F}_n$. Fix an ℓ-ideal J of $\mathcal{F}_n$, and suppose both $\mathbf{H}_{\Sigma_1}$ and $\mathbf{H}_{\Sigma_2}$ J-split. There exists another Schauder basis $\mathbf{H}_\Delta$ of $\mathcal{F}_n$ such that*

(i) $\Delta \preceq_2 \Sigma_1$,

(ii) $\Delta \leq \Sigma_2$,

(iii) $\mathbf{H}_\Delta$ *J-splits.*

Thus, we can 'construct' (in a non-effective and highly non-uniform manner) tailor-made splitting joint refinements for any given ideal $J \subseteq \mathcal{F}_n$. The implications of this fact are not well understood yet, but we trust they shall be fruitful.

CLOSING A CIRCLE OF IDEAS. Composing Grothendieck's functor with Γ, we return to AF C^*-algebras and close this section in a circular fashion.

Let M be a countable MV-algebra, and let H be the abelian ℓ-group with strong order unit such that $\Gamma(H) \cong M$. By Elliott's Classification Theorem, there is a unique (unital) AF C^*-algebra A such that $K_0(A) \cong H$. Hence, $M \cong \Gamma \circ K_0(A)$ — to every countable MV-algebra M there corresponds a unique unital AF C^*-algebra. This bridge between apparently remote mathematical worlds opens up many interesting possibilities. It can be proved that countable boolean algebras correspond precisely to commutative AF C^*-algebras. Hence MV-algebras (Lukasiewicz logic) may be seen as a noncommutative generalization of boolean algebras (classical propositional logic).[14] Further, since countable MV-algebras are, up to syntax, theories in Lukasiewicz infinite-valued logic, and any such theory is just a sequence of symbols from a fixed finite alphabet, one can naturally speak of computability and axiomatizability for their corresponding AF C^*-algebras. One can thus say that a certain AF C^*-algebra has an undecidable (resp., decidable, polynomial time decidable, recursively enumerable, Gödel incomplete) theory. Most AF C^*-algebras in the literature have polynomial time decidable theories. If A is an AF C^*-algebra and $K_0(A)$ is finitely generated and totally ordered, then A cannot have a Gödel incomplete theory. This is no longer true if we drop the hypothesis that $K_0(A)$ is finitely generated [40].

Space constraints do not allow us to further discuss the role of MV-algebras for various computability issues (notably, the isomorphism problem) concerning AF C^*-algebras and their associated groups.

References

[1] James Waddel Alexander, *Combinatorial analysis situs.* Trans. AMS **28** (1926), 301-329.

[2] James Waddel Alexander, *The combinatorial theory of complexes.* Ann. of Math. (2) **31** (1930), 292-320.

[3] Kirby A. Baker, *Free vector lattices.* Canad. J. Math. **20** (1968), 58-66.

[4] W. Meurig Beynon, *Duality theorems for finitely generated vector lattices.* Proc. London Math. Soc. (3) **31** (1975), 114-128.

[5] W. Meurig Beynon, *Applications of duality in the theory of finitely generated lattice-ordered abelian groups.* Canad. J. Math. **29** (2) (1977), 243-254.

[6] Garrett Birkhoff, *Lattices and their applications.* In *Lattice Theory and its Applications;* (Darmstadt, 1991); (1995) Heldermann, Lemgo, 7-25.

---

[14]With no less suggestive power than that offered by the traditional view of C^*-algebras as a noncommutative generalization of compact Hausdorff spaces.

[7] Bruce Blackadar, *K-Theory for Operator Algebras*. (1987) Springer Verlag, Berlin-Heidelberg-New York.

[8] Ola Bratteli, *Inductive limits of finite dimensional C^*-algebras*. Trans. AMS **171** (1972), 195-234.

[9] Constantin Carathéodory, *Algebraic theory of measure and integration*. (2nd Ed.; Edited and with a preface by P. Finsler, A. Rosenthal and R. Steuerwald; Translated from the German by F. E. J. Linton.) (1986) Chelsea Publ. Co., New York.

[10] Chen C. Chang, *Algebraic analysis of many valued logics*. Trans. AMS **88** (1958), 467-490.

[11] Chen C. Chang, *A new proof of the completeness of the Lukasiewicz axioms*. Trans. AMS **93** (1959), 74-80.

[12] Roberto Cignoli, *Free lattice-ordered abelian groups and varieties of MV-algebras*. In *Proceedings of the IX Latin American Symposium on Mathematical Logic, Part 1* (Bahía Blanca, 1992); (1993) Univ. Nacional del Sur, Bahía Blanca, 113-118.

[13] Roberto Cignoli and Daniele Mundici, *An elementary proof of Chang's completeness theorem for the infinite-valued calculus of Lukasiewicz*. Studia Logica **58** (1) (1997), 79-97.

[14] Roberto Cignoli, Itala D'Ottaviano, and Daniele Mundici, *Algebraic Foundations of Many-Valued Reasoning*. (2000) Kluwer Acad. Publ., Dordrecht.

[15] Paul F. Conrad, *Representation of partially ordered abelian groups as groups of real valued functions*. Acta Math. **116** (1966), 199-221.

[16] Paul F. Conrad and J. Roger Teller, *Abelian pseudo lattice ordered groups*. Publ. Math. Debrecen **17** (1971), 223-241.

[17] Corrado De Concini and Claudio Procesi, *Complete symmetric varieties, II. Intersection theory*. In *Algebraic groups and related topics* (Kyoto/Nagoya, 1983), (1985) North-Holland, Amsterdam, 481-513.

[18] Jacques Dixmier, *C^*-Algebras*. (1977) North-Holland, Amsterdam.

[19] Edward G. Effros, *Dimensions and C^*-algebras*. (1981) Conference Board of the Mathematical Sciences, Washington, D.C.; American Mathematical Society, Providence, RI.

[20] Edward G. Effros, David E. Handelman, and Chao Liang Shen, *Dimension groups and their affine representations*. Amer. J. Math. **102** (2) (1980), 385-407.

[21] George A. Elliott, *On the classification of inductive limits of sequences of semisimple finite-dimensional algebras.* J. Algebra **38** (1) (1976), 29-44.

[22] George A. Elliott, *On totally ordered groups, and K_0.* In *Ring theory* (Proc. Conf., Univ. Waterloo, Waterloo, 1978); Lecture Notes in Mathematics **734** Springer Verlag, Berlin-Heidelberg-New York, 1-49.

[23] Gérard G. Emch, *Mathematical and Conceptual Foundations of 20th Century Physics.* (1984) North-Holland, Amsterdam.

[24] Euclid, *The thirteen books of Euclid's Elements translated from the text of Heiberg.* Vol. I: Introduction and Books I, II; Vol. II: Books III-IX; Vol. III: Books X-XIII and Appendix. Transl. with introduction and commentary by Thomas L. Heath, 2nd ed.; (1956) Dover Publ. Inc., New York.

[25] Günter Ewald, *Combinatorial Convexity and Algebraic Geometry.* (1996) Springer Verlag, Berlin-Heidelberg-New York.

[26] Andrew M. W. Glass, *Partially Ordered Groups.* (1999) World Scientific Publ. Co. Inc., River Edge, NJ.

[27] Kenneth R. Goodearl, *Notes on Real and Complex C^*-Algebras.* (1982) Birkhäuser, Boston.

[28] Kenneth R. Goodearl, *Partially Ordered Abelian Groups with Interpolation.* (1986) AMS Providence, RI.

[29] David Handelman, *Ultrasimplicial dimension groups.* Arch. Math. (Basel) **40** (2) (1983) 109-115.

[30] Richard V. Kadison and John R. Ringrose, *Fundamentals of the theory of operator algebras*, Volume II. (1986) Academic Press, San Diego, CA.

[31] Vincenzo Marra, *Every abelian ℓ-group is ultrasimplicial.* J. Algebra, **225** (2000), 872-884.

[32] Robert McNaughton, *A theorem about infinite-valued sentential logic.* J. Symbolic Logic **16** (1951), 1-13.

[33] Daniele Mundici, *Interpretation of AF C^*-algebras in Lukasiewicz sentential calculus.* J. Funct. Anal. **65** (1) (1986), 15-63.

[34] Daniele Mundici, *The derivative of truth in Lukasiewicz sentential calculus.* In *Methods and applications of mathematical logic* (Campinas, 1985): Contemp. Math. (1988), AMS, Providence, RI, 209-227.

[35] Daniele Mundici, *Farey stellar subdivisions, ultrasimplicial groups, and K_0 of AF C^*-algebras.* Advances in Math. **68** (1) (1988), 23-39.

[36] Daniele Mundici, *A constructive proof of McNaughton's theorem in infinite-valued logic.* J. Symbolic Logic **59** (2) (1994), 596-602.

[37] Daniele Mundici, *Non-boolean partitions and their logic.* Soft Computing **2** (1988), 18-22.

[38] Daniele Mundici, *Classes of ultrasimplicial lattice-ordered abelian groups.* J. Algebra **213** (2) (1999), 596-603.

[39] Daniele Mundici and Giovanni Panti, *A constructive proof that every 3-generated l-group is ultrasimplicial.* In *Logic, Algebra, and Computer Science* (Warsaw, 1996); (1999) Banach Center Publ., Polish Acad. Sci., Warsaw, 169-178.

[40] Daniele Mundici and Giovanni Panti, *Decidable and undecidable prime theories in infinite-valued logic.* Annals of Pure and Applied Logic **108** (2001) 269-278.

[41] Tadao Oda, *Convex Bodies and Algebraic Geometry.* (1988) Springer Verlag, Berlin-Heidelberg-New York.

[42] Giovanni Panti, *A geometric proof of the completeness of the Lukasiewicz calculus.* J. Symbolic Logic **60** (2) (1995), 563-578.

[43] Alan Rose and J. Barkley Rosser, *Fragments of many-valued statement calculi.* Trans. AMS **87** (1958), 1-53.

[44] Alexander Schrijver, *Theory of Linear and Integer Programming.* Wiley-Interscience Publ. (1986). John Wiley & Sons Ltd., Chichester, 1986.

[45] Alfred Tarski, *Logic, Semantics, Metamathematics.* Papers from 1923 to 1938. (Oxford at the Clarendon Press, 1956;) Transl. by J. H. Woodger; (1983) Reprinted by Hackett, Indianapolis.

[46] J. Roger Teller, *On abelian pseudo lattice ordered groups.* Pacific J. Math. **27** (1968) 411-419.

[47] Mordchaj Wajsberg, *Beiträge zum Metaaussagenkalkül I.* Monatshefte für Mathematik und Physik **42** (1935), 221-242.

[48] Elliot Carl Weinberg, *Free lattice-ordered abelian groups.* Math. Ann. **151** (1963), 187-199.

[49] Ryszard Wójcicki, *On matrix representations of consequence operations of Lukasiewicz's sentential calculi.* Z. Math. Logik Grundl. Math. **19** (1973) 239-247. (Reprinted in [50].)

[50] Ryszard Wójcicki and Grzegorz Malinowski, editors. *Selected papers on Lukas-iewicz sentential calculi.* Zakład Narodowy imienia Ossolińskich, Wydawnictwo Polskiej Akademii Nauk, (1977); Ossolineum, Polish Academy of Sciences, Wrocław.

[51] Kôsaku Yosida, *On vector lattice with a unit.* Proc. Imp. Acad. Tokyo **17** (1941), 121-124.

[52] Günter M. Ziegler, *Lectures on Polytopes.* (1995) Springer-Verlag, Berlin-Heidel-berg-New York.

Computer Science Department, University of Milan, Via Comelico 39-41,
20135 Milan, Italy
marra@dsi.unimi.it
mundici@mailserver.unimi.it

Hull Classses of Archimedean Lattice-Ordered Groups with Unit: A Survey

Jorge Martínez

ABSTRACT. This is a survey of the literature on hull classes of archimedean lattice-ordered groups with a designated unit. There has been a substantial amount of activity in this specialty in the last decade, and the goal here is to put the subject in some perspective, with an account of some of the history of accomplishments, as well as of the most recent progress.

1 Introduction

The category that will frame this exposition is $\mathbf{W}$, the category whose objects are the archimedean lattice-ordered groups – henceforth ℓ-groups – with a designated weak order unit $u > 0$, and whose morphisms are the ℓ-homomorphisms which preserve the designated unit. As a category of discourse, $\mathbf{W}$ has been successfully promoted in the work of R. N. Ball and A. W. Hager, throughout the eighties and nineties. This is not to say that their research has neglected the category $\mathbf{Arch}$ of all archimedean ℓ-groups, and, indeed they have, in many instances, carried over results about $\mathbf{W}$ to $\mathbf{Arch}$. Many challenging questions remain concerning both these categories, questions which are likely to occupy researchers for some time. This article makes a choice: to focus on $\mathbf{W}$; there is plenty to think about in that category.

$\mathbf{W}$ has a certain appeal, which by now seems natural. This attraction stems, in part, from the interplay between algebra and topology, which in turn has its roots in Stone duality. Or one could put it more bluntly: $\mathbf{W}$-objects are naturally described in their Yosida representation. In this introduction the Yosida Representation Theorem will be stated in full generality (in 1.1). But in spite of the interplay with topology that this representation theorem gives rise to, the focus will remain on $\mathbf{W}$, and only incidental or, one might say, inevitable references to topology will be made. Inevitable, for example, in this sort of context, means reminding the reader that, for any topological space X, $C(X)$ is the ring of all continuous real valued functions defined on X; it is a

J. Martínez (ed.), Ordered Algebraic Structures, 89–121.
© 2002 *Kluwer Academic Publishers.*

ring under pointwise operations, and a commutative semiprime f-ring at that. It will also be regarded as an archimedean ℓ-group with designated unit 1.

For any **W**-object G with designated unit $u > 0$, G^* stands for the convex ℓ-subgroup of G generated by u. Specifically,

$$G^* = \{\, g \in G : \exists n \in \mathbb{N},\ |g| \leq nu \,\}.$$

As the title indicates, this is a survey on hulls and hull classes in **W**. Several of the hull classes discussed here may be considered in a wider context – such as, say, the category of all abelian ℓ-groups, or the category of semiprime rings (without any ordering). But, in this regard as well, the discussion will, to a large extent, be about hull classes in **W**.

Implicit in the stated goal is an effort to be thorough and accurate in giving references. On the other hand, it seems reasonable to assume that any serious and thoughtful reader will make the effort to supplement the references given here, as needed or desired. This applies, in particular, to background material on ℓ-groups. The concepts that seem most crucial to the discussion of the paper will be recalled. For any ℓ-group-theoretical terminology that remains unexplained, the reader is referred to either [BKW77] or [D95], or else [AF88]. The Yosida Representation Theorem is most efficiently discussed (in modern language) in [HR77]. Categorical and topological references will be recorded as needed, and the same applies to any other algebraic facts or constructs.

Unless otherwise noted, all groups in this paper are abelian, all rings commutative, and all topological spaces are Hausdorff, and when regarding $C(X)$ one might as well assume – see Chapter 3 of [GJ76] for the reasons – that the space X is *Tychonoff*; that is, Hausdorff and such that for any point $p \in X$ and any closed set K not containing X, there is an $f \in C(X)$ such that $f(p) = 1$ and $F(K) = \{0\}$. βX will stand for the Stone-Čech compactification of the Tychonoff space X.

Definition & Remarks 1.1. To begin, we recall that a *value* in an ℓ-group G is a convex ℓ-subgroup L which is maximal with respect to not containing some element a of G. We then say that L is a *value of* a. It is well established that each value is a prime convex ℓ-subgroup, and, evidently, any two values of the same element are incomparable.

YG stands for the *Yosida space* of the **W**-object G; that is to say, the space of values of the designated unit, with the hull-kernel topology. It is well known that YG is a compact Hausdorff space.

Now for any Tychonoff space X, $D(X)$ shall denote the set of all continuous functions $f : X \longrightarrow \mathbb{R} \cup \{\pm\infty\}$, where the range is the extended reals with the usual topology, such that $f^{-1}\mathbb{R}$ is a dense subset of X. $D(X)$ is a lattice under pointwise operations, but not a group or ring under the obvious pointwise operations, unless some assumptions are made about X.

It does make sense to talk about an "ℓ-group in $D(X)$": a subset $H \subseteq D(X)$ which is an ℓ-group such that under the lattice operations it is a sublattice of $D(X)$, and, for each $h, k \in H$, $(h + k)(x) = h(x) + k(x)$ on a dense subset of X. An ℓ-group H in $D(X)$ *separates the points of* X if, for each pair of points $x \neq y$ in X, there is an $h \in H$ such that $h(x) \neq h(y)$.

With all this in mind, here is the Yosida Representation Theorem:

For each **W**-object G with designated unit $u > 0$, there is an ℓ-isomorphism ϕ of G onto an ℓ-group G' in $D(YG)$ such that $\phi(u) = 1$, and G' separates the points of YG.

The Yosida space is unique (up to homeomorphism) in the following sense:

Suppose that Z is a compact space and that there is an ℓ-isomorphism θ of G onto an ℓ-group H in $D(Z)$ which separates the points of Z and such that $\theta(u) = 1$, then there is a homeomorphism $t : Z \longrightarrow YG$ such that, for each $g \in G$ and $z \in Z$, $\phi(g)(tz) = \theta(g)(z)$.

Finally, if the unit u of G is strong, that is, if $G = G^*$, the image of the Yosida Representation lies in $C(YG)$.

Here is the definition of a hull class. The reader will note the passing references to essential extensions and the essential closure. The reader may refer to these concepts in [BKW77], or else stay tuned for their discussion in the next section.

Definition & Remarks 1.2. Consider a class H of **W**-objects closed under formation of ℓ-isomorphic copies. We call the extension $G \leq hG$ in **W** an H-*hull* of G if it is an essential extension, $hG \in$ H, and

$$H \in \text{H} \Rightarrow \exists g : hG \longrightarrow H, \quad \text{an embedding in } \textbf{W}, \text{ extending the identity on } G.$$

Proposition 2.4 of [HM99b] tells us that each **W**-object G has an H-hull if and only if H is *essentially intersective*; that is, for each essentially closed **W**-object E, and each collection $\mathcal{B}$ of subobjects of E, such that $\mathcal{B} \subseteq$ H and $\bigcap \mathcal{B}$ is essential in E, then $\bigcap \mathcal{B} \in$ H.

A class H with these features is called a *hull class*. The reader will easily verify that every hull class contains all the essentially closed **W**-objects, and is therefore nontrivial. The operator h is a *hull (operator) associated with* H.

The simplest example of a hull class of **W**-objects is perhaps **D**, the class of divisible **W**-objects. The accompanying hull is the usual divisible hull, which will be denoted by d. For an ℓ-group G, dG is ordered as follows: $0 < x \in dG$ if and only if $nx > 0$ in G, for a suitable natural number n. It is well known that $G \leq dG$ is an essential extension, and that the map $L \mapsto dL$ defines a lattice isomorphism from the lattice

$C(G)$ of all convex ℓ-subgroups of G, onto $C(dG)$. (The latter is dismissively dealt with in [BKW77, Lemma 5.5.1].) d is, in fact, a reflection; see §4.

In the remarks which follow on uniform convergence we shall be very brief, while counting on the reader to either have some intuition about those ideas, or else to review the appropriate references.

Examples 1.3. **U** and **RU** stand, respectively, for the subclasses of uniformly complete **W**-objects and the *relatively* uniformly conplete ones. We shall not dwell, for example, on the distinction between these two strains of convergence, referring the reader to [LZ71] or [HM96] instead. It is easy to see, and well known, that both of these are hull classes.

In the case of **RU** the associated hull ru must be constructed by transfinite iteration. Starting with a **W**-object G, $(ru)^1 G$ collects the relative uniform limits in G^e (the essential closure of G) of sequences in G, and this process is carried forward by transfinite induction up to the ordinal ω_1. For details the reader is referred to [LZ71], [V69] and [HM99b]. There is also a discussion of relative uniform convergence in Chapter XV, §13, of [Bi67].

There is more to **RU** and **U**. They are monoreflective subcategories of **W**. We take these up in §4; the hull operator u associated with **U** is also a reflection, and we prefer to revisit it in that context.

Historically, the following have been some of the most prominent examples of hull classes in **W**. One could simply argue that these are the most natural hulls of ℓ-groups, having to do, after all, with one of the most basic instincts of anyone who deals with lattices; namely, the desire to enlarge a structure so as to create certain suprema and infima, and to do it in a minimal way.

For a more detailed discussion we refer the reader to [HM99b]. We shall try, here and in the next two sections, to be as accurate as possible with attributions; in that respect too the reader may rely on [HM99b]. Prudence and experience cautions us, however, to apologize in advance for misattributions or omissions.

Examples 1.4. In this article α will denote an infinite cardinal or else the symbol ∞, which may be thought of as exceeding all cardinals. The reader may prefer to think of ∞ in this situation as the one for which *no* cardinality restrictions are made.

(a) Here are the main definitions; explanations of terms follow in (b):

- **P**(α) is the class of all α-projectable objects in **W**. That is, $G \in$ **P**(α) if and only if $G = S \oplus S^\perp$, for any α-generated polar S.

- **C**(α) is the class of all conditionally α-complete objects.

- **L**(α) is the class of all laterally α-complete objects.

(b) Now let us briefly explain the terms employed in the above definitions.

(i) First, recall that if G is any ℓ-group and $S \subseteq G$ then

$$S^\perp = \{ x \in G \ : \ |x| \wedge |s| = 0, \ \forall s \in S \}.$$

A convex ℓ-subgroup of the form $S^\perp$ is called a *polar* of G. If $L \in C(G)$ then $L^\perp$ is the largest convex ℓ-subgroup which satisfies $L \cap L^\perp = \{0\}$. It is well known that $\mathcal{P}(G)$, the set of all polars of G, is a complete boolean algebra under inclusion. This is essentially a result about Brouwerian lattices, which goes back to Glivenko [Gv29] for lattices; it appears in this form in [Bi67], Theorem 26, p. 130. For ℓ-groups it is due to Šik [Si62].

(ii) An α-*generated* polar K is one for which $K = S^{\perp\perp}$, for some subset S of G with $|S| < \alpha$. Let $\mathcal{P}_\alpha(G)$ stand for the subalgebra of $\mathcal{P}(G)$ generated by all α-generated polars of G.

(iii) G is *conditionally α-complete* if every bounded set $S \subseteq G$ with $|S| < \alpha$ has a supremum in G.

(iv) G is *(boundedly) laterally α-complete* if every (bounded) subset S of G of elements which are pairwise disjoint, and with $|S| < \alpha$, has a supremum in G.

(c) For each α, $\mathbf{P}(\alpha)$, $\mathbf{C}(\alpha)$ and $\mathbf{L}(\alpha)$ are hull classes; this is summarized in Theorem 2.9 of [HM99b].

2 Classical Results

In this section we give a brief account of the development in the literature of the hulls and hull classes introduced in 1.4.

By introducing the auxiliary $\mathbf{bL}(\alpha)$, the class of all boundedly laterally α-complete $\mathbf{W}$-objects, one is able to show (see [HM96, Theorem 3.3]) that

$$\mathbf{bL}(\alpha) \subseteq \mathbf{P}(\alpha),$$

for each cardinal $\alpha > \omega_0$, and from there it is easy to see that

$$\mathbf{C}(\alpha), \mathbf{L}(\alpha) \subseteq \mathbf{bL}(\alpha) \subseteq \mathbf{P}(\alpha),$$

for all such α.

The history of this type of result, for the two extreme cases, $\alpha = \infty$ and $\alpha = \omega_1$, should be highlighted. Before doing that, one should reflect on what these classes are for $\alpha = \omega_0$. A moment's thought will make it clear that

$$\mathbf{W} = \mathbf{C}(\omega_0) = \mathbf{L}(\omega_0),$$

As to $\mathbf{P}(\omega_0)$, this is the class of all *projectable ℓ-groups*, and we will have more to say about this class in 3.3(a).

Traditionally, the objects in $\mathbf{P}(\infty)$, $\mathbf{C}(\infty)$ and $\mathbf{L}(\infty)$ have been referred to as *strongly projectable*, *conditionally complete* and *laterally complete*, respectively. The statement "$\mathbf{C}(\infty) \subseteq \mathbf{P}(\infty)$" was, apparently, first shown by F. Riesz, in [Ri40]. This citation is also made by Fuchs, in Theorem 16, p. 91, [Fu63]. The statement "$\mathbf{L}(\infty) \subseteq \mathbf{P}(\infty)$" is usually attributed to S. Bernau [Be75a].

The existence of the lateral completion $G \leq l(\infty)G$ was established by Conrad, in [C69], for the class of all representable ℓ-groups, which include all abelian ℓ-groups. For arbitrary ℓ-groups the result on existence followed a few years later, in Bernau's [Be75a].

Now $\alpha = \omega_1$: the objects in $\mathbf{P}(\omega_1)$, $\mathbf{C}(\omega_1)$ and $\mathbf{L}(\omega_1)$ have been called *σ-projectable*, *conditionally σ-complete* and *laterally σ-complete*, respectively. Veksler and Geiler [VG72] prove, actually, that $\mathbf{bL}(\omega_1) \subseteq \mathbf{P}(\omega_1)$, for vector lattices.

Until now we have said nothing about the hull operators associated with the hull classes discussed in this section. We take that up next. This is also the appropriate place to review the essential closure.

Definition & Remarks 2.1. (a) Suppose that G is an ℓ-group and H is an ℓ-subgroup of G. H is *essential* or *large* in G, or that G is an *essential extension* of H, if for each $0 < g \in G$ there is an element $0 < h \in H$ and a positive integer n such that $g \leq nh$. For archimedean ℓ-groups, H is essential in G if and only if the trace map $P \mapsto P \cap H$, from $\mathcal{P}(G)$ to $\mathcal{P}(H)$ is an isomorphism of Boolean algebras. (See [BKW77, Theorem 11.1.15]; the sufficiency of this result, which is the least obvious, is first proved in [C71, Theorem 3.5].)

G is *essentially closed* (in $\mathbf{W}$) if it has no proper essential extensions. (This is one place where the universe looks very different if one does not stick to archimedean ℓ-groups. In the class of all abelian ℓ-groups, for example, essentially closed objects do not exist, because one can always extend an ℓ-group essentially, by forming a lexicographic product with the totally ordered group of integers, $\mathbb{Z}$.)

(b) In [C71] Conrad showed that every archimedean ℓ-group G can be embedded in an essentially closed archimedean ℓ-group G^e, so that $G \leq G^e$ is essential. This is the *essential closure of G*.

Conrad goes on to characterize the essentially closed archimedean ℓ-groups; the following are equivalent:

(i) G is essentially closed.

(ii) G is divisible, laterally and conditionally complete.

(iii) $G \cong D(X)$, for some compact extremally disconnected space X. (Recall that a space X is *extremally disconnected* if the closure of any open set is open. We shall have more to say about this topological connection in the section on epireflections.)

(c) Using direct limits one can give a description of the essential closure of the ring $C(X)$; we refer the reader to [FGL65] for details. Recall, first, that if $\mathcal{U}$ is any filter base of dense subsets of a space X, then $C[\mathcal{U}]$ denotes the direct limit of the rings $C(V)$ (for all $V \in \mathcal{U}$), with domain restrictions being the bonding maps.

Let $\mathcal{G}_\delta(X)$ denote the filter base of dense G_δ-sets of the Tychonoff space X. Then (Theorem 4.6, [FGL65])

$$C(X)^e = C[\mathcal{G}_\delta(\beta X)].$$

Remarks 2.2. (a) As before, α denotes an infinite cardinal number, or else ∞. For the same reasons as in [HM99b], we must assume on occasion that α is also a regular cardinal; this is clearly indicated below. We re-introduce, from [HM99b], the following hull operators; for each **W**-object G, $p(\alpha)G$, $c(\alpha)G$ and $l(\alpha)G$ are, respectively, the least α-projectable, conditionally α-complete, and laterally α-complete **W**-object contained in G^e, containing G.

(b) Now the α-projectable hull $p(\alpha)G$ of G is constructed via a transfinite induction process, which is described in [HM99b, §3]; the transfinite construction is generalized in §3 of this article. The reader should refer to [M02], elsewhere in this volume, for a more elaborate and amply illustrated account of this subject. In the two extreme cases, $\alpha = \infty$ and $\alpha = \omega_0$, the respective projectable hulls are created in one step of this induction; see [M02] and also §3.

If G is already α-projectable one has the following descriptions of the positive cones of the conditionally and laterally α-complete hulls, respectively; the suprema that appear below are to be formed in G^e:

$$c(\alpha)G^+ = \{ \vee_i x_i : 0 \le x_i \in G, \ \{x_i : i \in I\} \ G\text{-bounded}, \ |I| < \alpha \};$$

$$l(\alpha)G^+ = \{ \vee_i x_i : 0 \le x_i \in G, \ i \ne i' \ \Rightarrow \ x_i \wedge x_{i'}0 = 0, \ i \in I, \ |I| < \alpha \}.$$

Note that a set S in G^e is *G-bounded* if there is an $x \in G$ such that $S \le x$.

(c) In [HM99b] one may find some useful hull identities, which we record here: suppose that $\omega_1 \le \alpha \le \infty$. If G is a **W**-object we have:

(i) $c(\alpha) \cdot l(\alpha) = l(\alpha) \cdot c(\alpha)$, for each regular cardinal α [HM99b, Theorem 4.1(a)].

(ii) For each $\alpha \ge \omega_1$,

$$((ru)^1 \cdot d \cdot p(\alpha))G = (u \cdot d \cdot bl(\alpha))G = ((ru)^1 \cdot d \cdot bl(\alpha))G = (c(\alpha) \cdot d)G,$$

and $(c(\alpha) \cdot d)G$ is the convex ℓ-subgroup of $D(Y(p(\alpha)G))$ generated by G [HM99b, Theorem 5.5(a)].

(iii) For each $\alpha \geq \omega_1$,

$$(u \cdot d \cdot l(\alpha))G = ((ru)^1 \cdot d \cdot l(\alpha))G = e(\alpha)G = D(Y(p(\alpha)G)).$$

[HM99b, Theorem 5.5(b)]

In (ii), $bl(\alpha)$ denotes the hull operator associated with the class $\mathbf{bL}(\alpha)$. Note also that $e(\alpha)$ is the composite $c(\alpha) \cdot l(\alpha) \cdot d$, and that it is the hull operator for the class of divisible objects, which are also both laterally and conditionally α-complete; we shall have occasion to briefly refer to this class again; we fix the notation $\mathbf{E}(\alpha)$ to do that.

We remind the reader as well that u denotes the uniform completion operator (see §4), and $(ru)^1 G$ stands for the adjunction of all relative uniform limits in G^e of sequences in G.

(d) Some comments on the development of the identities in (c) are in order.

Item (i) appears, for $\alpha = \infty$, in [C71, Proposition 4.2], applied to subdirect products of reals in which all finitely nonzero functions are present. It is our recollection, nonetheless, that Conrad, in a number of conversations over time, has always attributed this property of lateral and conditional completions to Bernau. The reader is referred to Bernau's [Be75b]. It is stated as Theorem 8.2.6 in [AF88], and a straightforward proof is given there.

Note that $e(\infty)$ is the essential closure.

$c(\infty)$ is none other than the familiar Dedekind cut completion operator, which is described in [Fu63], Chapter V, §10. Fuchs attributes the result that, for an archimedean ℓ-group G the Dedekind cut completion of G yields a conditionally complete ℓ-group to a host of authors; see Theorem [Fu63]. [Bi67], unfortunately, refers one to the aforementioned spot in [Fu63]. A description of $c(\infty)$, like the one in (ii) and for subdirect products of reals, may also be found in the article [CMc69] of Conrad and McAlister; (see, for example, Theorem 3.3.)

Finally, and in the context of vector lattices, reference should once again be made to the work of Veksler [V69], where it is shown that $(ru)^1 \cdot bl(\infty) = c(\infty)$, and to [VG72], where the role of relative uniform convergence in this context is discussed.

The reader will doubtless notice the prominence of the divisible hull in the preceding remarks, and, in particular, in part (b). It seems reasonable to ask what happens if divisibility is discarded. The comments that follow address some aspects of this problem.

Definition & Remarks 2.3. In [HM98b] the subcategory $\mathbf{W_s}$ of $\mathbf{W}$ is the focus of attention. First, here is some background.

(a) Let G be an ℓ-group and $0 < s \in G$. Then s is *singular* if $g \wedge (s - g) = 0$, for each $0 \leq g \leq s$, with $g \in G$. An object in $\mathbf{W}$ is said to be *singular* if the designated unit is singular. As is explained in [HM98b], G is singular if and only if the Yosida representation of G carries it into

$$D(YG, \mathbb{Z}) = \{ f \in D(Y) : f(y) \text{ real} \Rightarrow f(y) \in \mathbb{Z} \}.$$

$\mathbf{W_s}$ is the full subcategory of all singular $\mathbf{W}$-objects.

(b) The following remarks conflate several of the results in [HM98b], and apply 2.2(b). We phrase things using the current notation, as opposed to that of [HM98b], which is slightly different. We note, as it represents a curious omission, that items (iii) and (iv) below are not explicitly stated in [HM98b], in general, but only for $\alpha = \infty$, for lateral completion, and not at all for conditional completions.

(i) Every singular $\mathbf{W}$-object which is laterally α-complete is also conditionally α-complete. Indeed, if G is a $\mathbf{W_s}$-object, then $G \in \mathbf{L}(\alpha)$ if and only if $G \cong D(X, \mathbb{Z})$, where X is a compact α-disconnected space [HM98b, Proposition 6.6].

(ii) The lateral α-completion of a singular $\mathbf{W}$-object is singular; in fact, $l(\alpha)G = l(\alpha)S(G)$, where $S(G)$ denotes the subgroup generated by the characteristic functions of G. [HM98b, Proposition 6.2]

(iii) The lateral α-completion of a singular $\mathbf{W}$-object G is $D(Y(p(\alpha)G, \mathbb{Z})$. See [HM98b, Theorem 6.10], for the $\alpha = \infty$ case.

(iv) Suppose that G is in $\mathbf{W_s}$. Then $c(\alpha)G \leq l(\alpha)G$. Now $c(\alpha)G$ is the convex ℓ-subgroup of $l(\alpha)G$ generated by G. In particular, $c(\alpha)G$ is also singular. If G is a subdirect product of integers then so is $c(\alpha)G$. (The first statement follows from (i); the second is then obvious, or one may refer to [CMc69, Lemma 2.3(2)] for $\alpha = \infty$. The third is also clear. The latter is implicit, for $\alpha = \infty$, in [CMc69, Theorem 4.8].

(c) In the above statements, there remains only to explain the term "α-disconnected". (This is one of those instances where bringing up topological concerns is practically inevitable.)

Let X be a Tychonoff space. We assume that the reader is familiar with the notion of a cozeroset ([GJ76]). Now for α a cardinal as in the preceding discussion, an open subset U of X is called an α-cozeroset if it is the union of fewer than α cozerosets. (Note that every open set is an ∞-cozeroset, and "ω_1-cozeroset" simply means "cozeroset".

Next, X is α-disconnected if each α-cozeroset has open (and therefore clopen) closure. If $\alpha = \infty$ we have the extremally disconnected spaces, whereas if $\alpha = \omega_1$ we get the basically disconnected spaces. It is part of Corollary 2.4 in [HM96] that the Yosida space of any α-projectable $\mathbf{W}$-object is α-disconnected.

Mention of the Stone-Nakano Theorem may be somewhat anticlimactic here. The assertion – generalized for each α – is that, for X Tychonoff, $C(X)$ is conditionally

α-complete if and only if X is α-disconnected. For $\alpha = \infty$ this appears in [FGL65], Theorem 4.12. The reader may also find it as an exercise (3N, [GJ76]) for the extreme cases ω_1 and ∞.

Now in [FGL65] it is also shown (Theorem 4.11) that $c(\infty)C(X)$ is the convex ℓ-subgroup generated by $C(X)$ in $C(X)^e = C[\mathcal{G}_\delta(X)]$. Compare, for $\alpha = \infty$, with the hull identity in 2.2(c)(ii).

In the classical completion of the rational numbers, one has the option of completing by cuts, and we have already considered generalizations of that process with the operators $c(\alpha)$, or completing by introducing limits for Cauchy sequences. The latter might be considered as the topological/analytic approach. This approach has also been generalized for archimedean ℓ groups.

Definition & Remarks 2.4. o-completeness.

(a) In a **W**-object A a sequence $(s_n)_{n\in\mathbb{N}}$ is said to be o-*Cauchy* if there is a decreasing sequence $(v_n)_{n\in\mathbb{N}}$ in A, such that $\wedge_n v_n = 0$ and $|s_n - s_{n+p}| \leq v_n$, for all $n, p \in \mathbb{N}$. The sequence $(s_n)_{n\in\mathbb{N}}$ o-*converges* if there is an $s \in A$ and $(v_n)_{n\in\mathbb{N}}$ in A, such that $\wedge_n v_n = 0$ and $|s_n - s| \leq v_n$, for all $n, p \in \mathbb{N}$. A is o-*complete* if every o-Cauchy sequence o-converges.

The earliest account of a generalization of the construction of the real numbers from Cauchy sequences is in Everett's [Ev44]. Here is an account of the principal contributions on o-completeness and o-completeness from the literature:

(i) Every o-complete divisible **W**-object is uniformly complete. [LZ71, Theorem 16.2(i)])

(ii) G is o-complete if and only if for each pair of sequences in G

$$a_1 \leq a_2 \leq \ldots \leq \ldots \leq b_2 \leq b_1,$$

such that $\wedge_n b_n - a_n = 0$, there is a $c \in G$ such that $\vee_n a_n = c = \wedge_n b_n$. ([Pa64]) Evidently, when such a c exists it is unique.

(iii) Obvious from (ii): If G is conditionally ω_1-complete then it is also o-complete.

(iv) If G is a divisible **W**-object, then G is o-complete if and only if it is uniformly complete and
$$(a + b)^{\perp\perp} = a^\perp + b^\perp,$$
for each $a, b \in G$. [dP81, §11]

(v) If X is any Tychonoff space, then $C(X)$ is o-complete if and only if X is a quasi F-space. See [DHH80, Theorem 3.7]. Recall that X is *quasi F* if each dense cozeroset of X is C^*-embedded.

(b) We let **O** stand for the class of all o-complete **W**-objects. It is a hull class, and the corresponding hull operator $G \leq o_C G$ is the traditional extension of G by considering all o-Cauchy sequences in G modulo the ideal of sequences which o-converge to zero. [1]

(c) In view of (i) and (iii) in (a), we have, for divisible **W**-objects, the following strict inclusions:

$$\mathbf{C}(\infty) \subset \mathbf{C}(\omega_1) \subset \mathbf{O} \subset \mathbf{U}.$$

In the comments of 3.5 further on we briefly take up the subject of covering classes of compact spaces. Now a substantial amount of information on quasi F-spaces as a covering class may be found in the following references: [DHH80], [HVW87], [HP83], [dP81].

In closing, here are some comments on the o-completion of a ring of continuous functions.

It is shown in [DHH80, Corollary 3.5] that, for any Tychonoff space X, $o_C C(X)$ can be constructed as the uniform completion of the convex ℓ-subgroup generated by $C(X)$ in the direct limit $C[\mathcal{C}(\beta X)]$, where $\mathcal{C}(X)$ denotes the filter base of dense cozerosets of X. Note that $o_C C(X)$ may also be obtained as the convex ℓ-subgroup of $C[\mathcal{C}_\delta(\beta X)]$ generated by $C(X)$, where $\mathcal{C}_\delta(X)$ stands for the filter base of dense sets which are also countable intersections of cozerosets of X. [2]

3   Polar Functions: Propaganda

The projectable and α-projectable hulls discussed in the previous section are examples of a more general class of hull operators. This gets us into a discussion of polar functions, of which a great deal is said in [M02], in this volume, continuing in [HM∞b], and with ongoing work in [HM∞c]. We recommend [M02] to the reader; here we limit ourselves to a review of the basic concepts.

We let $\mathcal{S}(G)$ denote the subalgebra of $\mathcal{P}(G)$ consisting of all summands of G.

Definition & Remarks 3.1. (a) Suppose that $\mathcal{X}$ denotes a function which assigns to a **W**-object G a subalgebra $\mathcal{X}(G)$ of $\mathcal{P}(G)$ such that $\mathcal{S}(G) \subseteq \mathcal{X}(G)$; we then call $\mathcal{X}$ a *polar function*. We say that $\mathcal{X}$ is *invariant* if for each essential extension $G \leq H$, the assignment $K \mapsto K^{\perp_H \perp_H}$ carries $\mathcal{X}(G)$ into $\mathcal{X}(H)$. We use the abbreviation *ipf*, short for "invariant polar function".

Suppose $G \leq H \leq G^e$. We say that H is an $\mathcal{X}$-*splitting extension* of G, if $K^{\perp_H \perp_H} \in \mathcal{S}(H)$, for each $K \in \mathcal{X}(G)$. The idea of an $\mathcal{X}$-splitting extension is already

---

[1]The construction itself is due to Everett ([Ev44]), although the fact that $o_C G$ is actually o-complete, in general, was not settled until the work of Papangelou ([Pa64]).

[2]For the record, the Yosida space of $o_C C(X)$ is none other than the quasi F-cover of βX [DHH80, Theorem 3.9].

evident in the exposition of [Bl74], dealing with projectable and strongly projectable hulls.

(b) Suppose that $\mathcal{X}$ denotes a polar function. Consider a **W**-object G and the embedding $G \leq G^e$ in its esential closure. Since the latter is laterally complete, the polar $K^{\perp_{G^e} \perp_{G^e}}$ is a summand of G^e, for each $K \in \mathcal{P}(G)$. The following therefore makes sense: for each $g \in G$ let $g[K]$ denote the projection of g on $K^{\perp_{G^e} \perp_{G^e}}$.

Lemma 2.2, [M02], insures that for each $K \in \mathcal{P}(G)$ and each extension H of G in G^e, $K^{\perp_H \perp_H}$ is a summand of H if and only if, for each $h \in H$, $h[K] \in H$. This essentially yields:

> [M02, Theorem 2.4] For each **W**-object G and each polar function $\mathcal{X}$, there is a least $\mathcal{X}$-splitting extension of G, denoted $G[\mathcal{X}]$. The elements of $G[\mathcal{X}]$ are the finite sums of the form
>
> $$\sum_{i=1}^{n} g_i[K_i],$$
>
> where each $g_i \in G$ and each $K_i \in \mathcal{X}$ and $K_i \cap K_j = \{0\}$, whenever $i \neq j$.

To consider hull classes in this context, we need the notion of the idempotent closure of a polar function. It is defined transfinitely in §5 of [M02]; let us summarize it briefly here.[3]

Definition & Remarks 3.2. (a) Suppose that $\mathcal{X}$ is a polar function. Define, at the first step of the induction, $\mathcal{X}^1 \equiv \mathcal{X}$. Assume now that λ is an ordinal, and that for each ordinal $\gamma < \lambda$, $\mathcal{X}^\gamma$ is defined, such that for each **W**-object G, and ordinals $\gamma < \delta < \lambda$ then $\mathcal{X}^\gamma(G)$ is a subalgebra of $\mathcal{X}^\delta(G)$. If λ is a limit ordinal,

$$\mathcal{X}^\lambda(G) \equiv \bigcup_{\gamma < \lambda} \mathcal{X}^\gamma(G), \quad \text{for each } \mathbf{W}\text{-object } G.$$

On the other hand, if κ precedes λ, then set

$$\mathcal{X}^\lambda(G) \equiv \{G \cap P : P \in \mathcal{X}(G^\kappa[\mathcal{X}])\}.$$

We then have a transfinite sequence of polar functions

(†) $$\mathcal{X} = \mathcal{X}^1 \leq \cdots \leq \mathcal{X}^\lambda \leq \cdots \leq \mathcal{P}.$$

(Note: for polar functions $\mathcal{X}$ and $\mathcal{Y}$, $\mathcal{X} \leq \mathcal{Y}$ means that $\mathcal{X}(G) \leq \mathcal{Y}(G)$, for each **W**-object G.) If $\mathcal{X}$ is an ipf, then each member of the sequence $\mathcal{X}^\lambda$ is also an ipf [M02,

---

[3]The reader of [M02] will find there the topological "dual" of this construction, and, indeed, also the notion dual to invariant polar functions, which attaches to a compact space X the least cover which makes clopen sets out of the members of a selected subalgebra of regular closed sets of X.

Proposition 5.2]. Then (†) must stabilize; that is, for each **W**-object G there is an ordinal λ such that each $\mu > \lambda$, $\mathcal{X}^\lambda(G) = \mathcal{X}^\mu(G)$. Finally, for each **W**-object G, let

$$\mathcal{X}^b(G) \equiv \mathcal{X}^\tau(G),$$

where τ is the least ordinal such that $G^{\tau^0}[\mathcal{X}] = G^\tau[\mathcal{X}]$, for each ordinal $\tau' > \tau$. This is the *idempotent closure* of $\mathcal{X}$. $\mathcal{X}$ is *idempotent* if $\mathcal{X}^b = \mathcal{X}$. If $\mathcal{X}$ is invariant then so is $\mathcal{X}^b$ [M02, Proposition 5.2].

[M02, Proposition 5.2] also records, for the least splitting extensions corresponding to the members of the sequence $\mathcal{X}^\lambda$ that, for each ordinal λ and each **W**-object G,

$$G[\mathcal{X}^{\lambda+1}] = G[\mathcal{X}^\lambda][\mathcal{X}],$$

and for each limit ordinal λ,

$$G[\mathcal{X}^\lambda] = \bigcup_{\gamma < \lambda} G[\mathcal{X}^\gamma].$$

In the limit we have that $G[\mathcal{X}^b] = G^\tau[\mathcal{X}]$, for suitable τ.

(b) For any ipf $\mathcal{X}$, the **W**-objects of the form $G[\mathcal{X}^b]$ form a hull class, denoted $\mathbb{H}(\mathcal{X})$. Furthermore, $G[\mathcal{X}^b]$ is the hull of G in $\mathbb{H}(\mathcal{X})$ [M02, Theorem 5.13(a)].

Let us now consider examples; some old, and at least one new.

Examples 3.3. (a) As explained in [M02, 2.7(b)], the polar function $\mathcal{P}_{w_0}$ which selects the finitely generated (or, equivalently, the principal) polars of an ℓ-group is an idempotent ipf; $G[\mathcal{P}_{w_0}]$ is the projectable hull of G. This way of realizing the projectable hull is already found in [Bl74]. In disguise, perhaps, it is also to be found in [Ch71] and [C73].

(b) $\mathcal{P}$ itself is an idempotent ipf, and $G \leq G[\mathcal{P}]$ is the embedding of G in its strongly projectable hull, as demonstrated in [C73] and [Bl74].

(c) For any uncountable cardinal α we have $\mathcal{P}_\alpha(G)$, the subalgebra generated by the α-generated polars of G. Note that $\mathcal{P}_\alpha$ is an ipf, but it is not idempotent. The transfinite construction alluded to in 2.2(b), carried out in [HM99b], is essentially the construction outlined in 3.2(a). That is to say, $p(\alpha)G = G[\mathcal{P}_\alpha^b]$, for each **W**-object G.

The following example has, very recently, received a good deal of attention, from a number of viewpoints.

Example 3.4. For each **W**-object G, let $\mathcal{P}_\alpha^\alpha(G)$ stand for the α-generated polars of G having an α-generated complement. It is easy to see that $\mathcal{P}_\alpha^\alpha(G)$ is a subalgebra of

$\mathcal{P}(G)$. As is demonstrated in [M02], 5.5, $\mathcal{P}_\alpha^\alpha$ is an idempotent ipf, for each uncountable, regular cardinal α.

The hull class associated with $\mathcal{P}_\alpha^\alpha$ consists of the so-called "α-splitting" ℓ-groups. The hull $G \leq G[\mathcal{P}_\alpha^\alpha]$ surfaces in [HM01d] in connection with the ring of α-quotients.

We should like the reader to have some idea of the further developments involving polar functions. We conclude this section with some remarks on the content of [HM∞b] and [HM∞c]. We shall also have occasion to single out [HM∞d] in this respect, but that reference belongs in §5.

Remarks 3.5. (a) First, this is another one of those instances where an incursion of topology is inevitable.

We consider **KTop₂**, the category of compact spaces with all continuous maps. A continuous surjection $f : Y \longrightarrow X$ is said to be *irreducible* if $X = f(A)$, for a closed subset A of Y, implies that $A = Y$. Let us regard $f \leq g$, for irreducible surjections $f : Y \longrightarrow X$ and $g : Z \longrightarrow X$, provided there is a continuous map $h : Z \longrightarrow Y$ (necessarily an irreducible surjection), such that $f \cdot h = g$. This defines a quasi-ordering, under which the relation $f \sim g$ defined by $f \leq g$ and $g \leq f$ defines an equivalence relation. Modulo that we have a poset of equivalence classes of compact spaces; this poset is denoted $\mathrm{Cov}(X)$. Note that if $f : Y \longrightarrow X$ and $g : Z \longrightarrow X$ are equivalent irreducible maps, then $Y \cong Z$.

Now suppose that $\mathcal{T}$ is a class of compact spaces which is invariant under formation of homeomorphic copies. It is called a *covering class* if for each X, $\mathrm{Cov}(X) \cap \mathcal{T}$ has a minimum element. In view of the comment at the conclusion of the preceding paragraph, and since $\mathcal{T}$ is closed under formation of homeomorphic copies, this concept is unambiguously defined.

There is a considerable literature on this subject, beginning with the work of Glea-son [Gl58], which is nicely presented in Chapter 10 of [W74]. Very readable accounts of irreducible maps and covers may be found in [H89] and [PW89]. We single out [V84] and [HVW89] as well; the former discusses the intractable basically disconnected cover, whereas the latter discusses a topological counterpart of the notion of a polar function.

By a *completion class* A of **W**-objects we mean a hull class with the additional feature that if G is uniformly complete then $G \in$ A if and only if $G^* \in$ A. §3 of [HM∞b] lays out the particulars of an inclusion preserving bijection between the class of completion classes of uniformly complete real f-algebras with identity and the class of covering classes of compact spaces.

For each ipf $\mathcal{X}$, the discussion in [HM∞b], describes a modification of the transfinite construction of 3.2(a), to produce the hull of a **W**-object in the completion class $\mathbf{U}^f \cap \mathbb{H}(\mathcal{X}^\flat)$, of all uniformly complete real f-algebras which have no proper $\mathcal{X}$-splitting extensions [HM∞b]. (Note: $\mathbf{U}^f$ stands for the class of uniformly complete real f-algebras.)

(b) We summarize the ideas of [HM∞c] briefly. This investigation extends the notion of a polar function: we consider functions which assign to each **W**-object G a *sublattice* $\mathcal{L}(G)$ of $\mathcal{P}(G)$ containing all the summands of G. "Invariance" of the polar function $\mathcal{L}$ has the same meaning as before.

Suppose that $\mathcal{L}$ is an ipf. An essential extension $G \leq H$ is said to be $\mathcal{L}$-*essential* if the induced $K \mapsto K^{\perp\perp}$, from $\mathcal{P}(G)$ to $\mathcal{P}(H)$, induces a lattice isomorphism from $\mathcal{L}(G)$ onto $\mathcal{L}(H)$. $\mathcal{L}$-essential extension is well behaved: if $G \leq H \leq H'$ are essential extensions, then $G \leq H'$ is $\mathcal{L}$-essential if and only if both $G \leq H$ and $H \leq H'$ are $\mathcal{L}$-essential.

Let $\widehat{\mathcal{L}}$ denote the polar function defined by

$$\widehat{\mathcal{L}}(G) \equiv \{\, K \in \mathcal{L}(G) \,:\, K^{\perp} \in \mathcal{L}(G) \,\},$$

for each **W**-object G. Then $\widehat{\mathcal{L}}(G)$ is a subalgebra, and, indeed, it is the largest subalgebra of $\mathcal{P}(G)$ contained in $\mathcal{L}(G)$. Moreover, $\widehat{\mathcal{L}}$ is a (subalgebra) ipf. This is the *boolean center* of $\mathcal{L}$. It is further shown that, for ipfs $\mathcal{L}$ with a certain "local" property, $G[\widehat{\mathcal{L}}]$ is an $\mathcal{L}$-essential extension of G, and that $\widehat{\mathcal{L}}$ is idempotent.

Note, for example, that if $\mathcal{P}^{\alpha}(G)$ is the sublattice of all α-generated polars of G, then $\widehat{\mathcal{P}^{\alpha}} = \mathcal{P}^{\alpha}_{\alpha}$.

[HM∞c] also discusses conditions on $\mathcal{L}$ under which each **W**-object G has a largest $\mathcal{L}$-essential extension, denoted $G^{\mathcal{L}}$. These include the so-called local condition mentioned above, so that $G[\widehat{\mathcal{L}}] \leq G^{\mathcal{L}}$; we do not yet understand these phenomena well enough, and cannot say, in general, when these two extensions agree.

As the subtitle of this section already indicates, the exposition of it is meant to convey the obvious: that this is a rather new cottage industry of research, which is very much in flux, and which has connections that extend (into the theory of covers of compact spaces, and thus) beyond the scope of this article.

4   Monoreflections

As we have already seen in a few of the examples, sometimes a hull class is also reflective, in the categorical sense (which we shall review below in 4.1). On occasion a hull is even functorial, although that happens more infrequently. In this section we review some of the recent work on monoreflections in **W**. Consideration of monoreflections involving rings, *per sé* are postponed until the next section.

Our basic reference for the background material in category theory is [HS79]. All subcategories are here assumed to be full; that is, they contain all the morphisms of the parent category. We assume that the reader is familiar with terms such as "monomorphism" and "epimorphism", and that the reader is at least sensitized to the notion that, in general, a monomorphism which is also an epimorphism need not be an isomorphism.

Definition & Remarks 4.1. Suppose that **A** is a category and **B** is a subcategory. We say that **B** is *reflective in* **A**, if for each **A**-object A there is a **B**-object ρA, and a morphism $\rho_A : A \longrightarrow \rho A$ such that, for each morphism $f : A \longrightarrow B$ with B an object in **B**, there is a unique morphism $\bar{f} : \rho A \longrightarrow B$ such that $\bar{f} \cdot \rho_A = f$.

We also say that **B** is a *reflective subcategory* of **A**. The assignment ρ turns out to be a functor **A** $\longrightarrow$ **B**. It is, in fact, the left adjoint of the forgetful functor from **B** to **A**. We shall refer to ρ_A as the *reflection map on A*. The functor ρ itself is called the *reflection*.

If ρ_A is always an epimorphism (resp. a monomorphism) we say that ρ is an *epireflection* (resp. *monoreflection*). We also say then that **B** is an *epireflective* (resp. *monoreflective*) subcategory. It is shown in [HS79, 36.3] that every monoreflection is necessarily an epireflection.

Assuming the category is complete and is subject to certain foundational assumptions, all of which are met in **W**, the discussion of §36 and §37, [HS79], leads up to a characterization of the epireflective subcategories as those which are closed under products and extremal subobjects. We shall let the enterprising reader take up the details of this characterization. In particular, any epireflective subcategory is closed under intersections. Thus, it becomes clear why the object class of any epireflective subcategory of **W** is a hull class.

It is an easy exercise to show that d, the divisible hull operator, is a monoreflection in **W**. Indeed, this is true more widely, for example, in the category of all abelian ℓ-groups.

Ball and Hager are responsible for the first major advances in this research with several publications in the nineties. Here we highlight six which are in some measure connected to the topic at hand. First, [BH90a], [BH90b] and [BH93] contain most of the information we wish to report on here concerning **W**; [BH99a] applies mostly to the category of all archimedean ℓ-groups, building, as one might expect, on the work of [BH93]. Further, since one deals with epimorphisms in any discussion of epireflections, [BH89] and [BH99b] take on considerable importance; the first of these contains the characterization of epimorphisms in **W** in terms of the so-called "epi-indicators" (which we shall not recall here). [BH99b] shows that an epimorphic extension of a divisible object into an epicomplete one is relatively uniformly dense; the authors point out that the assumption of epicompleteness in this result cannot be dropped.

From a categorical vantage point, [HM94b] and [HM98a] consider the existence of the largest monoreflection beneath a so-called extension operator. There are some conditions placed on the category in those papers, and it is reasonably clear that **W** satisfies them, and so the existence of a maximum monoreflection does not come as a surprise. However, these very general considerations clearly go beyond our scope in this survey.

Our first remarks concern the maximum monoreflection in **W**.

Remarks 4.2. The Maximum Reflection in **W**.

(a) In [BH90a] the authors describe β, the maximum monoreflection of a **W**-object G as the ℓ-group of Baire functions on YG, modulo certain "null" functions. (For $C(X)$, with X compact, $\beta C(X)$ is the algebra of Baire functions on X.) They show that the associated monoreflective subcategory consists of the objects which are isomorphic to a $D(X)$, where X is compact and basically disconnected. In [BH90b, Theorem 3.9], this class is identified as

$$\mathbf{D} \cap \mathbf{C}(\omega_1) \cap \mathbf{L}(\omega_1).$$

This is also the class of epicomplete **W**-objects. (The **W**-object G is *epicomplete* if each monomorphism $m : G \longrightarrow H$ in **W** which is also an epimorphism is necessarily an isomorphism. Thus, in ge, every monoreflective subcategory contains all the epicomplete objects. The class of epicomplete objects need not be epireflective, however; in **W** it is, yet this property of **W**, while aesthetically satisfying seems not to have any dramatic consequences.)

(b) The role of laterally ω_1-complete ℓ-groups is interesting in this context. In [HM97] and [HM99a] it is shown that $\mathbf{L}(\omega_1)$ is monoreflective, and the associated reflection, σ is discussed in some detail. For example, [HM99a, Theorem 4.9] describes σG, inside βG, as Baire-functions-modulo-null-functions. In particular, if $G = C(X)$, with X compact, then $\sigma C(X)$ consists of all Baire functions f on X for which a countable partition of X by Baire sets $Y_1, Y_2, \ldots$ exists so that $f|_{Y_i}$ extends to a continuous function on X, for each i [HM99a, Corollary 4.10(a)].

For each **W**-object G, $\sigma G \le \beta G$ is essential and these two have the same Yosida space. In fact, $u \cdot d \cdot \sigma = \beta$, where u denotes the uniform completion [HM97, Corollary 3.4.1].

(c) Something which is both curious and fascinating: as is shown in [HM97, §7], $\mathbf{L}(\alpha)$ is epireflective if and only if $\alpha = \omega_1$. Moreover, $\mathbf{C}(\alpha)$, $\mathbf{bL}(\alpha)$ and $\mathbf{P}(\alpha)$ fail to be epireflective for all uncountable cardinals α and for ∞. Nor is $\mathbf{P}(\omega_0)$, the class of projectable ℓ-groups, epireflective.

Let us indicate, for $\alpha > \omega_1$, how one proves such claims as these. All the classes in question are closed under products, so, in order to show that $\mathbf{L}(\alpha)$, $\mathbf{C}(\alpha)$ or $\mathbf{P}(\alpha)$ is not epireflective, on must show that these are not closed under taking extremal subobjects. Recall that $A \le B$ is *extremal* if $A \le E \le B$, with an epic first factor, implies that $A = E$. Now take any compact basically disconnected space X which is not extremally disconnected. Then $A = D(X)$ is epicomplete, and therefore extremal in any **W**-object that contains it. Now, one may choose X so that A fails to be in $\mathbf{L}(\alpha)$, $\mathbf{C}(\alpha)$ or $\mathbf{P}(\alpha)$. Consider then the inclusion $A \le A^e$, and reflect that A^e, being essentially closed, lies in $\mathbf{L}(\infty) \cap \mathbf{C}(\infty) \subseteq \mathbf{P}(\infty)$. This suffices to prove the claim that, for $\alpha > \omega_1$ these three classes fail to be epireflective.

It is unkown whether the class $\mathbf{L}(\omega_1) \cap \mathbf{C}(\omega_1)$ is epireflective. This question is especially intriguing, since if this class is intersected with the class of divisible **W**-

objects, then one gets the least monoreflective subcategory of **W**.

With regard to $\mathbf{L}(\infty)$, it is shown in [HM01a] that the least epireflective subcategory containing it is $\mathbf{L}(\omega_1)$.

The techniques of [HM94b] and [HM98a] which predict that in **W** there must be a maximum monoreflection also promise that there is a maximum essential monoreflection. We discuss that reflection next.

Remarks 4.3. The Maximum Essential Monoreflection in **W**.

Among the accomplishments of [BH93] is the identification of the maximum essential monoreflection in **W**. First, let us be clear with a definition: in a category **A**, a monomorphism $m : A \longrightarrow B$ is *essential* if for each morphism $f : B \longrightarrow C$ such that $f \cdot m$ is monic, it follows that f too must be monic. A monoreflection ρ is essential if the reflection map ρ_A is always essential; if ρ reflects in the subcategory **B** we say that **B** is *essentially reflective*.

Suppose that G is a **W**-object, identified with its image in $D(YG)$ under the Yosida representation. Thus, $g^{-1}\mathbb{R}$ is dense in YG, for each $g \in G$. Denote by $G^{-1}\mathbb{R}$ the set of all these dense sets, and by $(G^{-1}\mathbb{R})_\delta$ the set of all countable intersections of sets in $G^{-1}\mathbb{R}$. The members of $(G^{-1}\mathbb{R})_\delta$ are also dense sets, by the Baire Category Theorem. Next, let

$$C[(G^{-1}\mathbb{R})_\delta] \equiv \lim C(U), \quad U \in (G^{-1}\mathbb{R})_\delta,$$

where the bonding maps of the direct limit are the restrictions induced by containment of sets. This construct is usually abbreviated $c^3 G = C[(G^{-1}\mathbb{R})_\delta]$; c^3 stands for "closure under countable composition". This closure was first introduced in [Ar71]; that it is reflective is shown in [ArH81]. The reader interested in closure under countable composition, *per sé*, should look at [HIJ61, 2.4] and also at [BH91] and [H85].

We shall refer to the subcategory in which c^3 reflects, as the class of c^3-*objects*. To get an idea of the prominence of this subcategory in localic terms, the reader should have a look at [MV90].

At any rate, it is shown in [BH93, Theorem 9.2] that $G \leq c^3 G$ is the largest essential monoreflection. More precisely, it is shown in [BH93] that $D(X)$ is a c^3-object (Theorems 4.1 and 5.1, [BH93]) if X is a compact quasi F-space. (X is a *quasi F-space* if every dense cozeroset of X is C^*-embedded. These are precisely the spaces which make $D(X)$ a group under pointwise operations.) Furthermore, every c^3-object is an extremal subobject of some $D(Y)$, for Y compact and quasi F [BH93, Theorem 9.4]. The latter is a notable achievement, not only because it is a sophisticated piece of mathematics, but because such tidy results involving extremal subobjects are in themselves remarkable.

Our next set of remarks involve the uniform completion operator u, in particular, and essential monoreflections, in general.

Remarks 4.4. [HR77, Theorem 5.5] establishes that there is a monoreflection of **W** in the subcategory **U** of all uniformly complete **W**-objects. The reflection map, $G \leq uG$ simply computes the uniform metric completion of G (or its image under the Yosida representation) in $D(YG)$, which is uniformly complete. In addition, it is shown there that $Y(uG) = YG$. The divisible hull d has this feature as well.

Unknown (and presumably also largely unexplored) is the answer to the question: what can be said about the monoreflections ρ for which $Y(\rho G) = YG$, for each **W**-object G?

One imagines that the machinery of [HM94b] and [HM98a] can be tuned to prove that a maximum monoreflection exists with this property. It is not u, however. It is easily shown that any $\mathbf{W_s}$-object G is uniformly complete. However, dG is larger. Thus, a presumptive maximum monoreflection which preserves the Yosida spaces must reflect on a subcategory consisting of divisible, uniformly complete archimedean ℓ-groups.

Nor is c^3 itself that maximum reflection. We shall be indicate why in the next section (Example 5.1).

As a follow-up, here is an interesting example, which ties in with projectable hulls.

Example 4.5. Suppose that G is a **W**-object, identified with its image under the Yosida representation. A function $f \in D(YG)$ is *locally* in G if for each $p \in YG$ there is a neighhborhood V of p and also a $g \in G$ such that $f|_V = g|_V$. G is said to be *local* if each $f \in D(YG)$ which is locally in G in fact belongs to G. Moreover, if one defines

$$\mathrm{loc}\, G \equiv \{ f \in D(YG) : f \text{ is locally in } G \},$$

then, as shown in [HR78], loc is an (essential) monoreflection.

Now if G is projectable then YG is zero-dimensional and local [HM96, Theorem 2.2]. However, the converse is false, as the example in [HM96, 2.3] demonstrates. Put differently, $G \leq \mathrm{loc}\, G \leq p(\omega_0)G$, but the two operators are different, and the projectable hull is not even a reflection. [4].

Referring back to the discussion in 4.4, let us add that loc too preserves the Yosida space of the ℓ-group. Thus, if there is a maximum monoreflection which preserves Yosida spaces, it must reflect in a class consisting of divisible, uniformly complete objects which are also local.[5]

Next, we take up the problem of tracing a monoreflection on an essential one; we shall explain how "tracing" is to be interpreted. And we shall trace more generally as well. For further discussion of this topic we refer the reader to [HM01b] and [HM01a].

---

[4]By contrast, returning to $\mathbf{W_s}$ for a moment, it is easily verified that if G is a $\mathbf{W_s}$-object, then so is loc G, and $p(\omega_0)G = \mathrm{loc}\, G$ [HM98b, Proposition 5.5]

[5]And a divisible, uniformly complete **W**-object need not be local. A. W. Hager reminds me of the following example, which is an extension of $C(\alpha N)$. Recall that αN denotes the one-point compactification of the discrete natural numbers, so that $C(\alpha N)$ is the ring of all convergent sequences. Letting j stand for the identity function, G denotes the set of all functions jf, for all $f \in C(\alpha N)$. It is, in fact relatively uniformly complete, but not local. This example appears in [BH99a].

Remarks 4.6. We assume throughout this discussion that **A** is a category which has products; a maximum monoreflection β and a maximum essential monoreflection ε, reflecting in $\mathbf{A_e}$; and one must have some set-basing axioms, for which we refer the reader to [HM01b]. **W** satisfies these assumptions.

We note that if **X** is any object class in **A** then there is a least epireflective subcategory, denoted $\mathcal{R}(\mathbf{X})$, containing **X**, and in establishing this, the unnamed set-based axioms come into play, albeit indirectly.

(a) The following is a paraphrase of Theorem 2 in [HM01b]:

(i) Suppose that **X** is a class of objects in $\mathcal{A}$. Then $\mathcal{R}(\mathbf{X} \cup \mathbf{A_e})$ is the least essentially reflective subcategory containing **X**.

(ii) Suppose that δ monoreflects on **X**. Then

$$(\delta \wedge \varepsilon)(G) \equiv \delta G \cap \varepsilon G$$

reflects on $\mathcal{R}(\mathbf{X} \cup \mathbf{A_e})$. Moreover, $\delta \wedge \varepsilon$ is the largest essential monoreflection beneath δ.

A more technical part (2(c)) of Theorem 2, [HM01b], yields that if h is a hull operator and **H** is the associated hull class, then, denoting the monoreflection on $\mathcal{R}(\mathbf{H})$ by $\overline{h}$, we have that $\mu(h) = \overline{h} \wedge \varepsilon$ is the largest monoreflection beneath h.

Note that since the subcategory of all epicomplete **W**-objects, $\mathbf{E}(\omega_1)$, is epireflective, we have that $\mathbf{E}(\omega_1) = \mathcal{R}(\mathbf{E}(\infty))$. The argument of 4.2(c) is more specific, however; it shows that every $\mathbf{E}(\omega_1)$-object is an extremal subobject of one that is essentially closed.

(b) Here's the application of the above which takes up a good part of [HM01a]. Consider $\mathbf{L}(\infty)$; as already noted, $\mathcal{R}(\mathbf{L}(\infty)) = \mathbf{L}(\omega_1)$. Thus, by our preceding remarks, the largest monoreflection under $l(\infty)$, is $\mu(l(\infty)) = \sigma \wedge c^3$. [HM01a, Theorem 5.5] gives a direct limit rendition of $\mu(l(\infty))$. This applies to the following observations (topology being, if not inevitable at this point, at least irresistible.)

(i) If G has a strong order unit then $\mu(l(\infty))G$ consists of all the continuous functions f on YG for which there is a countable partition by Baire sets $Y_1, Y_2, \ldots$ and a sequence $g_1, g_2, \ldots \in G$ such that $f|_{Y_i} = g_i|_{Y_i}$, for each i.

(ii) Suppose now that X is compact. Let $S(X)$ denote the subalgebra of all continuous step functions. Then (see [HM99a])

$$\sigma S(X) = \{\, f \in C(X) : f(X) \text{ is countable} \,\}.$$

It then follows that $\mu(l(\infty))S(X) = \sigma S(X)$. [6]

---

[6]It can then be established easily (with a little topological information) that $\mu(l(\infty))S(X) = C(X)$ if and only if X is a scattered space; that is, each nonvoid closed subspace has an isolated point.

Hulls are sometimes functorial. Let us conclude this section with some thoughts about that.

Remarks 4.7. (a) Unlike **D** and **U**, the hull operator $l(\omega_1)$ for $\mathbf{L}(\omega_1)$ and the reflection in this subcategory do not coincide. It is not difficult to see that if ρ is an essential monoreflection, then $G \leq \rho G$ is the hull in the subcategory in which ρ reflects. It is obvious, conversely, that if hull and reflection coincide then the reflection in question must be essential. But how does one go about recognizing that a monoreflective subcategory **R** of **W** is essentially reflective without knowing the reflection involved?

What is glaringly obvious about a subcategory like $\mathbf{L}(\omega_1)$ is that to belong to it the Yosida space of an object G must have some particular features; here, YG must be basically disconnected. So it is, perhaps, tempting to conjecture that if **X** is a reflective subcategory with the feature that any compact space is the Yosida space of some **W**-object in **X**, then **X** must be essentially reflective.

(b) Note that each ring of continuous functions $C(X)$ is a c^3-object. Thus, if ρ is an essential monoreflection then $\rho C(X) = C(X)$, which makes it clear that if a hull operator is reflective (in **X**) then each compact space is the Yosida space of an **X**-object. Our tentative conjecture in (a), likely, is not so far-fetched. One should pause to consider that what is assumed in the conjecture is that, given a compact X, there is some $G \in \mathbf{X}$ so that $YG = X$, not that G is necessarily $C(X)$ or even uniformly complete.

5 f-Rings

In this section we consider hulls and reflections in the category of archimedean f-rings with identity, with all ℓ-homomorphisms. This category is denoted **Arf**. It is a full subcategory of **W**. This is most easily seen (at least for divisible objects) by first realizing that the kernel of a **W**-morphism is uniformly closed [LZ71, Theorem 60.2]. Then using [HP82, Proposition 3.1], it follows that such a kernel is also a ring ideal.

However, before embarking on that discussion, let us first observe that **Arf** itself is a monoreflective subcategory of **W**. The monoreflection which witnesses this is denoted r, for "ringification"; see [HR77, §6]. It is shown there that r is an essential reflection. To follow up on the comments made about "Yosida-space preserving" reflections in 4.4 above, we should like to point out the following example; it is [HR77, 4.6], and may also be found in [HJ61, 3.6].

Example 5.1. On the half-ray $[0, \infty)$ consider the continuous functions which are eventually polynomials; this is G. Note that this is an f-ring. It is observed in [HR77, 5.8] that G cannot be embedded in a uniformly closed f-algebra H, such that $YG = YH$. In particular, uG is not a ring. (One may apply r, and again u, etc.; Aron in

[Ar71] has noted examples of situations in which this iteration of u and r must be continued to any preselected countable ordinal γ.)

Returning to the example at hand, some iteration of r and u monoreflects G in the class **Arf** $\cap$ **U**, and by the comment in the preceding paragraph, the reflected object has a different Yosida space. From this one concludes that r "enlarges" the Yosida space, and therefore that c^3 must too.

The above comments suggest several questions, none of which have received much attention. The reader will certainly appreciate, reading the list that follows, that the question posed in each item can be asked more generally.

Questions 5.2. (1) (Deliberately posed in this open-ended manner.) Describe (characterize, give information about) monoreflections ρ having the property that if G is in **Arf** then ρG is as well. Same question, more generally, for hull operators.

(2) More restrictively, when does a hull operator commute with r? Put differently, suppose **H** is a hull class with hull operator h; if $G \in$ **H**, when does it follow that $rG \in$ **H**, and if A is an f-ring, when does one also get that hA is a ring?

The operators $p(\alpha)$, $l(\alpha)$ and $c(\alpha)$ all have the second of these two properties. Refer, respectively, to [HM99b, §3]; [HM99a, §2], for the $\alpha = \omega_1$ case; and for $c(\alpha)$ use the description in 2.2(c).

However, there appears to be very little in print regarding the question of when rG is α-projectable (resp. conditionally α-complete, resp. laterally α-complete), assuming G has the corresponding property. The lone reference in this regard seems to be [HR79], in which an example is given of a laterally complete object G for which rG is not laterally complete.

(3) Characterize the objects G for which $YG = Y(rG)$.

It is implicit in the discussion of [HR77, §6] that if either G has a strong order unit or YG is a quasi F-space, then $YG = Y(rG)$.

One of the driving forces behind the joint work of this author with A. W. Hager on monoreflections in **Arf** has been curiosity about the functorial properties of rings of quotients. Our approach in [HM93] and [HM94a] was to start quite generally, and there are applications in these to abstract commutative rings and f-rings, as well as in [HM98a]; on the other hand, the first two of these references contain substantial results in **Arf**, some of which we will summarize presently. In [HM01c] we finally succeeded in identifying the maximum functorial ring of quotients in **Arf**.

As in **W**, it is convenient to view an **Arf**-object A as a ring of functions, under the Henriksen-Johnson Representation. [HJ61] Suppose that A is an archimedean f-ring. Let $m(A)$ denote the space of all maximal ℓ-ideals of of A, with the hull-kernel topology; this is a compact Hausdorff space. There is a ring ℓ-isomorphism ϕ from A onto an f-ring A' in $D(m(A))$, carrying the identity to the constant function 1, so that A' separates the points of $m(A)$.

Remarks 5.3. (a) Throughout this commentary it is assumed that all rings are commutative with an identity and semiprime; that is, that there are no nonzero nilpotent elements. Since in **Arf** all rings are semiprime anyway, this seems like a reasonable assumption.

The reader ought to be familiar with two constructions; with qA, the classical (or total) ring of quotients of a ring A, obtained in the usual manner via formal fractions; and with QA, the maximum ring of quotients. For information regarding the latter, we refer the reader to any of a number of articles in the literature. One could go to the source, Utumi's paper [U56]. The construction of the maximum ring of quotients via Hom-sets is carried out in [L86], whereas Banaschewski's approach (in [Ba65]) is to describe it from a functional point of view, for semiprime rings. [Wi89] characterizes semiprime rings which are their own maximum ring of quotients as the self-injective ones, and, equivalently, as those rings satisfying a strengthened von Neumann regularity. (We take that up ahead, in 5.4(b).) [An65] and [M95] discuss the maximum ring of quotients of an f-ring. In [FGL65], Theorem 2.6, one finds a direct limit description of both $qC(X)$ and $QC(X)$. For any Tychonoff space X, recall that $\mathcal{G}(X)$ and $\mathcal{C}(X)$ stand for the filter bases of dense open sets and dense cozerosets, respectively. Then

$$qC(X) = C[\mathcal{C}(X)] \quad \text{and} \quad QC(X) = C[\mathcal{G}(X)].$$

Note also [DHH80, Corollary 3.11]) that

(i) $(u \cdot q)C(X) = C[\mathcal{C}_\delta(X)]$, and

(ii) $(u \cdot q)C(X)^* = o_C C(X)$.

More generally, if $A \in$ **Arf** then so are qA and QA, with natural orderings that extend the given lattice ordering on A.

(b) Both q and Q are hull operators. We shall denote by **q** the class of all archimedean f-rings in which every regular element is invertible, while **Q** stands for the class of **Arf**-objects A for which $A = QA$. We have [HM01c, Theorem 1.5], which in **Arf** reduces to

$$\mathbf{Q} = \mathbf{q} \cap \mathbf{L}(\infty).$$

Furthermore, that theorem, together with Theorem 1.3 of [HM01c] gives a number of characterizations of **Q**-objects; for example, $A \in \mathbf{Q}$ if and only if A is laterally complete and von Neumann regular; if and only if A is laterally complete and has the bounded inversion property.

Recall that an f-ring A *has the bounded inversion property* if each $a \geq 1$ in A is invertible. It is well known (and easy to see) that each ring of continuous functions $C(X)$ has the bounded inversion property. A is *von Neumann regular* if for each $a \in A$ there is a $b \in A$ such that $a^2 b = a$. Already in Lambek's book, [L86], the reader may find a proof that every semiprime ring A for which $A = QA$ is necessarily von Neumann regular.

The role of von Neumann regularity is interesting, because it is a hull property on its own, and, indeed, the category **vN** of all von Neumann regular archimedean f-rings is a monoreflective subcategory of **Arf**. We will return to this aspect of von Neumann regular rings.

(c) For any semiprime commutative f-ring A, bA is the subring of qA consisting of all fractions a/s, in which $s \geq 1$. b is the monoreflection (of **Arf**, for our purposes) in the subcategory **BI** of all f-rings having the bounded inversion property [HM93, Proposition 7.4]. Putting the remarks of (b) another way,

$$\mathbf{Q} = \mathbf{vN} \cap \mathbf{L}(\infty) = \mathbf{BI} \cap \mathbf{L}(\infty).$$

It should be pointed out – see [HM01c, Example 1.4] – that the archimedean property is needed for the latter identity.

The reflection b is well behaved, in the sense that, for archimedean f-rings and for each α,

$$l(\alpha) \cdot b = b \cdot l(\alpha).$$

This is explicitly stated in [HM01c, Corollary 3.6] for $\alpha = \infty$, and it is implicit in the discussion of [HM01c, §4] for $\alpha = \omega_1$; in general, it will appear in [HM∞a].

Now to the functoriality of rings of quotients. The references already mentioned contain much more information than we have room for in this article. The following comments speak to the maximality issues of [HM94b] and [HM98a].

Remarks 5.4. (a) We discuss first the maximum monoreflection on **Arf** beneath q. For a ring A we consider *multiplicative systems* in A: subsets D of A consisting of regular elements and closed under multiplication.

Following the notation in [HM93], and identifying A with its image under the Henriksen-Johnson representation, we denote, for a multiplicative system D,

$$\phi(A, D) \equiv \{\, \operatorname{coz}(d) \cap a^{-1}\mathbb{R} : d \in D \text{ and } a \in A \,\},$$

where $\operatorname{coz}(d) = \{\, x \in \mathrm{m}(A) : d(x) \neq 0 \,\}$ and $a^{-1}\mathbb{R} = \{\, x \in \mathrm{m}(A) : a(x) \in \mathbb{R} \,\}$. $\phi(A, D)$ is a filter base of dense cozerosets.

§7 of [HM93] is concerned with the largest monoreflection under q, although the issue is not quite framed in that manner there. This functor, $\mu(q)$ according to the notation of 4.6(a) – but in [HM93] denoted $(\cdot)_H$ – has the following description in Corollary 7.6.1, [HM93]: for any divisible and uniformly complete **Arf**-object A,

$$\mu(q)A = C[\phi(A, h_q(A))]$$

where $h_q(A)$ is defined by

$$h_q(A) \equiv \{\, a \in A : Z(a) \cap F = \emptyset, \text{ for some } F \in A^{-1}\mathbb{R} \,\},$$

and $Z(a) = \{\, x \in \mathrm{m}(A) \ : \ a(x) = 0 \,\}$.

Proposition 7.8 of [HM93] adds that, if A is divisible and uniformly complete, then $\phi(A, D)$ consists of the dense cozerosets in $\mathrm{m}(A)$ which contain a member of $(A^{-1}\mathbb{R})_\delta$; this implies that $\mu(q)A$ is uniformly dense in $C[(A^{-1}\mathbb{R})_\delta] = c^3 A$.

In general terms we refer the reader to Theorems 5.3 and 5.5 of [HM93], which record that $\mu(q)$ reflects in the subcategory of all sections of objects in $\mathbf{q}$. That is, $A = \mu(q)A$ if and only if there is an **Arf**-object B, with $B = qB$, which is an extension of A, such that there is a morphism $\pi : B \longrightarrow A$ whose restriction to A is the identity.

On the other hand, we note that a description of $\mu(q)$ as $\overline{q} \wedge c^3$ (see 4.6(a)) has not been worked out; this seems to have been overlooked. From a different vantage point: we know nothing about $\mathcal{R}(\mathbf{q})$.

(b) Next, a summary of the maximum monoreflection under Q in **Arf**; that is to say, of the maximum functorial ring of quotients.

As with the discussion of the largest monoreflection beneath $l(\omega_1)$ (4.6(b)), it is helpful to view this maximum under Q against the backdrop of another reflection. First, a definition.

A ring A is *strongly α-regular* – following the lead of Wickstead in [Wi89], albeit not faithfully – if for each subset D of A, with $|D| < \alpha$, of pairwise annihilating elements there is an $s \in A$ so that $d^2 s = d$, for each $d \in D$ and $xs = 0$ whenever $xD = 0$. We abbreviate "strongly ∞-regular" to plain "strongly regular"; this is the regularity condition of [Wi89], already referred to in 5.3(a). It should be evident that strong regularity implies strong α-regularity, for each uncountable cardinal α. In turn, strong ω_1-regularity implies von Neumann regularity. Strong α-regularity is currently being investigated in the forthcoming [HM∞a].

[HM01c, Theorem 3.2] shows that the following are equivalent for an **Arf**-object A.

(i) A is strongly ω_1-regular.

(ii) $A \in \mathbf{vN} \cap \mathbf{L}(\omega_1)$.

(iii) $A \in \mathbf{BI} \cap \mathbf{L}(\omega_1)$.

[HM01c, Theorem 4.4] shows that the subcategory $\mathbf{SR}(\omega_1)$ of all strongly ω_1-regular **Arf**-objects is monoreflective, and in fact, that $r_{\omega_1} = b \cdot \sigma$ is the associated reflection. Moreover, $b \cdot \sigma = \sigma \cdot b$.

[HM01c, Theorem 5.3] shows that $\mathcal{R}(\mathbf{Q}) = \mathbf{SR}(\omega_1)$. Then, bringing to bear the full weight of the comments in 4.6(a), one obtains that $\mu(Q) = r_{\omega_1} \wedge c^3$ is the maximum monoreflection beneath Q [HM01c, Theorem 6.4]. We observe as well, in concordance with part (a) of these comments, that each strongly ω_1-regular **Arf**-object is an extremal subobject of one in $\mathbf{Q}$ [HM01c, Theorem 5.3]. Furthermore, one has the following decomposition: $\mu(Q)A = A$ if and only if A is an extremal subobject of $B \times C$, where $c^3 B = B$ and $C \in \mathbf{Q}$ [HM01c, Corollary 6.6].

Curiously, $\mu(Q) = \mu(l(\infty)) \cdot b$, but $b \cdot \mu(l(\infty)) \neq \mu(Q)$ [HM01c, Example 6.5].

We conclude these remarks about hull classes of archimedean f-rings with a (biased) discussion of the monoreflective subcategory of von Neumann regular f-rings.

Remarks 5.5. The first part of this discussion concerns the regular hull, and the second part is about the regular reflection in semiprime f-rings.

(a) All rings are commutative, semiprime and with identity 1. Historically, the regular hull HA of a ring A first appears in Storrer's [St68], as the maximum essential epimorphic extension in the category of all commutative semiprime rings with 1. Much more recently, [RW99] takes up the regular hull of a ring of continuous functions. At any rate, $A \leq HA$ is the hull in the subcategory of all von Neumann regular rings. It is shown in [HMood] that the class of von Neumann regular rings is essentially inter-sective, which means that it is a hull class. (Caution! The sense of being "essentially intersective" here has to be interpreted relative to taking intersections of rings all of which have the same maximum ring of quotients.)

In a semiprime f-ring "polar" and "annihilator" mean the same thing. One can therefore take advantage of this to generalize the notion of projectability to semiprime rings, via annihilator ideals, and then one is able to show that $HA = q(pA)$ [HMood], where pA denotes the projectable hull, and q is the classical ring of quotients hull. This helps explain why (and how) the regular hull of a semiprime f-ring A is (naturally) an f-ring – it is an ℓ-subring of QA – and also why the regular hull of an **Arf**-object lies in that category. The result displayed above is implicit in [RW99]. One important consequence of the identity $HA = q(pA)$ in **Arf** is that pA and HA have the same Yosida space.

(b) The most significant news from [HMood] concerns the behavior of prime d-ideals. Recall (in any semiprime f-ring A) that an ideal r of A is a *d-ideal* if $a \in r$ implies that $a^{\perp\perp} \subseteq r$. By "prime d-ideal" we mean a d-ideal which happens also to be a prime ideal. It is well known that every minimal prime ideal is a d-ideal. $\mathrm{Spec}_d(A)$ denotes the poset of prime d-ideals. As usual, $\mathrm{Min}(A)$ stands for the set of all minimal prime ideals. Unless the contrary is expressly noted, all topological references regarding structure spaces in this discussion are to the hull-kernel topology.

Since HA is the classical ring of quotients of pA, the first item of the following assertions is easy to see; the second is one of the main results of [HMood]:

(i) The trace map $p \mapsto p \cap pA$ is a homeomorphism of $\mathrm{Max}(HA) = \mathrm{Min}(HA)$ onto $\mathrm{Min}(pA) = \mathrm{Spec}_d(pA)$.

     The coincidence of the first two spectra is due to the fact that HA is von Neumann regular, whereas the latter pair of spectra coincide because in a projectable ring every prime d-ideal is minimal.

(ii) The trace map $p \mapsto p \cap A$ is a bijection of $\mathrm{Min}(pA)$ onto $\mathrm{Spec}_d(A)$.

There is also a description of the inverse to the bijection of (ii) above, and it is shown there that these bijections are homeomorphisms provided $\text{Spec}_d(A)$ is endowed with the *patch topology*, namely the least refinement of the hull-kernel topology in which the basic open sets

$$u(a) = \{\, p \in \text{Spec}_d(A) \,:\, a \notin p \,\}$$

are closed.

§6 of [HM∞d] also addresses the issue of when HA is representable as a ring of real valued functions. The main result is Theorem 6.9. Recall that $m(A)$ denotes the space of all maximal ℓ-ideals. $\text{Max}_d(A)$ denotes the space of all maximal d-ideals.

> Suppose A is a uniformly complete archimedean f-algebra. Let $m \in m(A)$ be the trace of a real maximal ideal of HA. Then m is an almost P-point of $m(A)$. Conversely, any almost P-point of $m(A)$ is a real ideal as well as a d-ideal, and therefore it is a real ideal in $\text{Max}_d(A)$. Consequently, it is the trace of a real maximal ideal of HA.

(A point x in the topological space X is an *almost P-point* if any continuous real valued function which vanishes at x vanishes also on a nonempty open set of X (not necessarily containing x).) An immediate consequence of this theorem is that if A is a uniformly complete archimedean f-algebra, then HA can be represented as a ring of real valued functions if and only if $m(A)$ has a dense subset of almost P-points.

(c) A proof of the fact that the subcategory of von Neumann regular f-rings is epireflective in the category of all semiprime f-rings appears in [HM98a, Proposition 9.6]; the proof is easily adapted to the abstract ring case. It is immediate from this then that the subcategory **vN** is monoreflective in **Arf**. Now [HM∞e] is principally concerned with the regular reflection in semiprime f-rings. We denote this monoreflection by ϱ.

Madden and Schwartz, in [SM99], 9C.2 and 9C.3, describe ϱ; their technique and language are too complex to recall here. On the other hand, in correspondence dated 23rd March, 1993, Schwartz gives the following description of ϱ in **Spf**, the category of all semiprime f-rings. Let $\text{Spec}_\ell(A)$ denote the set of prime ℓ-ideals of A. Put

$$P(A) \equiv \prod \{\, q(A/p) \,:\, p \in \text{Spec}_\ell(A) \,\}.$$

Observe that $P(A)$ is a von Neumann regular f-ring, relative to the coordinatewise lattice-ordering of $P(A)$. Each residue field $q(A/p)$ carries the natural total ordering induced by that on the domain A/p. Then ϱA is the intersection of all the von Neumann ℓ-subrings of $P(A)$.

Here are the main accomplishments of [HM∞e]:

(i) The trace map $p \mapsto p \cap A$ is a homeomorphism of $\text{Max}(\varrho A)$ onto $\text{Spec}_\ell(A)$, when the latter carries the patch topology.

(ii) $\varrho A = HA$ if and only if every prime ℓ-ideal of A is a d-ideal. This happens precisely when $A \leq \varrho A$ is essential.

(iii) For any infinite compact space X, $\varrho C(X)$ is not archimedean.

The latter condition says, loudly and clearly, that the restriction of ϱ to **Arf** cannot be the von Neuman regular reflection in that category. We are just beginning to consider the latter reflection.

6 In Conclusion ...

In this account of hulls and monoreflections the aim has been, principally, to sketch the development of the subject, starting with F. Riesz (because that seems like a good place to start) and leading up to the present and beyond, with citations of several "works in progress" with A. W. Hager. There are over 70 references in the bibliography, reaching back – somewhat awkwardly, perhaps – to Glivenko. Eighteen of those references are to collaborations with Hager. The reader might get the following impressions: that we are excited about the subject, that we believe there are plenty of questions to remain excited about, and that we appear to work well together. These impressions are correct. What is perhaps not so clear from the numbers is that my years of collaboration with Hager have also been among the most satisfying.

References

[An65]   F. W. Anderson, *Lattice-ordered rings of quotients.* Canad. Jour. Math. **17** (1965), 434-448.

[AF88]   M. Anderson & T. Feil, *Lattice-Ordered Groups, an Introduction.* (1988) Reidel Texts in the Math. Sci.; Kluwer, Dordrecht.

[Ar71]   E. R. Aron, *Embedding lattice-ordered algebras in uniformly closed algebras.* (1971) Thesis, University of Rochester.

[ArH81] E. R. Aron & A. W. Hager, *Convex vector lattices and ℓ-algebras.* Topology and its Appl. **12** (1981), 1-10.

[BH89]   R. N. Ball & A. W. Hager, *Epimorphisms in archimedean lattice-ordered groups and vector lattices.* In *Lattice-Ordered Groups, Advances and Techniques,* (A. M. W. Glass & W. C. Holland, Eds.); Math. and its Appl. (1989); Kluwer Acad. Publ., Dordrecht.

[BH90a]   R. N. Ball & A. W. Hager, *Epimorphisms in archimedean ℓ-groups and vector lattices with weak unit (and Baire functions)*. J. Austral. Math. Soc. (Ser. A) **48** (1990), 351-368.

[BH90b]   R. N. Ball & A. W. Hager, *Epicomplete archimedean ℓ-groups and vector lattices*. Trans. AMS **322** (No. 2) (1990), 459-478.

[BH91]   R. N. Ball & A. W. Hager, *On the localic Yosida representation of an archimedean lattice-ordered group with weak unit*. J. of Pure & Appl. Alg. **70** (1991), 17-43.

[BH93]   R. N. Ball & A. W. Hager, *Algebraic extensions of an archimedean lattice-ordered group, I*. J. of Pure & Appl. Algebra **85** (1993), 1-20.

[BH99a]   R. N. Ball & A. W. Hager, *Algebraic extensions of an archimedean lattice-ordered group, II*. J. of Pure & Appl. Algebra **138** (1999), 197-204.

[BH99b]   R. N. Ball & A. W. Hager, *The relative uniform density of the continuous functions in the Baire functions, and of a divisible archimedean ℓ-group in any epicompletion*. Topology and its Appl. **97** (1999), 109-126.

[Ba65]   B. Banaschewski, *Maximal rings of quotients of semi-simple commutative rings*. Archiv. Math. **XVI** (1965), 414-420.

[Be75a]   S. J. Bernau, *The lateral completion of a lattice ordered group*. J. Austral. Math. Soc. **19** (1975), 263-289.

[Be75b]   S. J. Bernau, *Lateral and Dedekind completion of archimedean lattice groups*. J. London Math. Soc. **12** (1975/76), 320-322.

[BKW77]   A. Bigard, K. Keimel & S. Wolfenstein, *Groupes et Anneaux Réticulés*. Lecture Notes in Math **608**, Springer Verlag (1977); Berlin-Heidelberg-New York.

[Bi67]   G. Birkhoff, *Lattice Theory* (3rd Ed.) AMS Colloq. Publ. **XXV** (1967), Providence, RI.

[Bl74]   R. D. Bleier, *The SP-hull of a lattice-ordered group*. Canad. Jour. Math. **XXVI**, No. 4 (1974), 866-878.

[Ch71]   D. Chambless, *The Representation and Structure of Lattice-Ordered Groups an f-Rings*. Tulane University Dissertation (1971), New Orleans.

[C69]   P. F. Conrad, *The lateral completion of a lattice-ordered group*. Proc. London Math. Soc. 3rd Series **XIX** (July 1969), 444-480.

[C71]   P. F. Conrad, *The essential closure of an archimedean lattice-ordered group.*
          Duke Math. Jour. **38** (1971), 151-160.

[C73]   P. F. Conrad, *The hulls of representable ℓ-groups and f-rings.* J. Austral.
          Math. Soc. **26** (1973), 385-415.

[CMc69]   P. F. Conrad & D. McAlister, *The completion of a lattice-ordered group.*
          J. Austral. Math. Soc. **9** (1969), 182-208.

[D95]   M. R. Darnel, *The Theory of Lattice-Ordered Groups.* Pure & Appl. Math.
          187, Marcel Dekker (1995); Basel-Hong Kong-New York.

[DHH80]   F. Dashiell, A. W. Hager & M. Henriksen, *Order-Cauchy completions of*
          rings and vector lattices of continuous functions. Canad. Jour. Math. **32**
          (1980), 657-685.

[Ev44]   C. J. Everett, *Sequence completions of lattice modules.* Duke Math. J. **11**
          (1944), 109-119.

[FGL65]   N. Fine, L. Gillman & J. Lambek, *Rings of Quotients of Rings of Functions.*
          (1965) McGill University.

[Fu63]   L. Fuchs, *Partially Ordered Algebraic Systems.* (1963) Pergamon Press,
          Oxford-New York-London-Paris.

[GJ76]   L. Gillman & M. Jerison, *Rings of Continuous Functions.* Grad. Texts in
          Math. **43**, Springer Verlag (1976); Berlin-Heidelberg-New York.

[G158]   A. M. Gleason, *Projective topological spaces.* Illinois J. Math. **7** (1958), 482-
          489.

[Gv29]   V. Glivenko, Bull. Acad. Sci. Belg. **15** (1929), 183-88.

[H85]   A. W. Hager, *Algebraic closures of ℓ-groups of continuous functions.* In *Rings*
          of Continuous Functions; C. Aull, Ed.; Lecture Notes in Pure & Appl. Math.
          95 (1985), Marcel Dekker, New York, 165-194.

[H89]   A. W. Hager, *Minimal covers of topological spaces.* In Papers on General
          Topology and Related Category Theory and Topological Algebra; Annals of
          the N. Y. Acad. Sci. **552** March 15, 1989, 44-59.

[HM93]   A. W. Hager & J. Martínez, *Functorial rings of quotients, I.* In *Proc. Conf.*
          Ord. Alg. Struc.; (W. C. Holland & J. Martínez, Eds.); Gainesville, 1991;
          (1993) Kluwer Acad. Publ., Dordrecht, 133-157.

[HM94a]   A. W. Hager & J. Martínez, *Functorial rings of quotients, II.* Forum Math.
          6 (1994), 597-616.

[HM94b]   A. W. Hager & J. Martínez, *Maximum monoreflections*. Appl. Categ. Struc. **2** (1994), 315-329.

[HM96]   A. W. Hager & J. Martínez, *α-Projectable and laterally α complete Archimedean lattice-ordered groups*. In *Proc. Conf. in Memory of T. Retta*; S. Bernahu, ed.; Temple U. (1995); Ethiopian J. Sci. (1996), 73-84.

[HM97]   A. W. Hager & J. Martínez, *The laterally σ-complete reflection of an archimedean lattice-ordered group*. In *Ord. Alg. Struc.*; Curacao (1995), (W. C. Holland & J. Martínez, Eds.); (1997) Kluwer Acad. Publ., Dordrecht, 217-236.

[HM98a]   A. W. Hager & J. Martínez, *Pushout-invariant extensions and monoreflections*. J. of Pure and Appl. Alg. **129** (1998), 263-295.

[HM98b]   A. W. Hager & J. Martínez, *Singular archimedean lattice-ordered groups*. Alg. Universalis **40** (1998), 119-147.

[HM99a]   A. W. Hager & J. Martínez, *More on the laterally σ-complete reflection of an archimedean lattice-ordered group*. Order **15** (1999), 247-260.

[HM99b]   A. W. Hager & J. Martínez, *Hulls for various kinds of α-completeness in archimedean lattice-ordered groups*. Order **16** (1999), 89-103.

[HM01a]   A. W. Hager & J. Martínez, *Functorial approximation to the lateral completion in archimedean lattice-ordered groups with unit*. Rend. Sem. Mat. Univ. Padova **105** (2001), 87-110.

[HM01b]   A. W. Hager & J. Martínez, *Maximum monoreflections and essential extensions*. Appl. Categ. Struc. **9** (2001), 517-523.

[HM01c]   A. W. Hager & J. Martínez, Functorial rings of quotients, III: the maximum in archimedean *f*-rings. To appear; J. of Pure & Appl. Alg.

[HM01d]   A. W. Hager & J. Martínez, *The ring of α-quotients*. To appear, Algebra Universalis.

[HM∞a]   A. W. Hager & J. Martínez, *On strong α-regularity*. Work in progress.

[HM∞b]   A. W. Hager & J. Martínez, *Polar functions, II: Completion classes of archimedean f-algebras vs. covers of compact spaces*. Preprint.

[HM∞c]   A. W. Hager & J. Martínez, *Polar functions, III: On irreducible maps vs essential extensions of archimedean ℓ-groups with unit*. Work in progress.

[HM∞d]   A. W. Hager & J. Martínez, *The projectable and regular hulls of a semiprime ring*. Preprint.

[HM∞e]   A. W. Hager & J. Martínez, *The regular reflection of a semiprime f-ring.* Work in progress.

[HR77]   A. W. Hager & L. C. Robertson, *Representing and ringifying a Riesz space.* Symp. Math. **21** (1977), 411-431.

[HR78]   A. W. Hager & L. C. Robertson, *Extremal units in an archimedean Riesz space.* Rend. Sem. Mat. Univ. Padova **59** (1978), 97-115.

[HR79]   A. W. Hager & L. C. Robertson, *On the embedding into a ring of an archimedean ℓ-group.* Canad. J. Math. **31** (1979), 1-8.

[HIJ61] M. Henriksen, J. R. Isbell & D. G. Johnson, *Residue class fields of lattice-ordered algebras.* Fund. Math. **50** (1961), 107-117.

[HJ61]   M. Henriksen & D. G. Johnson, *On the structure of a class of lattice-ordered algebras.* Fund. Math. **50** (1961), 73-94.

[HVW87] M. Henriksen, J. Vermeer & R. G. Woods, *The quasi F-cover of Tychonoff spaces.* Trans. AMS **303** (1987), 779-803.

[HVW89] M. Henriksen, J. Vermeer & R. G. Woods, *Wallman covers of compact spaces.* Diss. Math. **CCLXXX** (1989), Warsaw.

[HS79]   H. Herrlich & G. E. Strecker, *Category Theory.* Sigma Series in Pure Math. **1** (1979), Heldermann Verlag, Berlin.

[HP82]   C. B. Huismans & B. de Pagter, *Ideal theory in f-algebras.* Trans. AMS **269** (No. 1) (January, 1982), 225-245.

[HP83]   C. B. Huijsmans & B. de Pagter, *Maximal d-ideals in a Riesz space.* Canad. Jour. Math. **35** (1983), 1010-1029.

[L86]   J. Lambek, *Lectures on Rings and Modules.* (3rd Ed.) (1986) Chelsea Publ. Co., New York.

[LZ71]   W. A. J. Luxemburg & A. C. Zaanen, *Riesz Spaces, I.* (1971) North Holland, Amsterdam.

[MV90]   J. J. Madden & J. Vermeer, *Epicomplete archimedean ℓ-groups via a localic Yosida theorem.* J. of Pure & Appl. Alg. **68** (1990), 243-252.

[M95]   J. Martínez, *The maximal ring of quotients of an f-ring.* Alg. Universalis **33** (1995), 335-369.

[M02]   J. Martínez, *Polar functions, I: The summand-inducing hull of an archimedean lattice-ordered group with unit.* In these proceedings.

[dP81]   B. de Pagter, *On z-ideals and d-ideals in Riesz spaces, III.* Proc. Kon. Nederl. Akad. Wetensch., Series A, **84** (4) (1981), 409-422.

[Pa64]   F. Papangelou, *Order convergence and topological completion of commutative lattice groups.* Math. Annalen **155** (1964), 81-107.

[PW89]   J. R. Porter & R. G. Woods, *Extensions and Absolutes of Hausdorff Spaces.* Springer Verlag (1989); Berlin-Heidelberg-New York.

[RW99]   R. M. Raphael & R. G. Woods, *The epimorphic hull of C(X).* To appear.

[Ri40]   F. Riesz, *Sur quelques notions fondamentales dans la théorie générale des opérations linéaires.* Annals Math. **41** (1940), 174-206.

[SM99]   N. Schwartz & J. J. Madden, *Semi-Algebraic Function Rings and Reflectors of Partially Ordered Rings.* Lecture Notes in Math. **1712** (1999), Springer Verlag; Berlin-Heidelberg- et. al.

[Si62]   F. Šik, *Über die Beziehungen zwischen eigenen Spitzen und minimalen Komponenten einer ℓ-Gruppe.* Acta Math. Acad. Sci. Hungar. **13** (1962), 171-178.

[St68]   H. H. Storrer, *Epimorphismen von kommutativen Ringen.* Comm. Math. Helv. **43** (1968), 378-401.

[U56]   Y. Utumi, *On quotient rings.* Osaka Math. Jour. **8** (1956), 1-18.

[V69]   A. I. Veksler, *A new construction of the Dedekind completion for vector lattices and divisible ℓ-groups.* Siberian Math. J. **10** (1969), 891-896.

[VG72]   A. I. Veksler & V. A. Geiler, *Order and disjoint completeness of linear partially ordered spaces.* Siberian Math. J. **13** (1972), 43-51.

[V84]   J. Vermeer, *The smallest basically disconnected preimage of a space.* Topology and its Appl. **17** (1984), 217-232.

[W74]   R. C. Walker, *The Stone-Čech Compactification.* Ergebnisse der Math. u. i. Grenzgeb. **83** (1974), Berlin-Heidelberg-New York.

[Wi89]   A. W. Wickstead, *An intrinsic characterization of self-injective semiprime commutative rings.* Proc. Royal Irish Acad., Section A, **90A** (1) (1989), 117-124.

Department of Mathematics, University of Florida, P. O. Box 118105,
Gainesville, FL 32611-8105, USA
martinez@math.ufl.edu

Surveying Lattice-Ordered Fields

R. H. Redfield

Dedicated to Paul Conrad
on the occasion of his 80th birthday.

ABSTRACT. The object of what follows is to give a brief overview of the theory of lattice-ordered fields. While I have included no proofs, I have tried to give ample references for anyone interested in seeing the details. Section 1 briefly sketches the history behind the subject and section 2 recalls some basic definitions. The remainder reviews what is presently known: section 3 describes methods of constructing lattice-ordered fields; section 4 concerns the maximal totally ordered subfield; section 5 considers lattice-ordered fields as vector lattices; section 6 describes representations of lattice-ordered fields by means of power series fields; section 7 discusses the number of different compatible lattice orderings; section 8 deals with extensions to total orders; section 9 investigates the lattice-ordering of simple algebraic extensions of totally ordered fields; and section 10 lists some open questions. Of course not all results are mentioned below. I have chosen those that seem to me to be the most fundamental and the most interesting.

I have cited specific results in one of two ways: if an author has given the nth result in the mth section the number $m.n$, then I have used that number; otherwise, I have used the number of the page on which the result occurs.

1 Background

This section assumes some familiarity with the nomenclature of the subject. A reader who wants definitions of the technical terms should refer to section 2.

Vector lattices over the real field $\mathbb{R}$ were the first ordered algebraic structures to be studied; Freudenthal in 1936 [18], Kantorovich in 1937 [23], and Riesz in 1940 [43] introduced them to help in the study of the ring $C(X)$ of continuous real-valued functions on a topological space X.

Soon after, in 1942, Birkhoff gave the general definition of lattice-ordered group in [5] (see also pages 287-318 of [6]), where on page 298, he listed the following special cases which the theory of lattice-ordered groups subsumed: "the additive and multiplicative groups of ordered fields, which have long been studied by Hahn, Artin, and

J. Martínez (ed.), Ordered Algebraic Structures, 123–153.
© 2002 *Kluwer Academic Publishers.*

others, and are now extensively used in valuation theory; the study of abstract number and ideal theory initiated by Dedekind, and recently amplified by Krull, Ward, Lorenzen, Clifford, Dilworth and others; and the semi-ordered function spaces studied very recently by Riesz, Freudenthal, Kantorovich, the author, Bohenblust, Stone, Kakutani, and others" (for example [23], [24], [25], [43]). On page 118 of [8], Bourbaki is more precise. He credits Dedekind's work on divisibility in [14] as the primary impetus for the study of lattice-ordered groups.

Some time later, in 1956 [7], Birkhoff and Pierce gave the first abstract definition of a lattice-ordered ring in general and of an f-ring in particular (see also pages 397-410 of [6]). They devote nearly half the paper to the study of f-rings. Not only does the extra condition make the multiplication more tractable but also f-rings occur in two important settings: the function spaces $C(X)$ are f-rings, as are products of totally ordered fields with addition, multiplication, and order defined coordinatewise. However, as Birkhoff and Pierce noted (on page 57), if an ℓ-field is an f-ring, it is totally ordered; so results about f-rings are of little use to us.

When the paper of Birkhoff and Pierce appeared, totally ordered fields were well known, having held a prominent place in the literature since 1926 when Artin [2] used the theory of formally real fields developed in [3] to solve Hilbert's seventeenth problem. Artin and Schreier had shown in [3] that the complex numbers have no compatible total order, and although Birkhoff and Pierce recognized that they had in passing given the abstract definition of a lattice-ordered field, their primary concern with fields was to ask (on page 68): "Can the real numbers be made into an ℓ-ring by some ordering other than the usual one? Can the complex numbers be made into an ℓ-ring?"

The next decade saw steady work on f-rings and vector lattices and the appearance of two papers on partially ordered fields. Both Dubois [17], in 1956, and DeMarr [15], in 1967, investigated po-fields whose positive cones are closed with respect to both addition and division. By Proposition 4.5 below, any such lattice-ordered field is totally ordered, and hence these fields are not of interest to us, except to note the following. If $(L, +, \cdot, \geq)$ is an ℓ-field in which all squares are positive, then for all $0 < f \in L$, $(f^{-1})^2 > 0$, and hence $f^{-1} = f(f^{-1})^2 > 0$. So such an ℓ-field is one of the sort studied by Dubois and DeMarr and hence is totally ordered. In summary, we have proved the following result.

Proposition 1.1 ([30], 1.1). *If L is a lattice-ordered field in which every square is positive, then L is totally ordered.*

It was in 1969, with the appearance of the paper of Conrad and Dauns [11] containing the proof of their representation theorem (Proposition 6.7 below), that the study of lattice-ordered fields began in earnest. They concluded their paper by asking four questions. Proposition 1.1 answers Question (b), and Question (c) is answered in section 4 below. Questions (a) and (d) are still unanswered (see Questions 2 and 16 in section 10).

Subsequent to the paper of Conrad and Dauns, results have appeared with more regularity, especially in the last decade. Sections 3-9 below describe the progress that has been made since the appearance of the paper of Birkhoff and Pierce nearly fifty years ago.

2   Terminology and Notation

In what follows, all fields are commutative and all rings are associative, but a ring need not have a unit element. We use e to denote the identity of a generic, multiplicatively written group. For terms left undefined, see [1], [4], [10], [13], [20], or any book on abstract algebra (e.g., [36]).

First recall that if A and B are sets and if $f : A \longrightarrow B$, then a *right inverse* ([26], p. 19) of f is a function $r : B \longrightarrow A$ such that for all $b \in B$, $f(r(b)) = b$. If S is a subgroup of an abelian group G, then we say that S is a *direct factor* of G if the canonical map from G to G/S has a right inverse that is a group-homomorphism (or equivalently whose image is a subgroup of G ([38], 1.1)).

Next recall that a *lattice order* on a set S is a partial order $\geq$ for which every two elements $x, y \in S$ have a unique *least upper bound* $x \vee y$:

$$x \vee y \geq x, \ x \vee y \geq y, \text{ and if } w \geq x \text{ and } w \geq y, \text{ then } w \geq x \vee y,$$

and a unique *greatest lower bound* $x \wedge y$:

$$x \geq x \wedge y, \ y \geq x \wedge y, \text{ and if } x \geq w \text{ and } y \geq w, \text{ then } x \wedge y \geq w;$$

that a *trivial order* on S is a partial order $\geq$ on S for which distinct elements $x, y \in S$ are not comparable (i.e., $x \neq y \implies x \not\geq y$ and $y \not\geq x$); and that a *total order* on S is a partial order $\geq$ for which every two elements $x, y \in S$ are comparable (i.e., either $x \geq y$ or $y \geq x$). A subset O of a totally ordered set is *inversely well-ordered* if every nonempty subset of O contains a maximal element.

An ordered algebraic structure is then an algebraic structure with an order that is compatible with its algebraic operations. In particular, if $(G, \cdot)$ is a group, then a partial order $\geq$ on G is *compatible* with $\cdot$ if

$$\forall \, a, b, c, d \in G, \ a \geq b \implies cad \geq cbd;$$

and if $(R, +, \cdot)$ is a ring, then a partial order $\geq$ is *compatible* with $+$ and $\cdot$ if

$$\forall \, r, s, t \in R, \ r \geq s \implies r + t \geq s + t, \text{ and}$$

$$\forall \, r, s \in R, \ \forall \, 0 < t \in R, \ r \geq s \implies [\, tr \geq ts \ \text{ and } \ rt \geq st \,].$$

A group G with a compatible partial order is called a *partially ordered group* (or *po-group*) and a ring R with a compatible partial order is called a *partially ordered ring* (or *po-ring*); so a *partially ordered field* (or *po-field*) is a partially ordered ring whose underlying ring is a field. (In [17], Dubois defined a "partly ordered field" to be a field containing a nonempty subset P that is closed under addition and division. If P is multiplicatively closed, then P is the positive cone of a compatible partial order $\geq$ with respect to which $1 > 0$ and $[x > 0 \implies x^{-1} > 0]$. The *po*-fields considered by DeMarr in [15] also satisfy both these conditions.) If the underlying order is a lattice order, then we call these structures *lattice-ordered groups* (or *ℓ-groups*), *lattice-ordered rings* (or *ℓ-rings*), and *lattice-ordered fields* (or *ℓ-fields*); in the totally ordered case, we use *totally ordered group* (or *o-group*), *totally ordered ring* (or *o-ring*), and *totally ordered field* (or *o-field*). An *f-ring* is an *ℓ*-ring R such that

$$\forall \, a, b \in R, \ \forall \, 0 < c \in R, \ a \wedge b = 0 \implies ac \wedge b = ca \wedge b = 0.$$

A *vector lattice* (or *Riesz space*) over an *o*-field $(T, +, \cdot, \succeq)$ is a vector space V over T with a lattice order $\geq$ such that $(V, +, \geq)$ is an *ℓ*-group and

$$\forall \, u, v \in V, \ \forall \, 0 \prec t \in T, \ u \geq v \implies tu \geq tv.$$

And finally a *lattice-ordered algebra* (or *ℓ-algebra*) over an *o*-field T is an *ℓ*-ring $(A, +, \cdot, \geq)$ for which $(A, +, \geq)$ is simultaneously a vector lattice over T.

A homomorphism between ordered algebraic structures is a function that preserves both the operations and the order. In particular, if A and B are *po*-groups, then a function $f : A \longrightarrow B$ is an an *order-preserving homomorphsim* (or *o-homomorphism*) if for all $x, y \in A$,

$$f(xy) = f(x)f(y) \quad \text{and} \quad x \geq y \implies f(x) \geq f(y);$$

and an *order-preserving isomorphism* (or *o-isomorphism*) from A to B is an *o*-homomorphism that has an inverse that is also an *o*-homomorphism. For *ℓ*-groups, an *ℓ-homomorphism* is an *o*-homomorphism that also preserves the lattice operations; and *ℓ-homomorphisms* between *ℓ*-rings are *ℓ*-homomorphisms between the underlying *ℓ*-groups that also preserve the multiplication. In each case, *ℓ-isomorphisms* are *ℓ*-homomorphisms with inverses that are also *ℓ*-homomorphisms. For vector lattices and *ℓ*-algebras, the corresponding functions are called *Riesz homomorphisms* (*Riesz isomorphisms*) and are linear transformations of the underlying vector spaces as well as *ℓ*-homomorphisms (*ℓ*-isomorphisms) of the underlying *ℓ*-groups or *ℓ*-rings.

For a *po*-group $(G, \cdot, \geq)$, the subset $G^{\geq} = \{g \in G \mid g \geq e\}$ is called the *positive cone* of G; so the positive cone of a *po*-ring $(R, +, \cdot, \geq)$ is $R^{\geq} = \{r \in R \mid r \geq 0\}$. Analogously, $G^{>} = \{g \in G \mid g > e\}$ and $R^{>} = \{r \in R \mid r > 0\}$. The positive cone is important because it is equivalent to the order:

$$x \geq y \text{ in } G \iff xy^{-1} \in G^{\geq} \quad \text{and} \quad x \geq y \text{ in } R \iff x - y \in R^{\geq}.$$

Furthermore, if, for a subset X of a ring R, we let $-X = \{-x \mid x \in X\}$, then we may use positive cones to characterize compatible partial orders as follows. (The positive cones of a *po*-group have an analogous characterization; for details, see page 13 of [19].)

Proposition 2.1 ([19], p. 105). *A subset P of a ring $(R, +, \cdot)$ is the positive cone of a compatible partial order on R if and only if P satisfies*

(1) $P \cap (-P) = \{0\}$,

(2) $P + P \subseteq P$,

(3) $PP \subseteq P$;

R is totally ordered if and only if as well

(4) $R = P \cup (-P)$.

Suppose that $(G, +, \geq)$ is a partially ordered abelian group. We say that G is *archimedean* if for all $a, b \in G^{>}$, there exists some positive integer n such that $a \not\geq nb$. If $S \subseteq G$, then S is *bounded* if there exists $g \in G$ such that for all $s \in S, g > s > -g$; and S is *convex* if $x \in S$ whenever $s \geq x \geq t$ for some $s, t \in S$. If S is a convex subgroup of G, then the relation

$$a + S \geq b + S \iff \text{ there exists } s \in S \text{ such that } a \geq b + s$$

defines a compatible partial order $\geq$ on G/S ([13], p. 44).

If $(G, +, \geq)$ is a lattice-ordered abelian group, then for any element $x \in G$, we let $|x| = (x \vee 0) - (x \wedge 0)$. If $S \subseteq G$, then S is *disjoint* if $0 \notin S$ and for all $x \neq y$ in S, $x \wedge y = 0$; the *polar* of S is the set $S^{\perp} = \{x \in G \mid |x| \wedge |s| = 0 \text{ for all } s \in S\}$; and for $g \in G$, we write $g^{\perp} = \{g\}^{\perp}$. The subset S is an *ℓ-subgroup* of G if $(S, +, \geq)$ is an ℓ-group; a *value* of an element $g \in G$ is then a convex ℓ-subgroup that is maximal with respect to not containing g. We let $\Gamma(G)$ denote the set of values of elements of G, and we say that an element that has a unique value is *special* (see [1], [13], [20], [33]).

If V is a vector lattice over an *o*-field T, then an *ℓ-subspace* of V is a subset of V which is a vector lattice over T with respect to the inherited operations and order.

Note that an element b in a lattice-ordered field L can be "basic" in two senses. It can be an element of a basis of L as a vector space over some subfield K of L, or it can be a positive element for which the interval $[0, b]$ is totally ordered. We distinguish between these two situations as follows. A *v-basis* of L is a basis of L as a vector space over K; an *ℓ-basic* element of L is a positive element b such that $[0, b]$ is totally ordered; an *ℓ-basis* of L is a disjoint subset B of ℓ-basic elements of L such that $B^{\perp} = \{0\}$; and a *vℓ-basis* of L is a subset that is simultaneously a v-basis and an ℓ-basis. We let $B(L)$ denote the set of ℓ-basic elements of L.

3 Constructing Lattice-ordered Fields

A. Power Series Fields. In 1907, before the work of Artin and Schreier, Hahn [21] had shown how to use power series fields over totally ordered groups to construct arbitrarily large totally ordered fields. We can use a generalization of his method to construct arbitrarily large lattice-ordered fields.

Specifically, let $(T, +, \cdot, \geq)$ be a totally ordered field, and let $(\Delta, \cdot, \succeq)$ be a partially ordered abelian group. Unlike Hahn, we do not assume that $\succeq$ is a total order. Instead, we assume is that $\succeq$ is *rooted*, i.e., that the positive cone $\Delta^{\succeq}$ is totally ordered.

Example 3.1. With respect to the trivial order, any abelian group with more than one element is a rooted abelian group that is not totally ordered. The integers $\mathbb{Z}$ also form a rooted abelian group that is not totally ordered with respect to the order whose positive cone consists of the positive even integers $\{2n \mid 0 \leq n \in \mathbb{Z}\}$.

If $(\Delta, \cdot, \succeq)$ is rooted, then $[\alpha, \infty) = \{\delta \in \Delta \mid \delta \geq \alpha\}$ is totally ordered for all $\alpha \in \Delta$, and we say that a subset A of Δ is *locally inversely well-ordered* if for all $\alpha \in A$, $[\alpha, \infty)$ is inversely well-ordered. Then a collection $\mathcal{X}$ of subsets of Δ is called a *supporting collection* ([32], [34], [35]) if it has the following six properties:

1. every $A \in \mathcal{X}$ is locally inversely well-ordered;
2. if $\delta \in \Delta$, then $\{\delta\} \in \mathcal{X}$;
3. if $B \subseteq A \in \mathcal{X}$, then $B \in \mathcal{X}$;
4. if $A, B \in \mathcal{X}$, then $A \cup B \in \mathcal{X}$;
5. if $A, B \in \mathcal{X}$ and $AB = \{\alpha\beta \mid \alpha \in A, \beta \in B\}$, then $AB \in \mathcal{X}$;
6. if $A, B \in \mathcal{X}$ and $\delta \in \Delta$, then $\{(\alpha, \beta) \in A \times B \mid \alpha\beta = \delta\}$ is finite.

(Properties 3 and 4 say that $\mathcal{X}$ is an ideal of the lattice $(2^{\Delta}, \supseteq)$; property 5 says that $\mathcal{X}$ is a subsemigroup of $(2^{\Delta}, \cdot)$.) A supporting collection $\mathcal{X}$ is *strongly supporting* if it satisfies as well

7. if $\emptyset \neq A \in \mathcal{X}$ is countable, then there exists $\alpha \in A$ such that $\bigcup_n (\alpha^{-1}A)^n \in \mathcal{X}$.

Example 3.2. Suppose that $(\Delta, \cdot, \succeq)$ is a rooted abelian group. If Δ is finite, then 2^{Δ} is strongly supporting and is the only supporting collection in 2^{Δ}. If $\succeq_o$ is a compatible total order on Δ such that $\Delta^{\succeq} \subseteq \Delta^{\succeq_o}$, then by Proposition 1.1 of [34] the subsets of Δ that are inversely well-ordered with respect to $\succeq_o$ form a strongly supporting collection of subsets of $(\Delta, \cdot, \succeq)$. The field-families that Raynor defined in [29] (see also [40], [41], and [42]) are all strongly supporting collections by Proposition 2.1 of [35]. And the set of all finite subsets of the integers with the usual total order is supporting but not strongly supporting.

Recall that a *factor set* is a subset $\{t_{\alpha,\beta}\}_{\alpha,\beta\in\Delta}$ of T such that for all $\alpha, \beta, \delta \in \Delta$,

(i) $t_{\alpha,\beta} > 0$, (ii) $t_{\alpha\beta,\delta}t_{\alpha,\beta} = t_{\alpha,\beta\delta}t_{\beta,\delta}$, (iii) $t_{\delta,e} = 1 = t_{e,\delta}$, and (iv) $t_{\alpha,\beta} = t_{\beta,\alpha}$, and for any f in the product $\prod_\Delta T$, let $\mathrm{Supp}(f) = \{\delta \in \Delta \mid f_\delta \neq 0\}$ denote the *support* of f and let $\mathrm{Max}(f)$ denote the set of maximal elements of $\mathrm{Supp}(f)$. Then ([9], [11], [27]) if $\mathcal{X}$ is a supporting collection in 2^Δ, the set

$$_\mathcal{X}\prod_\Delta T = \{f \in \prod_\Delta T \mid \mathrm{Supp}(f) \in \mathcal{X}\}.$$

is a lattice-ordered ring with respect to addition defined coordinatewise, multiplication defined via convolution, and order determined by the maximal elements of the support:

$$(f+g)_\delta = f_\delta + g_\delta, \quad (fg)_\delta = \sum_{\alpha\beta=\delta} t_{\alpha,\beta} f_\alpha g_\beta, \quad \text{and}$$
$$f > 0 \iff f \neq 0 \text{ and} f_\mu > 0 \text{ for all } \mu \in \mathrm{Max}(f).$$

(Properties 3 and 4 ensure additive closure; properties 3, 5, and 6 ensure multiplicative closure; property 1 ensures that the order is properly defined; property 2 ensures that every element of Δ occurs in the support of at least one element of $_\mathcal{X}\prod_\Delta T$; property 7 is used to ensure the existence of multiplicative inverses - see Theorem 3.3 below.)

The ℓ-basic elements of $_\mathcal{X}\prod_\Delta T$ are precisely the positive elements whose support has exactly one element, and if b is ℓ-basic and b^{-1} exists and is positive, then b^{-1} is also ℓ-basic ([33], 6.2). As well, if $\succeq$ is the trivial order on Δ, then $\geq$ is the coordinatewise order on $_\mathcal{X}\prod_\Delta T$.

The lattice-ordered ring $(_\mathcal{X}\prod_\Delta T, \{t_{\alpha,\beta}\}, +, \cdot, \geq)$ is a lattice-ordered field in the following situations.

Theorem 3.3 ([34], 3.5; [35], 1.1). *Suppose that T is a totally ordered field, that Δ is a rooted abelian group, that $\{t_{\alpha,\beta}\} \subseteq \Delta$ is a factor set, and that $\mathcal{X}$ is a strongly supporting collection in 2^Δ. If Δ_t denotes the set of torsion elements of Δ and $\mathrm{tor}(\mathcal{X}) = \{A \in \mathcal{X} \mid A \subseteq \Delta_t\}$, then $\mathrm{tor}(\mathcal{X})$ is a supporting collection in 2^{Δ_t}, $_{\mathrm{tor}(\mathcal{X})}\prod_\Delta T$ is a subring of $_\mathcal{X}\prod_\Delta T$, and the following statements are equivalent:*

(1) $_\mathcal{X}\prod_\Delta T$ *is a lattice-ordered field;*

(2) $_{\mathrm{tor}(\mathcal{X})}\prod_\Delta T$ *is a lattice-ordered field;*

(3) $_{\mathrm{tor}(\mathcal{X})}\prod_\Delta T$ *has no zero-divisors.*

If $t_{\alpha,\beta} = 1$ for all $\alpha, \beta \in \Delta$, then $\{t_{\alpha,\beta}\}$ is a factor set, called the *trivial factor set*, and we write $(_\mathcal{X}\prod_\Delta T, +, \cdot, \geq)$ instead of $(_\mathcal{X}\prod_\Delta T, \{t_{\alpha,\beta}\}, +, \cdot, \geq)$.

Example 3.4. If $(\Delta, \cdot, \succeq)$ is torsion-free and rooted, then $(_\mathcal{X}\prod_\Delta T, +, \cdot, \geq)$ is an ℓ-field for any strongly supporting collection $\mathcal{X}$ in 2^Δ. In particular, suppose that $(\Delta, \cdot, \succeq_o)$ is a nontrivial abelian o-group and that $\mathcal{X}$ is the set of inversely well-ordered subsets of $(\Delta, \succeq_o)$. Then $(\Delta, \cdot)$ is torsion-free, and for the trivial order $\succeq$, $(\Delta, \cdot, \succeq)$ is a rooted abelian group for which $\mathcal{X}$ is a strongly supporting collection. Since Δ is nontrivial, $({}_{\mathcal{X}} \prod_\Delta T, +, \cdot, \geq)$ is a lattice-ordered field that is not totally ordered and whose cardinality is at least that of Δ. (See [32].)

Example 3.5. If $\succeq$ is the trivial order on the group $\mathbb{Z}_2 = \{0, 1\}$ of integers mod 2, then $(\mathbb{Z}_2, +, \succeq)$ is a rooted abelian group, and $({}_{2^{\mathbb{Z}_2}} \prod_{\mathbb{Z}_2} \mathbb{Q}, +, \cdot, \geq)$ is the f-ring $\mathbb{Q} \times \mathbb{Q}$ with coordinatewise operations and order; so $({}_{2^{\mathbb{Z}_2}} \prod_{\mathbb{Z}_2} \mathbb{Q}, +, \cdot, \geq)$ is an ℓ-ring that is not an ℓ-field ([7], p. 57). On the other hand, if

$$t_{m,n} = \begin{cases} 3 & \text{if } m = 1 = n \\ 1 & \text{otherwise,} \end{cases}$$

then $({}_{2^{\mathbb{Z}_2}} \prod_{\mathbb{Z}_2} \mathbb{Q}, \{t_{m,n}\}, +, \cdot, \geq)$ is the ℓ-field $\mathbb{Q}(\sqrt{3})$ with the compatible coordinatewise order:

$$a + b\sqrt{3} \geq 0 \iff a \geq 0 \text{ and } b \geq 0.$$

(See Example 7.4 of [33].)

Recall that a lattice-ordered group is *laterally complete* if every disjoint set has a least upper bound. This property is not the correct one for lattice-ordered fields because many lattice-ordered power series rings will not be laterally complete in this sense ([10], p. 5.9). For instance, if $(T, +, \cdot, \geq)$ is an o-field, if $(\Delta, \cdot, \succeq)$ is a trivially ordered torsion-free abelian group, and if $\mathcal{X}$ is a strongly supporting collection in 2^Δ, then ${}_{\mathcal{X}} \prod_\Delta T$ contains a disjoint set with no least upper bound ([35], p. 92). So we consider ℓ-rings of the following sort. An ℓ-ring R is *weakly laterally complete* if every disjoint subset D for which there exists $0 < d \in R$ such that $d^\perp \subseteq D^\perp$ has a least upper bound in R. An ℓ-ring R may also be *multiplicatively dense* in the sense that for all $a, b \in R^>$ and all $ab \geq x > 0$, there exist $a \geq a_1 > 0$ and $b \geq b_1 > 0$ such that $x \geq a_1 b_1$.

Proposition 3.6 ([35], 3.1). *For any totally ordered field* $(T, +, \cdot, \geq)$, *any rooted abelian group* $(\Delta, \cdot, \succeq)$, *any factor set* $\{t_{\alpha,\beta}\} \subseteq T$, *and any strongly supporting collection* $\mathcal{X}$ *in* 2^Δ, $({}_{\mathcal{X}} \prod_\Delta T, \{t_{\alpha,\beta}\}, +, \cdot, \geq)$ *is weakly laterally complete and multiplicatively dense.*

B. Changing the Multiplication. In the lattice-ordered power series fields described above, $1 > 0$. There are also many ℓ-fields in which $1 \not> 0$, and one easy way to construct examples of this kind of ℓ-field is to change the multiplication on ℓ-fields in which $1 > 0$. Suppose in general that $(L, +, \cdot, \geq)$ is a lattice-ordered field in which $1 > 0$ and that $u \in L$ is such that $u \not> 0$ but $L^\geq L^\geq \subseteq uL^\geq$. Then with respect to the operation $x \odot_u y = xyu^{-1}$, $(L, +, \odot_u, \geq)$ is a lattice-ordered field whose multiplicative identity u does not exceed 0. (See [32] and section 6 of [37].) In particular, any $u \in L$ such that $u^{-1} > 0$ will generate a new compatible multiplication.

If $(\Delta, \cdot, \succeq)$ is a rooted abelian group that is not totally ordered, then such an element u always exists in a power series field $(\chi \prod_\Delta T, \{t_{\alpha,\beta}\}, +, \cdot, \geq)$. For if $\eta \in \Delta$ is not comparable to e, then the element $d \in \chi \prod_\Delta T$ defined by letting

$$d_\delta = \begin{cases} 1 & \text{if } \delta = \eta \text{ or } \delta = e \\ 0 & \text{otherwise,} \end{cases}$$

is a positive element whose inverse is not positive, and hence, if $u = d^{-1}$, then $(\chi \prod_\Delta T, \{t_{\alpha,\beta}\}, +, \odot_u, \geq)$ is a lattice-ordered field whose multiplicative identity u is not positive [32].

Example 3.7. We noted in Example 3.5 that $\mathbb{Q}(\sqrt{3})$ is an ℓ-field with respect to the coordinatewise order $\geq$. If $u = -1 + \sqrt{3}$, then $u^{-1} = \frac{1}{2} + \frac{1}{2}\sqrt{3} > 0$, and hence $(\mathbb{Q}(\sqrt{3}), +, \odot_u, \geq)$ is an ℓ-field whose multiplicative identity $u = -1 + \sqrt{3}$ does not exceed 0.

C. Wilson Orders. Another way of constructing lattice-ordered fields in which $1 \not> 0$ is to use a method first described by Wilson in 1975 ([50], [51]).

Let L be an extension field of a totally ordered field $(T, +, \cdot, \geq)$; let B be a v-basis for L over T; suppose that for all $c, d \in B$, if $cd = \beta_1 b_1 + \cdots + \beta_n b_n$ for $b_1, \ldots, b_n \in B$, then $\beta_i \geq 0$ for all i; and let $\geq_B$ be the binary relation:

$$\alpha_1 b_1 + \cdots + \alpha_n b_n \geq_B 0 \iff \alpha_i \geq 0 \text{ for all } i.$$

Wilson observed that $(L, +, \cdot, \geq_B)$ is an ℓ-field which has B as a $v\ell$-basis. We will refer to B as a *Wilson basis* and to $\geq_B$ as the *Wilson order* on L determined by B.

Example 3.8. Consider the field $\mathbb{Q}(\sqrt{3})$. For $b_1 = \frac{1}{2} + \frac{\sqrt{3}}{2}$ and $b_2 = \frac{3}{2} + \frac{\sqrt{3}}{2}$, we have $b_1 b_1 = \frac{1}{2} b_1 + \frac{1}{2} b_2$, $b_1 b_2 = b_2 b_1 = \frac{3}{2} b_1 + \frac{1}{2} b_2$, and $b_2 b_2 = \frac{3}{2} b_1 + \frac{3}{2} b_2$. So $B = \{b_1, b_2\}$ determines a Wilson order $\geq_B$ on L with respect to which $\mathbb{Q}(\sqrt{3})$ is a lattice-ordered field in which, since $1 = -b_1 + b_2$, $1 \not\geq_B 0$. Note that if $\odot_u$ is the multiplication on $\mathbb{Q}(\sqrt{3})$ described in Example 3.7, then the function

$$T : (\mathbb{Q}(\sqrt{3}), +, \odot_u, \geq) \longrightarrow (Q(\sqrt{3}), +, \cdot, \geq_B),$$

defined by letting

$$T(\alpha + \beta\sqrt{3}) = \alpha b_1 + \beta b_2,$$

is an ℓ-isomorphism of lattice-ordered fields.

4 The Maximal Totally Ordered Subfield

In a totally ordered field T, $T^>$ forms a subgroup of the multiplicative group of nonzero elements of T. On the other hand, if L is a lattice-ordered field in which $1 \not> 0$, then $L^>$ cannot contain any multiplicative subgroups. The intermediate case between these two extremes is that of an ℓ-field L in which $1 > 0$.

A. Lattice-ordered Fields in Which $1 > 0$. Suppose that $(L, +, \cdot, \geq)$ is a lattice-ordered field in which $1 > 0$ and recall that $B(L)$ denotes the set of ℓ-basic elements of L. If $L^{\neq} = \{x \in L \mid x \neq 0\}$, then we have the following.

Theorem 4.1 ([33], 1.5, 3.2, 3.4). *Suppose that L is an ℓ-field in which $1 > 0$.*

(1) *If a subset S of $L^>$ is a subgroup of $(L^{\neq}, \cdot)$, then $S \subseteq B(L)$.*

(2) *As well, the following statements are equivalent:*

(a) *each ℓ-basic element of L has a positive multiplicative inverse;*

(b) $B(L) = \{x \in L \mid x > 0 \text{ and } x^{-1} > 0\}$;

(c) $B(L)$ *is the largest subgroup of $(L^{\neq}, \cdot)$ in $L^{\geq}$;*

(d) $B(L)$ *is a subgroup of $(L^{\neq}, \cdot)$;*

(e) $(B(L), \cdot, \geq)$ *is a rooted abelian group.*

Note that by Theorem 4.1(1), 1 is ℓ-basic. In fact, 1 lives in a very special subfield of L, a subfield which is denoted by $M(L)$ and which has the following properties.

Theorem 4.2. *Any lattice-ordered field L in which $1 > 0$ has a subfield $M(L)$ that may be characterized in any of the following ways:*

(1) ([30], 2.1; [33], 2.3; [37], 2.4; [44]; [46], p. 181) *$M(L)$ is the unique maximal totally ordered subfield of L;*

(2) ([33], 2.3) *$M(L)$ is the only totally ordered subfield of L that is convex;*

(3) ([33], 2.3) *$M(L) = \{x \in 1^{\perp\perp} \mid x = 0 \text{ or } |x|^{-1} > 0\}$;*

(4) ([30], 2.1) *$M(L) = \{x \in L \mid \text{either } 1 \geq |x| \text{ or } 1 \geq |x^{-1}|\}$;*

(5) ([37], 2.4; [46], p. 183) *$M(L)$ is the largest subfield K of L such that for all $0 < a \in L$, aK is a totally ordered subgroup of $(L, +, \geq)$;*

(6) ([37], 2.4; [46], p. 183) *$M(L)$ is the largest subfield K of L that has a compatible total order $\succeq$ such that $K^{\succeq} L^{\geq} \subseteq L^{\geq}$;*

(7) ([37], 2.4; [46], p. 181) $M(L)^{\ge} = \{x \in L \mid \forall a, b \in L, a \wedge b = 0 \Longrightarrow xa \wedge b = 0\}$.

In [44], Schwartz proved the existence of $M(L)$ in the archimedean case; the author in [30] and [33] and Schwartz in [46] independently proved its existence in the general case; Steinberg also knew the general result in the early 1970's [49].

Note that such a subfield may not exist in a partially ordered field. In section 1 of [37], there is an example of a partially ordered field that has nonisomorphic maximal totally ordered subfields.

Example 4.3. Let $(T, +, \cdot, \ge)$ be a totally ordered field, let $(\Delta, \cdot, \succeq)$ be a rooted abelian group, let $\mathcal{X}$ be a strongly supporting collection in 2^Δ, and suppose that $(\chi \prod_\Delta T, +, \cdot, \ge)$ is a lattice-ordered field. If Δ^p is the subset of elements of Δ that are comparable to e, then Δ^p is the unique maximal totally ordered subgroup of Δ and $M(\chi \prod_\Delta T)$ is the set of elements whose support is contained in Δ^p. In particular, if $\succeq$ is the trivial order, then $M(\chi \prod_\Delta T) = \{v \mid \text{Supp}(v) \subseteq \{e\}\}$.

Since $M(L)$ is a convex totally ordered subgroup of $(L, +, \ge)$ that contains 1, $M(L) \subseteq 1^{\perp\perp}$. For the other containment, we have the following.

Proposition 4.4. *Suppose that L is a lattice-ordered field in which $1 > 0$.*

(1) ([30], 3.6) *If L is algebraic over $M(L)$, then $M(L) = 1^{\perp\perp}$.*

(2) ([33], 3.2) *Each of statements (a) - (e) of Theorem 4.1 implies that for all $b \in B(L)$, $b^{\perp\perp} = bM(L)$, and hence that $M(L) = 1^{\perp\perp}$.*

In [11], Conrad and Dauns observed that each positive element in a totally ordered field $(T, +, \cdot, \succeq)$ is special in the sense that it has a unique value (see section 2) and hence that the positive special elements form a multiplicative subgroup of the group $(T^{\neq}, \cdot)$. Conrad and Dauns then developed their representation theory (see Proposition 6.7 below) for ℓ-fields in which the positive special elements form a group (see also [48]). In such ℓ-fields, the positive special elements are precisely the ℓ-basic elements.

Proposition 4.5 ([10], p. 3.13; [11], p. 389; [33], 5.1; [48], p. 411). *Let $(L, +, \cdot, \ge)$ be a lattice-ordered field in which the positive special elements form a group. Then for any $0 < t \in L$, the following statements are equivalent:*

(1) $t^{-1} > 0$;

(2) *t is ℓ-basic;*

(3) *t is special.*

B. Lattice-ordered Fields in Which $1 \not> 0$. If it is not assumed that $1 > 0$ in a lattice-ordered field L, then the structure of L is more difficult to determine than if it is assumed that $1 > 0$. The constructions in section 3 show that such fields certainly exist and Schwartz ([45], [46]) has investigated them in the archimedean case (see Propositions 4.7 and 5.3 below). In general, we can at least determine those ℓ-fields that possess subfields that behave as $M(L)$ does in the case in which $1 > 0$.

Specifically, for any ℓ-field $(L, +, \cdot, \geq)$,

$$\overline{P}(L) = \{x \in L \mid a \geq 0 \Longrightarrow ax \geq 0\}$$

is the positive cone of a compatible partial order $\geq^*$ on L. Since $1 \in \overline{P}(L)$, $(L, +, \cdot, \geq^*)$ contains a copy Q of the rational numbers to which $\geq^*$ gives the usual total order ([37], p. 149). So $(L, +, \cdot, \geq^*)$ contains at least one maximal totally ordered subfield, and hence there is at least one element the set $\mathcal{M}_1(L)$ of maximal totally ordered subfields of $(L, +, \cdot, \geq^*)$. As well, Q is a subfield of $(L, +, \cdot)$ such that

$(*)$ for all $0 < a \in L$, aQ is a totally ordered subgroup of $(L, +, \geq)$,

and hence there is also at least one element in the set $\mathcal{M}_2(L)$ of subfields of $(L, +, \cdot)$ that are maximal with respect to $(*)$. Finally we let

$$\begin{aligned} P_3(L) &= \{x \in L \mid \text{ for all } a, b \in L, a \wedge b = 0 \Longrightarrow ax \wedge b = 0\}, \text{ and} \\ M_3(L) &= \{x - y \mid x, y \in P_3(L)\}. \end{aligned}$$

By Theorem 4.2, if $1 > 0$, then ([37], 2.4)

$$\mathcal{M}_1(L) = \mathcal{M}_2(L) = \{M_3(L)\} = \{M(L)\}.$$

In general, we can say the following.

Theorem 4.6 ([37], 2.3, 3.4, 4.4). *Suppose that $(L, +, \cdot, \geq)$ is a lattice-ordered field. Then*

$$\mathcal{M}_1(L) = \mathcal{M}_2(L), \text{ and for any } T \in \mathcal{M}_2(L), \ T \subseteq M_3(L).$$

Furthermore, the following statements are equivalent:

(1) $P_3(L)$ *is a totally ordered subset of $(L, \geq^*)$;*

(2) $M_3(L)$ *is a totally ordered subfield of $(L, +, \cdot, \geq^*)$;*

(3) $M_3(L)$ *is the unique maximal totally ordered subfield of $(L, +, \cdot, \geq^*)$;*

(4) $M_3(L)$ *is a subfield of $(L, +, \cdot)$ such that for all $0 < a \in L$, $aM_3(L)$ is a totally ordered subgroup of $(L, +, \geq)$;*

(5) $M_3(L)$ *is the unique maximal subfield K of $(L, +, \cdot)$ such that for all $0 < a \in L$, aK is a totally ordered subgroup of $(L, +, \geq)$;*

(6) $\mathcal{M}_1(L) = \mathcal{M}_2(L) = \{M_3(L)\}$.

So in any of the equivalent cases (1) - (6) of Theorem 4.6, $M_3(L)$ is the analogue in arbitrary ℓ-fields of $M(L)$ in ℓ-fields in which $1 > 0$.

For the archimedean case, we have the following.

Proposition 4.7 ([46], pp. 180-183). *If L is an archimedean lattice-ordered field, then*
$$\mathcal{M}_1(L) = \mathcal{M}_2(L) = \{M_3(L)\}.$$

The known examples of ℓ-fields in which $1 \not> 0$ arise from ℓ-fields in which $1 > 0$ by either changing the multiplication or using a Wilson order. Both cases are ones to which the six conditions of Theorem 4.6 apply.

Proposition 4.8 ([37], 5.5, 5.6). *Suppose that L is an extension field of a totally ordered field $(T, +\cdot, \geq)$, that B is a Wilson basis of L over T, and that $\geq_B$ is the corresponding Wilson order on L. Then $M_3(L) = T$ and*
$$\mathcal{M}_1(L) = \mathcal{M}_2(L) = \{M_3(L)\}.$$

Proposition 4.9 ([37], 6.3). *Suppose that $(L, +, \cdot, \geq)$ is a lattice-ordered field in which $1 > 0$ and that $u \in L$ is such that $L^\geq L^\geq \subseteq uL^\geq$, and form the ℓ-field $(L, +, \odot_u, \geq)$. Then $M_3(L, +, \odot_u, \geq) = uM(L, +, \cdot, \geq)$ and*
$$\mathcal{M}_1(L, +, \odot_u, \geq) = \mathcal{M}_2(L, +, \odot_u, \geq) = \{M_3(L, +, \odot_u, \geq)\}.$$

5 Lattice-Ordered Fields as Vector Lattices

If $(L, +, \cdot, \geq)$ is a lattice-ordered field in which $1 > 0$, then, since $M(L)$ is a totally ordered subfield of L, we may consider L to be a vector lattice over $M(L)$, and an easy way of constructing vector lattices over $M(L)$ is to form products or sums of totally ordered vector spaces. Specifically, if $\{T_\gamma\}_{\gamma \in \Gamma}$ is a set of totally ordered vector spaces over $M(L)$, and if $\prod_\Gamma T_\gamma$ denotes the product of the T_γ over Γ with addition, scalar multiplication, and order all defined coordinatewise, then $\prod_\Gamma T_\gamma$ is a vector lattice over $M(L)$; the subspace of $\prod_\Gamma T_\gamma$ consisting of all vectors with finite support is also a vector lattice over $M(L)$ and is denoted by $\sum_\Gamma T_\gamma$.

The simplest such products are those of the form $\prod_\Gamma M(L)$, and ℓ-fields L that live between such a product and its corresponding sum are among those considered by Conrad and Dauns in [11], i.e., among those ℓ-fields which have a finite ℓ-basis and in which the positive special elements form a group. Specifically, we have the following.

Theorem 5.1 ([10], p. 3.15; [33], 5.3). *For an ℓ-field L in which $1 > 0$, the following statements are equivalent.*

(1) *The positive special elements of L form a group and each positive element of L is the disjoint join of special elements.*

(2) *The ℓ-basic elements of L form a group and each positive element of L is the disjoint join of ℓ-basic elements.*

(3) *Each ℓ-basic element of L has a positive multiplicative inverse and L has an ℓ-basis B such that $b^{\perp\perp}$ is unbounded for each $b \in B$.*

(4) *Each ℓ-basic element of L has a positive multiplicative inverse and there exists a one-to-one Riesz homomorphism $\varphi : L \longrightarrow \prod_\Gamma M(L)$ of vector lattices over $M(L)$ such that $\varphi(L)$ contains $\sum_\Gamma M(L)$.*

If we drop the assumption that the positive special elements form a group, then there are several other assumptions that force the structure of L as a vector lattice over $M(L)$ to be simple.

For instance, if L is finite-dimensional over $M(L)$, then, although we cannot show that L is Riesz isomorphic to a product of the form $\prod_\Gamma M(L)$, we can characterize L as a product of the form $\prod_\Gamma V_i$, where the V_i are totally ordered vector spaces over $M(L)$ that contain no proper, nontrivial, convex subspaces, i.e., such that for all $0 \neq b \in V_i$,

$$\{v \in V_i \mid tb \geq |v| \text{ for some } t \in M(L)\} = V_i.$$

Specifically, we have the following.

Proposition 5.2 ([33], 3.2, 4.4, 4.6). *Suppose that L is an ℓ-field in which $1 > 0$ and that L is finite-dimensional over $M(L)$. Then*

(1) *for all $b \in B(L)$, $b^{\perp\perp}$ is unbounded;*

(2) *for all $b \in B(L)$, $b^{\perp\perp}$ contains no proper, nontrivial, convex subpaces;*

(3) *L has a finite ℓ-basis $\{b_1, \ldots, b_n\}$;*

(4) *L is Riesz isomorphic to $\prod_{i=1}^n b_i{}^{\perp\perp}$ as a vector lattice over $M(L)$.*

We can both sharpen and generalize Proposition 5.2 if we further assume that L is archimedean. For by Theorem 4.6, we can consider L as a vector lattice over $M_3(L)$, and, in the archimedean case, when we replace $M(L)$ by $M_3(L)$ in Proposition 5.2, we have the following.

Proposition 5.3 ([46], pp. 184-186). *Suppose that L is an archimedean ℓ-field that is finite-dimensional over $M_3(L)$. Then*

(1) *for all $b \in B(L)$, $b^{\perp\perp}$ is unbounded;*

(2) *for all $b \in B(L)$, $b^{\perp\perp} = bM_3(L)$;*

(3) *L has a finite ℓ-basis $\{b_1, \ldots, b_n\}$;*

(4) *L is Riesz isomorphic to $\prod_{i=1}^{n} b_i M_3(L)$ as a vector lattice over $M_3(L)$.*

6 Representations in Power Series Rings

A. General Theory. Hahn's Theorem [21] for totally ordered groups says that every o-group can be embedded in a lexicographically ordered product of subfields of the real numbers, and in [12], Conrad, Harvey, and Holland generalized this result to lattice-ordered groups. Conrad and Dauns [11] sketched a proof of the corresponding result for totally ordered fields, and Prieß-Crampe [28] gave several detailed proofs of this result. In [31], the author investigated the situation for totally ordered rings.

In general, we want to determine those lattice-ordered fields that can be embedded in power series fields of the kind described in section 3. We begin with the following notation. Suppose that $(T, +, \cdot, \geq)$ is a totally ordered field, that $(\Delta, \cdot, \succeq)$ is a rooted abelian group, and that $\mathcal{X}$ is a supporting collection in 2^Δ, and form the ℓ-ring $(\mathcal{X}\prod_\Delta T, +, \cdot, \geq)$. Then for $\alpha \in \Delta$ and $t \in T$, we let $\overline{\alpha}, \overline{t} \in \mathcal{X}\prod_\Delta T$ be the elements

$$\overline{\alpha}_\delta = \begin{cases} 1 & \text{if } \delta = \alpha \\ 0 & \text{otherwise,} \end{cases} \quad \text{and} \quad \overline{t}_\delta = \begin{cases} t & \text{if } \delta = e \\ 0 & \text{otherwise;} \end{cases}$$

and we let $\overline{\Delta} = \{\overline{\alpha} \mid \alpha \in \Delta\}$ and $\overline{T} = \{\overline{t} \mid t \in T\}$. In general, if L is an ℓ-field, if D is a maximal disjoint set of ℓ-basic elements of L, and if $u \in L$, then we let

$$D[u] = D \cap u^{\perp\perp} = \{d \in D \mid d \wedge |u| > 0\}.$$

For a guide to the general results, consider the following examples.

Example 6.1. Let $\mathcal{X}$ be the set of inversely well-ordered subsets of the integers $\mathbb{Z}$ with respect to its usual total order, and let $\succeq$ be the trivial order on $\mathbb{Z}$. Then $\mathcal{X}$ is a strongly supporting collection of $2^\mathbb{Z}$ with respect to $\succeq$, and $(\mathbb{Z}, +, \succeq)$ is a torsion-free rooted abelian group; so $(_\chi\prod_\mathbb{Z}\mathbb{Q}, +, \cdot, \geq)$ is a lattice-ordered field by Theorem 3.3. (One way to think of $(_\chi\prod_\mathbb{Z}\mathbb{Q}, +, \cdot, \geq)$ is as the field of all Maclaurin series $\sum_{n\in\mathbb{Z}} q_n x^n$, where $q_n \in \mathbb{Q}$ and $q_n = 0$ for all n greater than some N, with coordinatewise order.) For this ℓ-field, we have

$$M\left(_\chi\prod_\mathbb{Z}\mathbb{Q}\right) = 1^{\perp\perp} = \overline{\mathbb{Q}} \quad \text{and} \quad B\left(_\chi\prod_\mathbb{Z}\mathbb{Q}\right) = \{q\overline{n} \mid q \in \mathbb{Q}^>, n \in \mathbb{Z}\}.$$

Note that $B(_\chi\prod_\mathbb{Z}\mathbb{Q})$ is a rooted abelian group with respect to multiplication, that $M(_\chi\prod_\mathbb{Z}\mathbb{Q})^>$ is a convex subgroup of $B(_\chi\prod_\mathbb{Z}\mathbb{Q})$, and that the quotient group $B(_\chi\prod_\mathbb{Z}\mathbb{Q})/M(_\chi\prod_\mathbb{Z}\mathbb{Q})^>$ is a trivially ordered group that is o-isomorphic to $(\mathbb{Z}, +, \succeq)$. Furthermore, $\overline{\mathbb{Z}}$ is an ℓ-basis of $_\chi\prod_\mathbb{Z}\mathbb{Q}$, and, if a and b are strictly positive elements in $_\chi\prod_\mathbb{Z}\mathbb{Q}$, then

$$\overline{\mathbb{Z}}[ab] = \{\overline{\alpha\beta} \mid \alpha \in \text{Supp}(a),\ \beta \in \text{Supp}(b)\} \subseteq (\overline{\mathbb{Z}}[a]\overline{\mathbb{Z}}[b])^{\perp\perp}$$

and the set

$$\{(\overline{\alpha}, \overline{\beta}) \in \overline{\mathbb{Z}}[a] \times \overline{\mathbb{Z}}[b] \mid (\overline{\alpha\beta})^{\perp\perp} = 1^{\perp\perp}\} = \{(\overline{\alpha}, \overline{\beta}) \in \overline{\mathbb{Z}}[a] \times \overline{\mathbb{Z}}[b] \mid \alpha\beta = e\}$$

is finite.

Note finally that if, for $x \in B(_\chi\prod_\mathbb{Z}\mathbb{Q})$, we denote the single element in $\text{Supp}(x)$ by $\sigma(x)$, then the map $r(xM(_\chi\prod_\mathbb{Z}\mathbb{Q})^>) = \overline{\sigma(x)}$ is a right inverse of the canonical projection from $B(_\chi\prod_\mathbb{Z}\mathbb{Q})$ to $B(_\chi\prod_\mathbb{Z}\mathbb{Q})/M(_\chi\prod_\mathbb{Z}\mathbb{Q})^>$, and hence, since r is a group-homomorphism, that $M(_\chi\prod_\mathbb{Z}\mathbb{Q})^>$ is a direct factor of $B(_\chi\prod_\mathbb{Z}\mathbb{Q})$.

Example 6.2. Let $\succeq$ be the trivial order on the group $\mathbb{Z}_2 = \{0, 1\}$ of integers mod 2; let $\mathcal{X} = 2^{\mathbb{Z}_2}$; and let $t_{1,1} = 3$ and $t_{m,n} = 1$ otherwise. As noted in Example 3.5, $(\mathbb{Z}_2, +, \succeq)$ is a rooted abelian group, and $(_\chi\prod_{\mathbb{Z}_2}\mathbb{Q}, \{t_{\alpha,\beta}\}, +, \cdot, \geq)$ is the ℓ-field $\mathbb{Q}(\sqrt{3})$ with the coordinatewise order.

In this case as well, $M(_\chi\prod_{\mathbb{Z}_2}\mathbb{Q}) = \overline{\mathbb{Q}}$ and $B(_\chi\prod_{\mathbb{Z}_2}\mathbb{Q}) = \{q\overline{n} \mid q \in \mathbb{Q}^>, n \in \{0, 1\}\}$; $B(_\chi\prod_{\mathbb{Z}_2}\mathbb{Q})$ is an abelian group with respect to multiplication; $M(_\chi\prod_{\mathbb{Z}_2}\mathbb{Q})^>$ is a convex subgroup of $B(_\chi\prod_{\mathbb{Z}_2}\mathbb{Q})$; $B(_\chi\prod_{\mathbb{Z}_2}\mathbb{Q})/M(_\chi\prod_{\mathbb{Z}_2}Q)^>$ is o-isomorphic to $(\mathbb{Z}_2, +, \succeq)$; $\overline{\mathbb{Z}_2}$ is an ℓ-basis of $_\chi\prod_{\mathbb{Z}_2}\mathbb{Q}$; and for $0 < a, b \in {}_\chi\prod_{\mathbb{Z}_2}\mathbb{Q}$,

$$\overline{\mathbb{Z}_2}[ab] \subseteq (\overline{\mathbb{Z}_2}[a]\overline{\mathbb{Z}_2}[b])^{\perp\perp} \quad \text{and} \quad \{(\overline{\alpha}, \overline{\beta}) \in \overline{\mathbb{Z}_2}[a] \times \overline{\mathbb{Z}_2}[b] \mid (\overline{\alpha\beta})^{\perp\perp} = 1^{\perp\perp}\} \text{ is finite.}$$

In this case, however, $B(_\chi\prod_{\mathbb{Z}_2}\mathbb{Q})$ is torsion-free and hence, since

$$B\left(_\chi\prod_{\mathbb{Z}_2}\mathbb{Q}\right)\Big/M\left(_\chi\prod_{\mathbb{Z}_2}\mathbb{Q}\right)^>$$

is a torsion group, the canonical projection to it from $B(_\chi\prod_{\mathbb{Z}_2}\mathbb{Q})$ cannot have a right inverse that is a group-homomorphism. So $M(_\chi\prod_{\mathbb{Z}_2}\mathbb{Q})^>$ cannot be a direct factor of

$B(_\chi\prod_{\mathbb{Z}_2}\mathbb{Q})$. Note that nonetheless $r(xM(_\chi\prod_{\mathbb{Z}_2}\mathbb{Q})^>) = \overline{\sigma(x)}$ still defines a right inverse of the canonical projection, even though it is not a group-homomorphism.

The general case is remarkably similar to the two examples above.

Theorem 6.3 ([33], 7.1, 7.2). *If $(L, +, \cdot, \geq)$ is a lattice-ordered field in which $1 > 0$, then the following statements are equivalent.*

(1) *There exist a rooted abelian group $(\Delta, \cdot, \succeq)$, a supporting subset $\mathcal{X}$ of 2^Δ, a factor set $\{t_{\alpha,\beta}\} \subseteq M(L)$, and a one-to-one Riesz homomorphism*

$$\lambda : (L, +, \cdot, \geq) \longrightarrow \left(_\chi\prod_\Delta M(L), \{t_{\alpha,\beta}\}, +, \cdot, \geq\right)$$

of lattice-ordered algebras over $M(L)$ such that $\lambda(L) \supseteq \overline{\Delta}$.

(2) *Every ℓ-basic element of L has a positive multiplicative inverse; L has an ℓ-basis; and if D is any ℓ-basis of L, then $d^{\perp\perp}$ is unbounded for all $d \in D$, and for all $u, v \in L^>$, $D[uv] \subseteq (D[u]D[v])^{\perp\perp}$ and the set $\{(a, c) \in D[u] \times D[v] \mid (ac)^{\perp\perp} = 1^{\perp\perp}\}$ is finite.*

(3) *The ℓ-basic elements of L form a multiplicative group, and there exist a supporting subset $\mathcal{X}$ of $2^{B(L)/M(L)^>}$, a factor set $\{t_{\alpha,\beta}\} \subseteq M(L)$, and a one-to-one Riesz homomorphism*

$$\lambda : (L, +, \cdot, \geq) \longrightarrow \left(_\chi\prod_{B(L)/M(L)^>} M(L), \{t_{\alpha,\beta}\}, +, \cdot, \geq\right)$$

of lattice-ordered algebras over $M(L)$ such that $\lambda(L) \supseteq \overline{B(L)/M(L)^>}$.

In case (1), $(\Delta, \succeq)$ must be trivially ordered. If $M(L)^>$ is a direct factor of $B(L)$, then the factor set $\{t_{\alpha,\beta}\}$ may be assumed to be trivial.

Analogous results are true for representations based on index groups that are not assumed to be trivially ordered. In this case, the power series algebras that are used for the representations are over o-fields that are constructed as follows. Suppose that L is a lattice-ordered field in which $1 > 0$, that T is a subfield of $M(L)$, and that $M(L)^r$ is the real closure of $M(L)$. If

$$M(L)^r_e = \{u \in M(L)^r \mid |u| < t \text{ for some } t \in T^>\}, \text{ and}$$
$$M(L)^r_m = \{u \in M(L)^r \mid |u| < t \text{ for all } t \in T^>\},$$

then $M(L)^r_m$ is a maximal convex ideal of the totally ordered ring $M(L)^r_e$; so we may form the the totally ordered field $T^R = M(L)^r_e/M(L)^r_m$. The more general representation theorems, based on the o-fields T^R, are given in section 9 of [33].

B. Special Cases. Not surprisingly, Theorem 6.3 is simpler when it is applied to more restrictive classes of the lattice-ordered fields.

For instance, for archimedean ℓ-fields, we have the following.

Proposition 6.4 ([35], 5.3). *For an archimedean lattice-ordered field* $(L, +, \cdot, \geq)$, *the following statements are equivalent.*

(1) *There exist a rooted abelian group* $(\Delta, \cdot, \succeq)$, *a supporting collection* $\mathcal{X}$ *in* 2^Δ, *a factor set* $\{t_{\alpha,\beta}\} \subseteq M(L)$, *and an* ℓ*-isomorphism*

$$\lambda : (L, +, \cdot, \geq) \longrightarrow \left({}_\mathcal{X}\prod\nolimits_\Delta M(L), \{t_{\alpha,\beta}\}, +, \cdot, \geq \right).$$

(2) *The positive special elements of* L *form a multiplicative group and* L *is weakly laterally complete and multiplicatively dense.*

And we can characterize those ℓ-fields that can be represented by using torsion index groups as follows.

Proposition 6.5 ([34], 4.6; [35], 5.3). *If* $(L, +, \cdot, \geq)$ *is a lattice-ordered field in which* $1 > 0$, *then the following statements are equivalent.*

(1) *There exist a trivially ordered torsion abelian group* $(\Delta, \cdot, \succeq)$, *a factor set* $\{t_{\alpha,\beta}\} \subseteq M(L)$, *and an* ℓ*-isomorphism*

$$\lambda : (L, +, \cdot, \geq) \longrightarrow \left({}_\mathcal{F}\prod\nolimits_\Delta M(L), \{t_{\alpha,\beta}\}, +, \cdot, \geq \right),$$

where $\mathcal{F}$ *is the collection of finite subsets of* Δ.

(2) *There exist a factor set* $\{t_{\alpha,\beta}\} \subseteq M(L)$, *and an* ℓ*-isomorphism*

$$\lambda : (L, +, \cdot, \geq) \longrightarrow \left({}_\mathcal{F}\prod\nolimits_{B(L)/M(L)^>} M(L), \{t_{\alpha,\beta}\}, +, \cdot, \geq \right)$$

where $\mathcal{F}$ *is the set of finite subsets of* $B(L)/M(L)^>$.

(3) *Every* ℓ*-basic element of* L *has a positive multiplicative inverse;* L *has an* ℓ*-basis* D; *for all* $u, v \in L^>$, $\{(a, c) \in D[u] \times D[v] \mid (ac)^{\perp\perp} = 1^{\perp\perp}\}$ *is finite; and* L *is algebraic over* $M(L)$.

(4) *The positive special elements of* L *form a multiplicative group and no positive element of* L *exceeds an infinite disjoint set.*

If an ℓ-field L is finite-dimensional over $M(L)$, then no positive element of L exceeds an infinite disjoint set ([38], 0.1), and in this even more restrictive case, we can also describe the ℓ-basic elements.

Proposition 6.6 ([33], 10.1). *Let* $(L, +, \cdot, \geq)$ *be a lattice-ordered field in which* $1 > 0$, *and suppose that* L *is finite-dimensional over* $M(L)$. *Then the following statements are equivalent.*

(1) *There exist a finite trivially ordered abelian group* $(\Delta, \cdot, \succeq)$, *a factor set* $\{t_{\alpha,\beta}\} \subseteq M(L)$, *and an* ℓ-*isomorphism*

$$\lambda : (L, +, \cdot, \geq) \longrightarrow \left({}_{2^\Delta}\prod_\Delta M(L), \{t_{\alpha,\beta}\}, +, \cdot, \geq \right).$$

(2) *Every* ℓ-*basic element of* L *has a positive multiplicative inverse.*

(3) *Every* ℓ-*basic element of* L *is the root of an element in* $M(L)$.

In case (1), $B(L)$ *is a group and the group* Δ *may be taken to be* $B(L)/M(L)^>$.

The representation theorem proved by Conrad and Dauns in [11] also involved representing $M(L)$ in a totally ordered power series field over the real numbers $\mathbb{R}$. Specifically, they proved the following result. (This result can also be derived from the general representation theory in [33]. Stuart Steinberg pointed out an error in the proof given there on page 354, an error which was corrected in [39].)

Proposition 6.7 ([11], p. 393; [33], 10.7; [39]). *Suppose that* L *is a lattice-ordered field with a finite* ℓ-*basis and that the positive special elements of* L *form a group. Then the set of values* $\Gamma(L)$ *of* L *is a rooted abelian group, and there exists a one-to-one homomorphism of lattice-ordered rings from* L *into* $\left({}_W\prod_{\Gamma(L)} R, +, \cdot, \geq \right)$, *where* W *is the collection of locally inversely well-ordered subsets of* $\Gamma(L)$. *If* $\Gamma(L)$ *is torsion-free, then* $\left({}_W\prod_{\Gamma(L)} R, +, \cdot, \geq \right)$ *is a lattice-ordered field.*

7 The Number of Compatible Lattice Orderings

In their 1956 paper [7], Birkhoff and Pierce asked whether there were any nontotal compatible lattice orders of the real numbers. Wilson answered this question twenty years later in [51]. Not only did he find a nontotal compatible lattice order, he found infinitely many such orders. We may summarize his method as follows.

Let $\succeq$ denote the usual total order on $\mathbb{R}$; let L be a subfield of $\mathbb{R}$ with a compatible lattice order $\geq$. If $\{b_1, \ldots, b_k\} \subseteq \mathbb{R}$, then $L^{\geq}\langle b_1, \ldots, b_k \rangle$ denotes the set of finite sums of the form $m_1 b_1 + \cdots + m_k b_k$ for $m_i \in L^{\geq}$. And $\geq$ *quotient-represents* $L \cap \mathbb{R}^{\succeq}$ if for all $x \in L \cap \mathbb{R}^{\succeq}$, there exist $p, q \in L^{\geq}$ such that $q \neq 0$ and $x = pq^{-1}$. For such orders, we have the following. (Note that Lemma 2.2(2) of [51] states only that $a_0, \ldots, a_{n-1} \in L^{\geq}$; however, the proof on page 574 of [51] shows that in fact $a_0, \ldots, a_{n-1} \in L^>$.)

Lemma 7.1 ([51], 2.2). *Suppose that* K *is a subfield of* $\mathbb{R}$ *and that* $\geq$ *is a compatible lattice order on* K *that quotient-represents* $K \cap \mathbb{R}^{\succeq}$. *If* K' *is an extension field of* K *in* $\mathbb{R}$ *whose dimension* n *over* K *is finite, then there exists* $\alpha \in K'$ *such that*

(1) $K' = K[\alpha]$;

(2) *there exist* $a_0, \ldots, a_{n-1} \in K^>$ *such that* $\alpha^n = a_0 + \cdots + a_{n-1}\alpha^{n-1}$;

(3) $K^{\geq}\langle 1, \alpha, \ldots, \alpha^{n-1}\rangle \subseteq \mathbb{R}^{\geq}$;

(4) *the order on* K' *with positive cone* $K^{\geq}\langle 1, \alpha, \ldots, \alpha^{n-1}\rangle$ *quotient-represents* $K' \cap \mathbb{R}^{\geq}$.

Now suppose that K, J, and L are subfields of $\mathbb{R}$ such that

(i) $K \subseteq J \subseteq L$,

(ii) L is a proper algebraic extension of K,

(iii) J has a Wilson basis B over K satisfying properties (1)-(4) of Lemma 7.1, and

(iv) $\beta \in L \backslash J$.

Then by Lemma 7.1, there exists $\alpha \in J[\beta]$ such that $\{1, \alpha, \ldots, \alpha^{n-1}\}$ is a Wilson basis of $J[\beta]$ over J and hence such that

$$B' = \{b\alpha^i \mid b \in B, 0 \leq i \leq n-1\}$$

is a Wilson basis of $J[\beta]$ over K that also satisfies properties (1)-(4). Since $\{1\}$ is a Wilson basis of K over K, Zorn's Lemma implies that L has a Wilson basis over K and thus we have the following.

Proposition 7.2 ([51], p. 571). *Let L be a subfield of $\mathbb{R}$ which is a proper algebraic extension of a subfield K. Then there exists a compatible lattice order $\geq$ on L with respect to which $1 > 0$ and $M(L) = K$.*

So in general, we have:

Proposition 7.3 ([51], 1.1). *Let L be a subfield of $\mathbb{R}$ containing κ distinct subfields K such that L is algebraic over K. Then L admits at least κ distinct lattice orders.*

But if c is the cardinality of $\mathbb{R}$, then $\mathbb{R}$ is algebraic over 2^c distinct subfields, and the subfield A of algebraic numbers in $\mathbb{R}$ is algebraic over $2^{\aleph_0}$ distinct subfields. So we have the following consequence of Proposition 7.3, which answers the question of Birkhoff and Pierce mentioned at the beginning of the section.

Theorem 7.4 ([51], 1.2). *The field of real numbers $\mathbb{R}$ admits exactly 2^c distinct compatible lattice orders, and the subfield of algebraic numbers of $\mathbb{R}$ admits exactly $2^{\aleph_0}$ distinct lattice orders.*

By Proposition 7.2, the compatible lattice orders that we have constructed so far in this section all have $1 > 0$. Since $1 = a_0^{-1}(\alpha^n - a_{n-1}\alpha^{n-1} - \cdots - a_1\alpha)$, if we use

$$B'' = \{b\alpha^i \mid b \in B, 1 \leq i \leq n\}$$

instead of B', we can construct orders in which $1 \not> 0$, and thus we have the following companion to Proposition 7.2.

Proposition 7.5 ([51], p. 576). *Let L be a subfield of $\mathbb{R}$ which is a proper algebraic extension of a subfield K. Then there exists a compatible lattice order $\geq$ on L with respect to which $1 \not> 0$ and K is the largest trivially ordered subfield in $(L, +, \cdot, \geq)$.*

Now if L is a finite-dimensional extension of K, then there are finitely many intermediate subfields, and hence Proposition 7.3 guarantees only finitely many distinct lattice orders. We can construct more orders by using Proposition 7.5. In fact, since the minimum polynomial $f(x) = -a_0 - \cdots - a_{n-1}x^{n-1} + x^n$ of α has coefficients $-a_i < 0$, we can find $\aleph_0$ distinct $r \in \mathbb{Q}$ such that $\alpha - r$ also has a minimum polynomial with negative coefficients. And therefore we have the following companion to Proposition 7.3.

Proposition 7.6 ([51], 4.3). *If L is a subfield of $\mathbb{R}$ that is a finite-dimensional extension of $\mathbb{Q}$, then L admits at least $\aleph_0$ distinct lattice orders.*

To find an upper bound, we can use the next result which, as noted in [46], follows from a proof similar to that used to prove Proposition 3 in Chapter VIII of [19].

Proposition 7.7 ([19], p. 127; [46], p. 186). *Let $(L, +, \cdot, \geq)$ be a lattice-ordered field, and suppose that T is a subfield of L such that L is algebraic over T and T admits a compatible total order $\succeq$ such that $T^{\succeq}L^{\geq} \subseteq L^{\geq}$. If $(T, +, \cdot, \succeq)$ is archimedean, then $(L, +, \cdot, \geq)$ is also archimedean.*

So if L is a subfield of $\mathbb{R}$ that is a finite-dimensional extension of $\mathbb{Q}$, then any lattice order on L is archimedean, and Proposition 5.3 applies. But there are only finitely many candidates for $M_3(L)$ and there are at most $\aleph_0$ bases of L over each of these subfields. So there can be at most $\aleph_0$ compatible lattice orders.

Theorem 7.8 ([46], p. 186). *If L is a subfield of $\mathbb{R}$ that is a finite-dimensional extension of $\mathbb{Q}$, then L admits exactly $\aleph_0$ distinct lattice orders.*

8 Extending Lattice Orders to Partial Orders

Birkhoff and Pierce also asked in [7] whether the complex numbers admit a compatible lattice order. The answer to this question would be negative if it could be shown that

every lattice-ordered field has a compatible total order. Since such a field already has
a compatible lattice-order, there are two possible strategies to use to try to prove this:
first, find a compatible total order that may be unrelated to the given lattice-order, or
second, find a compatible total order that arises from the given lattice-order.

In [3], Artin and Schreier found a simple criterion for applying the first strategy;
they showed that a field has a compatible total order if and only if -1 is not the sum
of squares (i.e, if and only if it is *formally real*).

For the second strategy, recall that if $(F, +, \cdot, \geq)$ is a partially ordered field, then
a partial order $\succeq$ on F *extends* $\geq$ (or $\geq$ is *extendible* to $\succeq$) if $(F, +, \cdot, \succeq)$ is also a
partially ordered field and $F^{\geq} \subseteq F^{\succeq}$. Now determining whether a given lattice order is
extendible to a total order is not as simple as determining merely whether the field has
some compatible total order. Fuchs gave the following criterion for this more restrictive
situation.

Proposition 8.1 ([19], p. 117; [47]). *Suppose that $(F, +, \cdot, \geq)$ is a partially ordered field. Then $\geq$ is extendible to a compatible total order if and only if for all
$a_1, \ldots, a_n \in F^{\neq}$ and all $p_1, \ldots, p_n \in F^{>} \cup \{1\}$, $p_1 a_1{}^2 + \cdots + p_n a_n{}^2 \neq 0$.*

There are various classes of ℓ-fields whose orders are known to be extendible to
total orders. For instance, if $(L, +, \cdot, \geq)$ is an ℓ-field, then the set of quotients

$$Q(L^{>}) = \{pq^{-1} \mid p, q \in L^{>}\}$$

is the strictly positive cone of a compatible partial order, $\geq_Q$ on L by Proposition
2.1, and we can use this order to show that certain archimedean lattice orders can be
extended to total orders.

Proposition 8.2 ([45], p. 193). *Suppose that $(L, +, \cdot, \geq)$ is an archimedean
lattice-ordered field in which $1 > 0$ and which is algebraic over $M(L)$. Then the order
$\geq_Q$ is the intersection of total orders. Hence $\geq$ can be extended to a total order, and
$\geq$ is uniquely extendible to a total order if and only if $\geq_Q$ is already a total order.*

In the general case, when we do not assume that $1 > 0$, we nonetheless have the
following result.

Theorem 8.3 ([46], p. 189). *If $(L, +, \cdot, \geq)$ is an archimedean lattice-ordered field
that is algebraic over $M_3(L)$, then $\geq$ is extendible to a compatible total order.*

Combining Theorem 8.3 and Proposition 7.7, we can answer the question of Birkhoff
and Pierce mentioned at the beginning of the section in the case of the algebraic
numbers.

Corollary 8.4 ([46], p. 192). *Suppose that A is the field of all complex numbers
that are algebraic over $\mathbb{Q}$. Then there is no compatible lattice order definable on A.*

And for finite-dimensional ℓ-fields, we can sharpen Theorem 8.3 as follows.

Proposition 8.5 ([46], p. 191). *If $(L, +, \cdot, \geq)$ is an archimedean lattice-ordered field that is finite-dimensional over $M_3(L)$, then $\geq$ can be extended to a unique compatible total order.*

Since squares are always positive in a totally ordered field, the order on an ℓ-field that has a negative square cannot be extended to a compatible total order. So if we could find an ℓ-field with a negative square, we would have an ℓ-field whose order is not extendible to a total order. From this point of view, we can conclude the following from Theorem 8.3. (In the finite-dimensional case, this result follows from a generalization of Corollary 2 of [16] - see page 26 of [30].)

Proposition 8.6. *If $(L, +, \cdot, \geq)$ is an archimedean lattice-ordered field that is algebraic over $M_3(L)$, then L has no negative squares.*

9 Simple Algebraic Extensions

The lattice-ordered fields whose structure we know in the most detail are those that are finite-dimensional extensions of $M(L)$. In the archimedean case, we know the structure of these fields as vector lattices precisely (Proposition 5.3), we know exactly which of these fields are power series fields (Proposition 6.6), and we know all compatible total orders that extend the given order (Proposition 8.5). The nonarchimedean case, on the other hand, is more complicated.

One way to approach this more general case is to notice that simple algebraic extensions arise as quotient rings of the polynomial ring $M(L)[x]$, a ring that has a natural compatible lattice order. Surprisingly, if the quotient order is defined, it is a lattice-order, not when the principal ideals are lattice-ordered, but rather when they are trivially ordered! Specifically we have the following.

If $(T, +, \cdot, \succeq)$ is a totally ordered field and a is algebraic over T, then $T(a)$ is field-isomorphic to the quotient ring $T[x]/(p(x))$, where $p(x)$ is the minimum polynomial of a over T (i.e., $p(x) \in T[x]$ is monic and irreducible and $p(a) = 0$) and $(p(x))$ is the principal ideal generated by $p(x)$ (i.e., $(p(x)) = \{f(x)p(x) \mid f(x) \in T[x]\}$). Now the order defined by letting

$$\alpha_0 + \alpha_1 x + \cdots + \alpha_n x^n \geq 0 \iff \alpha_i \succeq 0 \text{ for all } i$$

is a compatible lattice order on the ring $T[x]$ and we can say the following about the structure of the ideal $(p(x))$ with respect to the order $\geq$.

Proposition 9.1 ([38], 4.2). *Suppose that T is a totally ordered field and that $p(x)$ is a polynomial in $T[x]$ of positive degree such that $p(0) \neq 0$. Then the following statements are equivalent:*

(1) $(p(x))$ is convex;

(2) $(p(x))$ is trivially ordered;

(3) $(p(x))^2 = \{0\}$.

When $(p(x))$ is convex, the quotient order on $T[x]/(p(x))$ is a compatible partial order, and, if n is the degree of $p(x)$, then $T[x]/(p(x))$ is isomorphic to T^n as a vector space over T. We would like to know when an isomorphism of vector spaces is an isomorphism of vector lattices, i.e, when $T[x]/(p(x))$ with the quotient order is Riesz isomorphic to the vector lattice $\prod_{i=1}^n T$ with the coordinatewise order. Since copies of $\prod_{i=1}^n T$ are abundant back in the ℓ-ring $T[x]$, we are led to consider right inverses of the canonical projection from $T[x]$ to $T[x]/(p(x))$. In particular, we can say the following.

Theorem 9.2 ([38], 4.5). *Suppose that T is a totally ordered field, that $p(x)$ is a polynomial in $T[x]$ of positive degree n such that $p(0) \neq 0$, and that $(p(x))^2 = \{0\}$. Let $\pi : T[x] \to T[x]/(p(x))$ be the canonical projection. Then, with respect to the quotient order on $T[x]/(p(x))$, the following statements are equivalent:*

(1) *π has an order-preserving right inverse whose image is a subspace of $T[x]$;*

(2) *$T[x]/(p(x))$ is an ℓ-algebra which has a $v\ell$-basis;*

(3) *$T[x]/(p(x))$ is a vector lattice with a $v\ell$-basis;*

(4) *$T[x]/(p(x))$ is a vector lattice in which every totally ordered subspace has dimension one over T;*

(5) *$T[x]/(p(x))$ is lattice-ordered and has n disjoint elements.*

If $p(x)$ is monic and irreducible, then each of the following statements is equivalent to each of the above statements:

(6) *$T[x]/(p(x))$ is a vector lattice in which every convex totally ordered subspace has dimension one over T;*

(7) *$T[x]/(p(x))$ is a vector lattice in which every convex totally ordered subspace that has no proper, nontrivial, convex subspaces has dimension one over T.*

In each case (1) - (7), $T[x]/(p(x))$ is isomorphic to an ℓ-subspace of $T[x]$ as a vector lattice over T.

Now there is always a natural right inverse of the canonical projection π from $T[x]$ to $T[x]/(p(x))$, viz., the function which takes a coset $f(x) + (p(x))$ to the remainder of $f(x)$ when divided by $p(x)$. This right inverse is a linear transformation and when it preserves order, we can be very specific about the polynomial $p(x)$ and the structure of $T[x]/(p(x))$.

Theorem 9.3 ([38], 4.6). *Suppose that T is a totally ordered field and that $p(x) = \lambda_0 + \lambda_1 x + \cdots + \lambda_{n-1} x^{n-1} + x^n \in T[x]$, where $n > 0$ and $\lambda_0 \neq 0$. Let $\pi : T[x] \longrightarrow T[x]/(p(x))$ be the canonical projection and let $r : T[x]/(p(x)) \longrightarrow T[x]$ be defined by letting $r(f(x) + (p(x)))$ be the remainder of $f(x)$ when it is divided by $p(x)$. Let $\tilde{r}$ denote r with codomain restricted to $r(T[x]/(p(x)))$. Then r is a well-defined linear transformation and a right inverse of π; and the following statements are equivalent:*

(1) *with respect to the quotient order, $T[x]/(p(x))$ is an ℓ-algebra over T in which $\{1 + (p(x)), \ldots, x^{n-1} + (p(x))\}$ is a $v\ell$-basis;*

(2) *$(p(x))^{\geq} = \{0\}$ and r preserves order;*

(3) *$\tilde{r}$ is an isomorphism of vector lattices;*

(4) *$(p(x))^{\geq} = \{0\}$ and for all distinct i and j in $\{0, \ldots, n-1\}$, $(x^i + (p(x))) \wedge (x^j + (p(x)))$ exists in $T[x]/(p(x))$ and equals $0 + (p(x))$;*

(5) *$\lambda_i \leq 0$ for all $0 \leq i \leq n-1$.*

The following example shows that there are cases in which the quotient order is defined and the remainder is not an order-preserving right inverse of π, but nonetheless π has a right inverse onto a subspace of $T[x]$ that is a vector lattice with respect to the inherited order.

Example 9.4 ([38], 4.7). Let $p(x) = x^2 + 2x - 2 \in Q[x]$. Then $(p(x))^{\geq} = \{0\}$; so the quotient order is defined on $\mathbb{Q}[x]/(p(x))$. Since $3x - 2 + (\frac{1}{2}x - 1)(x^2 + 2x - 2) = \frac{1}{2}x^3 > 0$, $3x - 2 + (p(x))$ is positive with respect to the quotient order, but in this case, the remainder, $3x - 2$, is not positive in $\mathbb{Q}[x]$; so the function determined by the remainder is not an order-preserving right inverse. (This also follows from statement (5) of Theorem 9.3.)

Let $L = \{c(x^3 + \frac{5}{2}x^2) + d \mid c, d \in \mathbb{Q}\}$ and note that L is a subspace of $\mathbb{Q}[x]$ which is a vector lattice with respect to the inherited order and which has a $v\ell$-basis $\{1, x^3 + \frac{5}{2}x^2\}$. Then the function $r : Q[x]/(p(x)) \longrightarrow L$, defined by letting

$$r(ax + b + (p(x))) = a\left(x^3 + \frac{5}{2}x^2\right) + (b - a),$$

is a right inverse of the canonical projection $\pi : \mathbb{Q}[x] \longrightarrow \mathbb{Q}[x]/(p(x))$. And r determines a lattice order on $\mathbb{Q}[x]/(p(x))$:

$$ax + b + (p(x)) \geq 0 + (p(x)) \iff r(ax + b + (p(x))) \geq 0.$$

In fact, $(\mathbb{Q}[x]/(p(x)), +, \cdot, \geq)$ is an ℓ-algebra over $\mathbb{Q}$ and hence a lattice-ordered field.

10   Unanswered Questions

The oldest and most famous question about lattice-ordered fields is the one that Birkhoff and Pierce first posed in [7] and that remains unanswered today.

Question 1.   Does there exist a compatible lattice order on the complex numbers $\mathbb{C}$?

In view of the work of Artin and Schreier [3], a positive answer to the following question would provide a negative answer to Question 1. This question is Question (a) of Conrad and Dauns [11].

Question 2.   Can the order on every lattice-ordered field be extended to a compatible total order?

More generally we can ask the following question.

Question 3.   Does every lattice-ordered field have a compatible total order?

If an ℓ-field has a negative square, then, as noted in section 8, its order cannot be extended to a compatible total order and hence such an ℓ-field shows that the answer to Question 2 is negative. Proposition 8.6 provided a class of ℓ-fields with no negative squares, but the general question remains unanswered.

Question 4.   Can a lattice-ordered field have a negative square?

The constructions described in section 3 have produced the only examples of lattice-ordered fields in which $1 \ngtr 0$. So the following question is unanswered.

Question 5.   Does every ℓ-field in which $1 \ngtr 0$ arise from an ℓ-field in which $1 > 0$ by one of the methods given in section 3?

Certainly a Wilson basis B is an ℓ-basis with respect to $\geq_B$; so every lattice-ordered field whose order is determined by a Wilson basis has an ℓ-basis. For a power series field $(_\chi \prod_\Delta T, \{t_{\alpha,\beta}\}, +, \cdot, \geq)$, any maximal set Ω of pairwise incomparable elements of Δ determines the ℓ-basis $\overline{\Omega} = \{\overline{\omega} \mid \omega \in \Omega\}$; in fact, such a set Ω, together with a set $\{s_\omega \mid \omega \in \Omega\}$ of positive elements of T, generates the ℓ-basis $\{s_\omega \overline{\omega} \mid \omega \in \Omega\}$. Nonetheless, there is no compelling reason to expect that every ℓ-field should have an ℓ-basis. So the following question is also unanswered.

Question 6.   Does every ℓ-field have an ℓ-basis?

Suppose that T is an o-field, that Δ is a rooted abelian group, and that χ is a strongly supporting collection in 2^Δ such that $_\chi \prod_\Delta T$ is an ℓ-field. If $\mathcal{F}$ is the set of finite subsets of Δ, then $\mathcal{F}$ is supporting, and we may form the ℓ-ring $_{\mathcal{F}}\prod_{\Delta}T$; certainly $_{\mathcal{F}}\prod_{\Delta}T \subseteq {}_{\mathcal{X}}\prod_{\Delta}T$. One way of approaching Question 6 would to find an o-field T, a rooted abelian group Δ, and a strongly supporting collection $\mathcal{X}$ such that some subfield of $_{\mathcal{X}}\prod_{\Delta}T$ does not contain $_{\mathcal{F}}\prod_{\Delta}T$.

Question 7. Must every subfield of a power series ℓ-field $_{\mathcal{X}}\prod_{\Delta}T$ that is lattice-ordered with respect to the inherited order contain $_{\mathcal{F}}\prod_{\Delta}T$?

Every ℓ-field in which $1 > 0$ has a unique maximal o-subfield $M(L)$. In the case of an ℓ-field L with a Wilson basis, Proposition 4.8 implies that $M(L) = T = 1^{\perp\perp}$. In the case of a power series field $\left(_{\mathcal{X}}\prod_{\Delta}T, \{t_{\alpha,\beta}\}, +, \cdot, \geq\right)$ over a rooted abelian group $(\Delta, \cdot, \succeq)$, if we let $\Delta^p = \{\alpha \in \Delta \mid \alpha \succeq e \text{ or } e \succeq \alpha\}$, then Δ^p is the maximal totally ordered subgroup of Δ and $M(_{\mathcal{X}}\prod_{\Delta}T) = \{f \mid \text{Supp}(f) \subseteq \Delta^p\} = 1^{\perp\perp}$. In general, however, the answer to the following question is unknown.

Question 8. Is there an ℓ-field L in which $1 > 0$ but $M(L) \neq 1^{\perp\perp}$?

If L is finite-dimensional over $M(L)$, then by part (1) of Proposition 4.4, $M(L) = 1^{\perp\perp}$. But there is still the possibility that L may contain an ℓ-basic element b such that $b^{\perp\perp} \neq bM(L)$. That is, the following question also remains unanswered.

Question 9. Does there exist an ℓ-field L in which $1 > 0$ and $1^{\perp\perp} = M(L)$ but which has an ℓ-basic element b such that $b^{\perp\perp} \neq bM(L)$?

According to Proposition 5.3, if L is an archimedean ℓ-field in which $1 > 0$ and if L is finite-dimensional over $M(L)$, then for any ℓ-basic $b \in L$, $b^{\perp\perp} = bM(L)$, and according to Proposition 4.4, the same can be said if the ℓ-basic elements of L form a multiplicative group. However, it may happen that L is finite-dimensional over $M(L)$, but it is not archimedean and its ℓ-basic element do not form a multiplicative group. In this case, we know from Proposition 5.2 only that $b^{\perp\perp}$ contains no proper nontrivial convex subspaces. That is, the answer is also unknown to the following more specific question.

Question 10. Does there exist an ℓ-field L in which $1 > 0$ and L is finite-dimensional over $M(L)$ but which has an ℓ-basic element b such that $b^{\perp\perp} \neq bM(L)$?

For an ℓ-field $(L, +, \cdot, \geq)$ in which $1 \not> 0$, then Theorem 4.6 implies that the analogue of $M(L)$ exists provided that the partially ordered field $(L, +, \cdot, \geq^*)$ has only one maximal totally ordered subfield.

Question 11. For every lattice-ordered field $(L, +, \cdot, \geq)$, does the partially ordered field $(L, +, \cdot, \geq^*)$ have a unique maximal totally ordered subfield?

According to Proposition 7.5, Wilson's construction in [51] gives ℓ-fields in which $M_3(L)$ is the largest trivially ordered subfield of L. In general, we can ask:

Question 12. If $1 \not> 0$ and if $(L, +, \cdot, \geq^*)$ has a maximal totally ordered subfield, is $M_3(L)$ the largest trivially ordered subfield of $(L, +, \cdot, \geq)$?

According to Proposition 4.5, if the positive special elements form a group, then the positive special elements are precisely the ℓ-basic elements. Since in general every ℓ-basic element is special ([4], p. 134; [10], p. 3.13), this leads to the following questions.

Question 13. Does there exist a lattice-ordered field in which the ℓ-basic elements form a group but in which there are positive special elements that are not ℓ-basic?

Or more generally,

Question 14. Does there exist a lattice-ordered field in which there are positive special elements that are not ℓ-basic?

There has been little study the field of quotients of a lattice-ordered integral domain. In particular, the following questions have not been answered (see [22]). Question 16 is Question (d) of Conrad and Dauns [11].

Question 15. Is every lattice-ordered integral domain a lattice-ordered subring of a lattice-ordered field?

Question 16. Does the field of quotients of a lattice-ordered integral domain $(D, +, \cdot, \geq)$ have a compatible lattice order that extends $\geq$?

References

[1] M. Anderson and T. Feil, *Lattice-Ordered Groups: An Introduction.* (1987) D. Reidel, Dordrecht, ISBN90-277-2643-4.

[2] E. Artin, *Über die Zerlegung definiter Funktionen in Quadrate.* Abh. Math. Sem. Hamb. Univ. **5** (1926), 100-115.

[3] E. Artin and O. Schreier, *Algebraische Konstruktion reeler Körper.* Abh. Math. Sem. Hamb. Univ. **5** (1926), 85-99.

[4] A. Bigard, K. Keimel and S. Wolfenstein, *Groupes et Anneaux Réticulés.* Lecture Notes in Mathematics **608** (1977), Springer-Verlag, Berlin, ISBN 3-540-08436-3.

[5] G. Birkhoff, *Lattice-Ordered Groups.* Annals of Math. **43** (1942), 298-331.

[6] G. Birkhoff, *Lattice Theory*. 3rd ed. (AMS Coll. Pub. 25) (1973), Amer. Math. Soc., Providence.

[7] G. Birkhoff and R. S. Pierce, *Lattice-ordered rings*. An. Acad. Brasil Ci. **28** (1956), 41-69.

[8] N. Bourbaki, *Eléments d'Histoire des Mathématiques*. (1969) Hermann, Paris.

[9] P. Conrad, *Generalized semigroup rings*. J. Indian Math. Soc. **21** (1957), 73-95.

[10] P. Conrad, *Lattice Ordered Groups*. Tulane University (1970), New Orleans.

[11] P. Conrad and J. Dauns, *An embedding theorem for lattice-ordered fields*. Pacific J. Math. **3** (1969), 385-398.

[12] P. Conrad, J. Harvey and W. C. Holland, *The Hahn embedding theorem for lattice-ordered groups*. Trans. Amer. Math. Soc. **108** (1963), 143-169.

[13] M. R. Darnel, *The Theory of Lattice-Ordered Groups*. (1995) Marcel Dekker, New York, ISBN 0-8247-9326-9.

[14] R. Dedekind, *Über Zerlegungen von Zahlen durch ihre grössen ten Gemein-samen Teiler*. Gesammelte Math. Werke, t. **II** (1931), 103-147; (originally appeared in Festschrift der Techn. Hochsch. zu Braunschweig bei Gelegenheit der 69 Versammlung Deutscher Naturforscher und Ärtze (1897), 1-40.)

[15] R. DeMarr, *Partially ordered fields*. Amer. Math. Monthly **74** (1967), 418-420.

[16] R. DeMarr, A. Steiger, *On elements with negative squares*. Proc. Amer. Math. Soc. **31** (1972), 57-60.

[17] D. W. Dubois, *On partly ordered fields*. Proc. Amer. Math. Soc. **7** (1956), 918-930.

[18] H. Freudenthal, *Teilweise geordnete moduln*. Proc. Ned. Akad. Wet. **39** (1936), 641-651.

[19] L. Fuchs, *Partially Ordered Algebraic Systems*. (1963) Pergamon Press, Oxford.

[20] A. M. W. Glass, *Partially Ordered Groups*. (1999) World Scientific, Singapore, ISBN 981-02-3493-7.

[21] H. Hahn, *Über die nichtarchimedean Grössensysteme*. Sitzungber. Kaiserlichen Akad. Wiss. Vienna Math. Nat. Klasse Abt. IIa **116** (1907), 601-653.

[22] M. Henriksen, *On the difficulties in embedding lattice-ordered integral domains in lattice-ordered fields*. General Topology and its Relations to Modern Analysis and Algebra III (Proc. Third Prague Topological Sympos., 1971) (1972) Academia, Prague, 183-185.

[23] S. Kakutani, *Concrete representation of abstract (L)-spaces and the mean ergodic theorem.* Annals of Math. **42** (1941), 523-537.

[24] S. Kakutani, *Concrete representation of abstract (M)-spaces (a characterization of the space of continuous functions).* Annals of Math. **42** (1941), 994-1024.

[25] L. V. Kantorovich, *Partially ordered linear spaces.* (Russian) Mat. Sbornik **2** (**44**) (1937), 121-168.

[26] S. MacLane, *Categories for the Working Mathematician.* (1971) Springer-Verlag, New York, ISBN 0-387-90035-7.

[27] B. Neumann, *On ordered division rings.* Trans. Amer. Math. Soc. **66** (1949), 202-252.

[28] S. Prieß-Crampe, *Angeordnete Strukturen: Gruppen, Körper, Projektive Ebenen.* Ergeb. der Math. u. i. Grenzgeb. **98** (1983) Springer-Verlag, Berlin.

[29] F. J. Raynor, *An algebraically closed field.* Glasgow Math. J. **9** (1968) 146-151.

[30] R. H. Redfield, *Algebraic extensions and lattice-ordered fields.* Analysis Paper **No. 15** (1975) Monash Univ., Clayton, Victoria, Australia.

[31] R. H. Redfield, *Embeddings into power series rings.* Manuscripta Math. **56** (1986), 247-268.

[32] R. H. Redfield, *Constructing lattice-ordered fields and division rings.* Bull. Austr. Math. Soc. **40** (1989), 365-369.

[33] R. H. Redfield, *Lattice-ordered fields as convolution algebras.* J. Algebra **153** (1992) 319-356.

[34] R. H. Redfield, *Lattice-ordered power series fields.* J. Austr. Math. Soc. (Series A) **52** (1992), 229-321.

[35] R. H. Redfield, *Internal characterizations of lattice-ordered power series fields.* Per. Math. Hungarica **32** (1996), 85-101.

[36] R. H. Redfield, *Abstract Algebra: A Concrete Introduction.* (2001) Addison Wesley Longman, Boston, ISBN 0-201-43721-X.

[37] R. H. Redfield, *Subfields of lattice-ordered fields that mimic maximal totally ordered subfields.* Czech. Math. J. **51** (**126**) (2001), 143-161.

[38] R. H. Redfield, *Unexpected lattice-ordered quotient structures.* Ordered Algebraic Structures: Nanjing (2001); W. C. Holland, ed.; Gordon and Breach, The Netherlands, ISBN 90-5699-325-9, 111-132.

[39] R. H. Redfield, *Letter to Stuart Steinberg*. (2001).

[40] P. Ribenboim, *Rings of generalized power series: nilpotent elements*. Abh. Math. Sem. Univ. Hamburg (Series A) **61** (1991), 15-33.

[41] P. Ribenboim, *Noetherian rings of generalized power series*. J. Pure Appl. Algebra **79** (1992), 293-312.

[42] P. Ribenboim, *Rings of generalized power series, II: units and zero-divisors*. J. Algebra **168** (1994), 71-89.

[43] F. Riesz, *Sur quelques notions fondamentales dans la théorie générale des opérations linéaires*. Annals of Math **41** (1940), 174-206.

[44] N. Schwartz, *Verbandsgeordnete Körper*. (1978) Dissertation, Universität München.

[45] N. Schwartz, *Archimedean lattice-ordered fields that are algebraic over their o-subfields*. Pacific J. Math. **89** (1980), 189-198.

[46] N. Schwartz, *Lattice-ordered fields*. Order **3** (1986), 179-194.

[47] J.-P. Serre, *Extensions des corps ordonnés*. C. R. Acad. Sci. Paris **229** (1949), 576-577.

[48] S. Steinberg, *An embedding theorem for commutative lattice-ordered domains*. Proc. Amer. Math. Soc. **31** (1972), 409-416.

[49] S. Steinberg, *Personal Communication*. (1990).

[50] R. R. Wilson, *Lattice orderings on fields and certain rings*. Symposia Mathematica (Convegno sulle Misure su Gruppi e su Spazi Vettoriali, Convegno sui Gruppi e Anelli Ordinati, INDAM, Rome) **21** (1975), 357-364.

[51] R. R. Wilson, *Lattice orderings on the real field*. Pacific J. Math. **63** (1976), 571-577.

Department of Mathematics, Hamilton College, Clinton, NY 13323, USA
rredfiel@hamilton.edu

Research Articles

Stone's Real Gelfand Duality
in Pointfree Topology

B. Banaschewski [1]

ABSTRACT.   The familiar 1940 result of M. H. Stone characterizing the
rings of real-valued continuous functions on compact Hausdorff spaces as certain
partially ordered rings is obtained here in its pointfree form, replacing the spaces
in question by appropriate frames, which avoids the classical recourse to the
choice-dependent existence of maximal ideals. The main tools for this are a direct
proof that the partially ordered rings involved are f-rings, the pointfree notion of
rings of real-valued continuous functions, and the representation of archimedean
bounded f-rings as rings of that kind.

Introduction

In 1940 M. H. Stone [13] gave the following characterization of the rings $C(X)$ of the
real-valued continuous functions on a compact Hausdorff space X: up to isomorphism,
they are exactly the partially ordered commutative $\mathbb{Q}$-algebras A with unit 1 such that

(i) for each $a \in A$, $a^2 \geq 0$,

(ii) for each $a \in A$, if all $na \leq b$ for some b then $a \leq 0$ (A is *archimedean*),

(iii) for each $a \in A$, there exist natural n such that $\pm a \leq n$ $(= n \cdot 1)$ (A is *bounded*),

and

(iv) A is *complete* in the ring topology given by the sets

$$W_n = \left\{ a \in A \mid -\frac{1}{n} \leq a \leq \frac{1}{n} \right\}, \quad n = 1, 2, \ldots,$$

[1] Thanks go to the Natural Sciences and Engineering Research Council of Canada for continuing
support in the form of a research grant; the University of Florida for financial assistance in connection
with the Conference on Lattice-Ordered Groups and f-Rings, February 28 to March 3, 2001; my late
friend and collaborator J. J. C. Vermeulen for some stimulating discussion of the problems dealt with
in Section 1; and the referees for some valuable criticism.

J. Martínez (ed.), Ordered Algebraic Structures, 157–177.
© 2002 *Kluwer Academic Publishers.*

as basic neighbourhoods of 0.

In the following, such partially ordered rings will be called *Stone rings*. We note that elsewhere they are also referred to as C^*-algebras (Johnstone [9]) but, given the well-established meaning of that term in connection with Banach algebras, we prefer not to follow that terminology.

Of course, the $C(X)$ in question have all the listed properties, and hence the only thing to show is the representation theorem stating the converse. Stone gave only a minimal outline of proof for this, saying "combine principles of algebra and topology with the existence and properties of square roots", besides suggesting that the latter could be obtained by means of a continued fraction algorithm. In addition, in a somewhat different vein, he noted that, for any partially ordered ring satisfying all the above conditions but (iv), its completion with respect to the topology involved here will be a partially ordered ring extension for which *all* these conditions hold.

There is a nice detailed proof of Stone's result due to Johnstone [9], and in order to put the present paper in perspective it may be useful to give a brief account of its main steps (with some slight modifications of our own).

(1) *Square roots and absolute values.* First one shows that, for any $a \geq 0$, there exist $b \geq 0$ such that $a = b^2$. Starting with $0 \leq a \leq 1$, one defines

$$s(a) = 1 - \sum_{n=1}^{\infty} c_n (1-a)^n \quad \text{where} \quad c_n = \left| \binom{\frac{1}{2}}{n} \right|,$$

using the given completeness of A and the convergence of $\sum_{n=1}^{\infty} c_n$, proves that $0 \leq s(a)$ and $s(a)^2 = a$ by the familiar properties of the power series $1 - \sum_{n=1}^{\infty} c_n x^n$, and then extends this to all $a \geq 0$ by boundedness. Of course, this is not the procedure Stone suggested but the power series argument seems to have become more popular in this kind of context over the years.

Next, one puts $|a| = s(a^2)$ for $a \in A$ such that $-1 \leq a \leq 1$ and again extends this definition by boundedness to all $a \in A$. Finally, one shows that $0, a \leq |a|$ where the first is quite obvious but the second somewhat more subtle.

(2) *A has bounded inversion*, that is, each $a \geq 1$ in A is invertible in A, with inverse given by the geometric series

$$\frac{1}{a} = \frac{1}{k} \sum_{0}^{\infty} \left(1 - \frac{a}{k} \right)^n \quad \text{for} \quad 0 \leq 1 - \frac{a}{k}.$$

In passing we note that Stone did not specifically mention this step even though it really plays quite an essential rôle.

(3) *Properties of the maximal ideals of A.* Any maximal ideal M of A is *closed* because the unit of A has a neighbourhood of invertible elements by (2). Further, M

is *convex*: its convex hull

$$\widetilde{M} = \{\, a \in A \mid b \le a \le c, \text{ for some } b, c \in M \,\}$$

is trivially an additive subgroup of A and stable under multiplication by elements $s \ge 0$ of A, hence an *ideal* since generally

$$s = \frac{1}{2}\Big(|s| + s\Big) - \frac{1}{2}\Big(|s| - s\Big)$$

where $0 \le |s| \pm s$ by (1), and *proper* again by (2) – showing $M = \widetilde{M}$.

As a result, A/M has a partial order, induced by that of A, in which it is archimedean because M is closed and totally ordered since $(|a|+a)(|a|-a) = 0 \in M$ and hence $|a| + a \in M$ or $|a| - a \in M$ for each $a \in A$. It follows by familiar results that A/M is isomorphic as totally ordered field to a subfield of $\mathbb{R}$, but then actually $A/M \cong \mathbb{R}$ since $\mathbb{R}$ is embedded in A by completeness.

(4) *The representation of A.* On the set of maximal ideals M of A, each $a \in A$ defines a real-valued function $\hat{a}$ by taking $\hat{a}(M) \in \mathbb{R}$ as the real number corresponding to $a + M \in A/M$, and the map $a \mapsto \hat{a}$ is then a *ring homomorphism*. Further, taking $\operatorname{Max} A$ as the *space* of maximal ideals of A with the Zariski topology, $\hat{a}$ is *continuous*: the cozero set

$$\operatorname{coz}(\hat{a}) = \{\, M \in \operatorname{Max} A \mid a \notin M \,\}$$

is open by definition while, for any rational $p < q$,

$$(\hat{a})^{-1}[(p, q)] = \operatorname{coz}(\hat{b}) \quad \text{for} \quad b = (a - p)(q - a) + |a - p||q - a|,$$

resulting from the fact that

$$(\lambda - p)(q - \lambda) + |\lambda - p|\,|q - \lambda| \ne 0 \quad \text{if and only if} \quad p < \lambda < q$$

in $\mathbb{R}$ and the observation that the homomorphism $A \longrightarrow A/M \longrightarrow \mathbb{R}$ preserves $|\cdot|$.

Note that, in particular, it now follows that $\operatorname{Max} A$ is Hausdorff: any two distinct points are in fact separated by a real-valued continuous function.

In the next two steps the Axiom of Choice, not used so far, enters into the argument.

(5) $\operatorname{Max} A$ *is compact*, as a familiar application of Zorn's Lemma.

(6) *The homomorphism $A \longrightarrow C(\operatorname{Max} A)$ taking a to $\hat{a}$ is an embedding.* Evidently, this amounts to saying that $\bigcap \operatorname{Max} A = \{0\}$, and one way to obtain this is to use the fact that, by the Axiom of Choice, this intersection is the *Jacobson radical* of A, meaning: the ideal J of all $a \in A$ such that $1 + ar$ is invertible for each $r \in A$. Now, the latter implies for any $a \in J$ that $a + b$ is invertible whenever b is, hence by (2)

$a - \frac{1}{n} - |a - \frac{1}{n}|$ is invertible for any $n = 1, 2, \ldots$, and since $(c - |c|)(c + |c|) = 0$ for all $c \in A$, it follows that $a - \frac{1}{n} + |a - \frac{1}{n}| = 0$. Consequently, $a \leq \frac{1}{n}$ for all natural $n \geq 1$, showing that $a \leq 0$ and then further $a = 0$, for any $a \in J$.

(7) *The embedding* $A \longrightarrow C(\text{Max } A)$ *is an isomorphism.* The image $\hat{A}$ of A obviously separates the points of Max A so that the Stone-Weierstrass Theorem makes it dense relative to the sup-norm topology of $C(\text{Max } A)$, and since the latter induces on $\hat{A}$ the topology corresponding to that considered on A in (iv) the given completeness of A implies $\hat{A} = C(\text{Max } A)$.

It should be added that, as Johnstone [9] shows, the characterization of Stone rings thus obtained can readily be augmented to a dual equivalence between the category of compact Hausdorff spaces and continuous maps and the category of Stone rings and ring homomorphisms: the only thing left to show is that the (contravariant) functor $X \mapsto C(X)$ from compact Hausdorff spaces to Stone rings is full and faithful, where the latter is completely obvious and the former fairly straightforward. Thus, the final result is a *real-valued counterpart* of the familiar Gelfand Duality which concerns the rings of complex-valued continuous functions on compact Hausdorff spaces with their sup-norm and the commutative C^*-algebras with unit over the complex number field.

Of course, there is also the alternative real-valued Gelfand Duality involving the $C(X)$ not as partially ordered rings but as *Banach algebras*, again with their sup-norm, on the one hand and the commutative Banach algebras with unit over $\mathbb{R}$ in which $\|a^2\| = \|a\|^2$ and all $1 + a^2$ are invertible on the other (Banaschewski [2]) – but that is clearly a different story.

Regarding the foundations required for Stone's result, we note first that it certainly cannot be proved in Zermelo-Fraenkel set theory (as usually understood: without the Axiom of Choice) because it implies the Prime Ideal Theorem that *every non-trivial Boolean algebra contains a prime ideal.* Indeed, given any such B one can construct a Stone ring A for which B is isomorphic to the Boolean algebra $Idp\, A$ of idempotents of A, and A has maximal ideals P if $A \simeq C(X)$ for some X; further, $P \cap Idp\, A$ is then a prime ideal of $Idp\, A$, showing that B contains a prime ideal. On the other hand, as a result of what will be shown here, the Prime Ideal Theorem turns out to be sufficient for the two crucial points in the arguments outlined above, that Max A is compact and $\bigcap \text{Max} A = \{0\}$, although the (strictly stronger) Axiom of Choice does provide rather shorter proofs for them.

This paper is concerned with the question whether the above representation theorem for Stone rings can be formulated in such a way that it becomes provable without recourse to any choice principle, and the aim here is to show this is indeed the case. The foundations assumed for this will be *Zermelo-Fraenkel set theory treated within classical logic.* We note, though, that it would ultimately be desirable to establish the result obtained here in a way which is constructively valid in the familiar sense of topos theory – for the obvious reason that this would extend its scope considerably. There is little doubt that appropriate modifications of the definitions and arguments used here will do this, but for the time being we were content with the present approach, leaving that aspect aside for another occasion.

We close with a brief outline of the paper.

Our first step will be to establish that any Stone ring is an f-ring (Section 1). Of course, this is a trivial consequence of Stone's original result but the point here is that this can be shown directly, by means of a more extensive analysis of the absolute value described above. Once this is done we are in the more familiar territory of archimedean bounded f-rings over $\mathbb{Q}$, and the task will then be to describe the pointfree version of their classical representation as ℓ-rings of real-valued continuous functions. For this we first provide a general review of the main aspects of function rings in pointfree topology, meaning: defined on frames rather than on spaces (Section 2), and then derive the representation in question (Section 3), leading to our form of Stone's characterization that, *up to isomorphism, the Stone rings are exactly the rings of real-valued continuous functions on compact completely regular frames*. In addition we show that, as in the classical case, this can be extended to a category equivalence.

1 Stone Rings as f-Rings

As already explained, the purpose here is to show that every Stone ring is an f-ring, but we begin with a more general result, which seems of independent interest. Below, all rings are taken to be commutative with unit.

Proposition 1.1. *An archimedean bounded partially ordered ring A over $\mathbb{Q}$ is an f-ring if and only if for each $a \in A$ there exists $b \geq 0$ such that $a^2 = b^2$.*

Proof. Since any f-ring clearly has the property in question ($a^2 = |a|^2$) only the "if" part has to be proved. We do this in several separate steps, using repeatedly, without specific mention, that A is archimedean and $a^2 \geq 0$, for all $a \in A$.

(1) *A is semiprime*, that is, $a^2 = 0$ implies $a = 0$. For any natural $n \geq 1$,

$$0 \leq (na - 1)^2 = -2na + 1,$$

hence $2na \leq 1$, so that $a \leq 0$, and the same conclusion for $-a$ shows $a = 0$.

(2) *If $a^2 = b^2$ and $b \geq 0$ then $a \leq b$.* We first note that $ab \leq 0$ and $0 < r \leq b$, for some $r \in \mathbb{Q}$, implies $a \leq 0$, where we may assume that $a, b \leq 1$ since A is bounded. For any natural $n \geq 1$,

$$0 \geq ab\left(\sum_{k=0}^{n-1}(1-b)^k\right) = a(1 - (1-b))\left(\sum_{k=0}^{n-1}(1-b)^k\right) = a(1 - (1-b)^n),$$

hence $a \leq a(1-b)^n \leq (1-r)^n$, and therefore $a \leq 0$ since $0 \leq 1-r < 1$.

Next, $c \leq 0$ whenever $c^3 \leq 0$. For any natural n,

$$(nc - 1)(n^2c^2 + 4) = n^3c^3 - n^2c^2 + 4nc - 4 \leq -(nc - 2)^2 \leq 0,$$

hence $nc \leq 1$ by what was just proved, and consequently $c \leq 0$.

Finally, $a - b \leq a + b$ because $b \geq 0$, hence

$$(a - b)^3 \leq (a + b)(a - b)^2 = 0,$$

since $a^2 = b^2$, and by the preceding result this shows that $a \leq b$.

(3) As an immediate consequence $a^2 = b^2$ for $a, b \geq 0$ implies $a = b$. In particular it follows that, for any $a \in A$, the $b \geq 0$ such that $a^2 = b^2$ is unique, to be denoted $|a|$, and by (2) we then have $a \leq |a|$.

(4) If $a^2 \leq b^2$ for $a, b \geq 0$ then $a \leq b$. In the following we use that $(|c|+c)(|c|-c) = 0$ and $|c| + c \geq 0$ for any $c \in A$. Now

$$0 \leq b^2 - a^2 = (b - a)(b + a) \leq (|b - a| + b - a)(b + a),$$

and since $0 \leq (a - b)^2 \leq (a + b)^2$, because $ab \geq 0$, we obtain

$$0 \leq (a - b)^4 \leq (a + b)^2(a - b)^2 = (b^2 - a^2)^2 \leq (|b - a| + b - a)^2(b + a)^2;$$

further, multiplying by $|b - a| + a - b$ yields

$$0 \leq (a - b)^4(|b - a| + a - b) \leq 0$$

and therefore $(a - b)(|b - a| + a - b) = 0$ by (1). Next, for any $c \in A$, if $c(|c| + c) = 0$ then $(|c| + c)^2 = 0$ so that $|c| + c = 0$, again by (1):

$$(|c| + c)^2 = |c|(|c| + c) = |c|^2 + |c|c = c^2 + |c|c = c(|c| + c) = 0.$$

Further, applying this to $c = a - b$ shows $|a - b| + a - b = 0$ and therefore $a \leq b$, as desired.

We are now ready to derive the two crucial properties of $|\cdot|$.

(5) $|ab| = |a| \, |b|$ by (3) since the squares of either side are equal.

(6) $|a + b| \leq |a| + |b|$ by (4) since

$$|a + b|^2 = (a + b)^2 = a^2 + 2ab + b^2 \leq |a|^2 + 2|a| \, |b| + |b|^2 = (|a| + |b|)^2,$$

the inequality because $2ab \leq |2ab| = 2|a| \, |b|$ by (3) and (5).

Our final step is

(7) $a \wedge b$ *exists and is given by the expected element* $\frac{1}{2}(a + b - |a - b|)$. This will further imply that the $|a|$ used here is indeed $a \vee (-a)$, and by the familiar fact that f-rings are characterized by the identity $|ab| = |a||b|$, it then follows from (5) that A is not just an ℓ-ring but indeed an f-ring.

First,

$$\frac{1}{2}\left(a + b - |a - b|\right) - b = \frac{1}{2}\left(a - b - |a - b|\right) \leq 0,$$

by (3), and hence $\frac{1}{2}(a + b - |a - b|) \leq a, b$. On the other hand, if $c \leq a, b$, and hence $0 \leq a - c, b - c$ then

$$\frac{1}{2}\left(a + b - |a - b|\right) - c = \frac{1}{2}\left((a - c) + (b - c) - |a - b|\right) \geq 0,$$

because

$$|a - b| = |(a - c) + (b - c)| \leq |a - c| + |b - c|$$

by (6), while $|a - c| = a - c$ and $|b - c| = b - c$. ∎

Note that A being over $\mathbb{Q}$ cannot be dropped from the above proposition, as is shown by the partially ordered ring of all ordered pairs (k, l) of integers which are both either odd or even. On the other hand, the boundedness of A only enters into the first part of the second step of the proof, where it can be replaced by the condition that $0 \leq ab$ and $1 \leq b$ imply $0 \leq a$ – again obviously satisfied by any f-ring. Hence we also have the following modification of Proposition 1.1.

Corollary 1.2. *An archimedean partially ordered ring A over $\mathbb{Q}$ is an f-ring if and only if for each $a \in A$ there is a $b \geq 0$ such that $a^2 = b^2$, and for any a and $b \geq 1$ in A, $0 \leq ab$ implies $0 \leq a$.*

Now, the argument described in the Introduction concerning square roots shows that any Stone ring is of the type considered in Proposition 1.1, and hence we conclude:

Proposition 1.3. *The Stone rings are exactly the uniformly complete archimedean bounded f-rings over $\mathbb{Q}$.*

As noted earlier, Stone [13] pointed out that, for any archimedean bounded partially ordered ring over $\mathbb{Q}$ in which all $a^2 \geq 0$, completion with respect to the topology involved retains these properties and hence produces a Stone ring, in our terminology. This observation has a natural extension as follows. Let A be any archimedean bounded partially ordered ring such that $a^2 \geq 0$ for all $a \in A$ and $na \geq 0$ implies $a \geq 0$ for all $a \in A$ and natural $n \geq 1$. Then the latter condition ensures that the ring of fractions of A given by inverting all natural $n \geq 1$ is partially ordered such that $0 \leq \frac{a}{n}$ if and only if $0 \leq a$, and as such it retains all the given properties of A. Hence we have the following consequence of the above proposition.

Corollary 1.4. *The archimedean bounded partially ordered rings A such that $a^2 \geq 0$ and $na \geq 0$ implies $a \geq 0$ for all $a \in A$ and all natural $n \geq 1$ are exactly the partially ordered subrings of archimedean bounded f-rings.*

Remark 1.5. In the context of Stone rings, the important step (2) in the above development, showing that $a \leq |a|$ for all $a \in A$, can alternatively be obtained by analyzing the decomposition of the rational polynomials involved in approximating $|a| - a$ for $-1 \leq a \leq 1$ into their irreducible factors in $\mathbb{R}[x]$, as indicated by Johnstone [9]. It seemed of interest to present a direct proof of this, without reference to any prior history of $|a|$. Moreover, from the point of view of possibly extending these results beyond the presently adopted foundations, the direct argument is decidedly preferable because the classical decomposition in $\mathbb{R}[x]$ is not constructively valid.

Remark 1.6. An analogous comment applies to step (4). Again, in a Stone ring this can be obtained from the definition $|a| = s(a^2)$ for $-1 \leq a \leq 1$, using that the coefficients in the power series involved are positive and that $s(a^2) \geq 0$ for all a involved.

Remark 1.7. Concerning the question of constructive validity, it may be worth pointing out that the arguments outlined in (1) of the Introduction, defining $s(a)$ for each $a \in A$ such that $0 \leq a \leq 1$ and proving that $0 \leq s(a)$ and $a = s(a)^2$, are valid in this sense for any Stone ring. In fact, there are two rather different ways of seeing this. On the one hand, the description of the remainder term in the Taylor expansion of the function $(1 - x)^{1/2}$ in Cauchy's integral form is valid in constructive analysis (Witt [16]), and this can then be used to obtain the required estimates (while the remainder term given by Bishop [7] does not seem to suffice for this purpose). Specifically, the crucial point is to establish that $\sum_{n=1}^{\infty} c_n = 1$ and $2c_n = \sum_{k+\ell=n} c_k c_\ell$ for all $n \geq 2$: once this is done the remaining calculations are routine and obviously constructively valid. On the other hand, Vermeulen [14] has given explicit numerical inequalities for the Taylor polynomials involved which serve the same purpose, but do not require the apparatus of constructive analysis.

2   Function Rings in Pointfree Topology

Recall that pointfree topology is the study of abstractly defined "lattices of open sets" called *frames* and their homomorphisms, where a frame is a complete lattice L which satisfies the distribution law

$$a \wedge \bigvee S = \bigvee \left\{ a \wedge t \mid t \in S \right\}$$

for all $a \in L$ and $S \subseteq L$, and a frame homomorphism is a map $h : L \longrightarrow M$ between (the underlying sets of) frames preserving finitary meets, including the unit (= top)

e of L, and arbitrary joins, including the zero (= bottom) 0 of L. **Frm** will then be the category defined by these notions. As a general reference to frames we suggest Johnstone [9] or Vickers [15].

Obvious examples of frames are the lattices $\mathfrak{O}X$ of open sets of a topological space X, the complete Boolean algebras, and the complete chains. A frame isomorphic to some $\mathfrak{O}X$ is called *spatial*. All finite frames and all complete chains are of this kind, but a complete Boolean algebra is spatial if and only if it is atomic.

The correspondence $X \mapsto \mathfrak{O}X$ provides a contravariant functor from the category **Top** of topological spaces and continuous maps to **Frm** which induces a dual equivalence between the full subcategories of sober spaces and spatial frames, respectively.

As a general comment concerning frames versus spaces we note that many topological spaces which are derived from other entities have their useful properties only on the basis of some foundational assumptions, as pointed out earlier in the case of Max A for a Stone ring A. On the other hand, the frames of open sets of these spaces tend to exist independent of such assumptions and may serve various purposes just as well as the spaces. Hence, if one aims for choice-independent results in such situations the proper setting is likely to be pointfree topology. For an account of several instances of this kind see Banaschewski [6].

The following notions involving frames and frame homomorphisms will be of specific interest here.

A frame L is called

compact if $\bigvee S = e$ for any $S \subseteq L$ implies that $\bigvee T = e$ for some finite $T \subseteq S$;

regular if, for all $a \in L$, $a = \bigvee\{x \in L \mid x \prec a\}$ where $x \prec a$ means that $a \vee x^* = e$ for the pseudocomplement $x^* = \bigvee\{y \in L \mid y \wedge x = 0\}$ of x;

completely regular if, for all $a \in L$, $a = \bigvee\{x \in L \mid x \prec\prec a\}$ where $x \prec\prec a$ means there exists a family $(c_{ik})_{i=0,1,2,\dots;k=0,1,\dots,2^i}$ such that

$$c_{00} = x, \quad c_{01} = a, \quad c_{ik} = c_{i+1\,2k}, \quad c_{ik} \prec c_{i\,k+1}.$$

We note for each of these conditions that it is satisfied by the frame $\mathfrak{O}X$ for a space X if and only if X has the same-named property in the usual sense. On the other hand, concerning spatiality, the compact regular frames are spatial if and only if the Prime Ideal Theorem holds.

Regarding homomorphisms, $h : L \longrightarrow M$ is called *dense* if $h(a) = 0$ implies $a = 0$, and any dense homomorphism with compact regular domain is one-one. Further, for any homomorphisms $f, g : L \longrightarrow M$, if L is *regular* and $f(a) \leq g(a)$ for all $a \in L$ then $f = g$.

Next we turn to the notion of function ring in this context. We begin with a brief outline of the concepts and results needed here; for a detailed treatment we refer to

Banaschewski [4], specifically sections 1, 2, 4, and 5, but some aspects are also dealt with in Johnstone [9].

The real numbers make their appearance in this setting as a frame, the *frame* $\mathcal{L}(\mathbb{R})$ *of reals*, originally due to Joyal [10], which is defined by generators and relations where the generators are the ordered pairs (p, q) of rational numbers and the relations are

(R1) $(p, q) \wedge (r, s) = (p \vee r, q \wedge s)$,
(R2) $(p, q) \vee (r, s) = (p, s)$ whenever $p \le r < q \le s$,
(R3) $(p, q) = \bigvee\{(r, s) \mid p < r < s < q\}$, and
(R4) $e = \bigvee\{(p, q) \mid$ all $p, q \in \mathbb{Q}\}$.

Note that sometimes the condition that $(p, q) = 0$ whenever $q \le p$ is added but (R3) actually makes that redundant. Also, for a somewhat different but equivalent description see Johnstone [9].

A basic property of $\mathcal{L}(\mathbb{R})$, important for present purposes, is its (complete) regularity resulting from (R3) and the fact that $(r, s) \prec (p, q)$ in $\mathcal{L}(\mathbb{R})$ whenever $p < r < s < q$, an easy consequence of (R2) and (R4).

Now, the *real-valued continuous functions on a frame* L are defined to be the homomorphisms $\mathcal{L}(\mathbb{R}) \longrightarrow L$, a natural choice given the following fact ([4], Proposition 4.4). For any space X, there is a one-one correspondence between the continuous maps $f : X \longrightarrow \mathbb{R}$ and the homomorphisms $\varphi : \mathcal{L}(\mathbb{R}) \longrightarrow \mathfrak{O}X$ such that

$$f \mapsto \varphi, \qquad \varphi(p, q) = f^{-1}[\{\lambda \in \mathbb{R} \mid p < \lambda < q\}],$$
$$\varphi \mapsto f, \qquad p < f(x) < q \text{ if and only if } x \in \varphi(p, q).$$

We indicate a proof which is somewhat more direct than that given in [4]. The first formula makes it immediately obvious that the correspondence $(p, q) \mapsto \varphi(p, q)$ turns the relations (R1)–(R4) into identities in $\mathfrak{O}X$ and hence defines a frame homomorphism $\varphi : \mathcal{L}(\mathbb{R}) \longrightarrow \mathfrak{O}X$, as claimed. On the other hand, for each $x \in X$, repeated application of (R2) shows that

$$\{\{\lambda \in \mathbb{R} \mid p < \lambda < q\} \mid p, q \in \mathbb{Q} \text{ such that } x \in \varphi(p, q)\}$$

is a Cauchy filter basis with limit $f(x)$, and the resulting map $f : X \longrightarrow \mathbb{R}$ is clearly continuous by its definition. Finally, the two correspondences are evidently inverse to each other.

Further, we note that there is an isomorphism $\mathcal{L}(\mathbb{R}) \longrightarrow \mathfrak{O}\mathbb{R}$ which takes each generator (p, q) to the corresponding open interval in $\mathbb{R}$ (Johnstone [9], IV, 1.3, Exercise; also [4], Proposition 2). Obviously, then, one could replace $\mathcal{L}(\mathbb{R})$ by $\mathfrak{O}\mathbb{R}$ in our context – an approach actually taken by some authors; on the other hand, though, the result that $\mathcal{L}(\mathbb{R})$ provides a presentation of $\mathfrak{O}\mathbb{R}$ by generators and relations is an important insight which does offer certain advantages, especially when it comes to defining specific real-valued continuous functions from certain types of given data. Apart from this, however, there is a more fundamental reason for preferring $\mathcal{L}(\mathbb{R})$ to $\mathfrak{O}\mathbb{R}$: constructively, it is $\mathcal{L}(\mathbb{R})$ which has the better properties. For instance, the frame derived from it which corresponds to (the frame of open sets of) the unit interval is constructively compact (Banaschewski [4], pp. 15-17), but this is not the case of the concrete unit interval determined by $\mathbb{R}$, as shown by Fourman-Hyland [8]. Consequently, as observed by Johnstone [9], if one is interested in constructively valid results, $\mathcal{L}(\mathbb{R})$ does seem to be the right frame to deal with.

Next we introduce algebraic operations on the set of real-valued continuous functions on a frame L, derived from the ℓ-ring operations of $\mathbb{Q}$ as follows.

(O1) For $\diamond = +, \cdot, \wedge, \vee$,

$$(\alpha \diamond \beta)(p, q) = \bigvee \left\{ \alpha(r, s) \wedge \beta(t, u) \mid \langle r, s \rangle \diamond \langle t, u \rangle \subseteq \langle p, q \rangle \right\}$$

where $\langle \cdot, \cdot \rangle$ stands for open interval in $\mathbb{Q}$ and

$$\langle r, s \rangle \diamond \langle t, u \rangle = \{ x \diamond y \mid x \in \langle r, s \rangle, \ y \in \langle t, u \rangle \}.$$

(O2) $(-\alpha)(p, q) = \alpha(-q, -p)$.

(O3) For each $r \in \mathbb{Q}$, the corresponding *constant* $\mathbf{r}$

$$\mathbf{r}(p, q) = e \ \text{ if } \ p < r < q \ \text{ and } \ \mathbf{r}(p, q) = 0 \text{ otherwise} .$$

One verifies that these specifications indeed define homomorphisms $\mathcal{L}(\mathbb{R}) \longrightarrow L$ by checking that they transform the defining relations of $\mathcal{L}(\mathbb{R})$ into identities in L, basically a straightforward procedure, even if somewhat lengthy. Once this is done it is easily shown further that the resulting algebraic system satisfies all the identities which hold for the corresponding operations in $\mathbb{Q}$, making it into a commutative f-ring with unit over $\mathbb{Q}$ which will be denoted $\mathcal{R}L$. For the details, see section 4 of Banaschewski [4].

$\mathcal{R}L$ has a number of additional properties the most important of which, for present purposes, are that in general it is archimedean and any $\alpha \geq 0$ is a square, and it is bounded whenever L is compact – the latter since (R4) and compactness implies, for any $\alpha \in \mathcal{R}L$, that there exist natural n such that $\alpha(-n, n) = e$, and this is the same as $|\alpha| \leq \mathbf{n}$.

Further we note that, for any space X, the correspondence between the continuous $X \to \mathbb{R}$ and the homomorphisms $\mathcal{L}(\mathbb{R}) \longrightarrow \mathfrak{O}X$ described earlier is an ℓ-ring isomorphism between the usual function ring $C(X)$ and the present $\mathcal{R}(\mathfrak{O}X)$, showing that the $\mathcal{R}L$ provide a natural extension of the classical notion of function ring to arbitrary frames.

It is clear that the correspondence $L \mapsto \mathcal{R}L$ is functorial such that, for any homomorphism $h : L \longrightarrow M$, $\mathcal{R}h : \mathcal{R}L \longrightarrow \mathcal{R}M$ takes $\alpha \in \mathcal{R}L$ to $h\alpha \in \mathcal{R}M$: by the definition of the algebraic structures involved, $\alpha \mapsto h\alpha$ is obviously an ℓ-ring homomorphism.

We close this section with a brief account of some points regarding the relation between a frame L and its function ring $\mathcal{R}L$. For the details we again refer to Banaschewski [4].

To begin with, note that any frame L has a largest completely regular subframe $C\mathrm{Reg}L$ for which the identical embedding $C\mathrm{Reg}L \longrightarrow L$ is the universal homomorphism to L from completely regular frames. Now, $\mathcal{L}(\mathbb{R})$ is quite clearly completely regular $((r, s) \prec\!\!\prec (p, q)$ whenever $p < r < s < q)$ so that any $\mathcal{L}(\mathbb{R}) \longrightarrow L$ factors through $C\mathrm{Reg}L$ and hence the embedding $C\mathrm{Reg}L \longrightarrow L$ determines an isomorphism $\mathcal{R}(C\mathrm{Reg}L) \longrightarrow \mathcal{R}L$. As a consequence, similar to the classical case, the study of the function rings $\mathcal{R}L$ may be restricted to the case of completely regular L without loss of generality. Moreover, these L have the property that they are generated by the set $\{\alpha(p, q) \mid \alpha \in \mathcal{R}L; \; p, q \in \mathbb{Q}\}$ and hence the functor $\mathcal{R}$ is faithful on completely regular frames: if $\mathcal{R}f = \mathcal{R}g$ for homomorphisms $f, g : L \longrightarrow M$ then $f\alpha = g\alpha$ for all $\alpha \in \mathcal{R}L$ so that f and g coincide on $\{\alpha(p, q) \mid \alpha \in \mathcal{R}L; \; p, q \in \mathbb{Q}\}$ which makes them equal if L is completely regular.

An important connection between $\mathcal{R}L$ and L is provided by the pointfree version of the classical cozero sets. Putting

$$\mathrm{coz}(\alpha) = \bigvee \{\, \alpha(p, 0) \vee \alpha(0, q) \mid p < 0 < q \text{ in } \mathbb{Q} \,\}$$

for $\alpha \in \mathcal{R}L$ defines a map $\mathrm{coz} : \mathcal{R}L \longrightarrow L$ which has the following properties:

(1) For all $\alpha, \beta \in \mathcal{R}L$,

$$\mathrm{coz}(\alpha\beta) = \mathrm{coz}(\alpha) \wedge \mathrm{coz}(\beta), \quad \mathrm{coz}(\mathbf{1}) = e,$$
$$\mathrm{coz}(\alpha + \beta) \leq \mathrm{coz}(\alpha) \vee \mathrm{coz}(\beta), \quad \mathrm{coz}(\mathbf{0}) = 0,$$
$$\mathrm{coz}(\alpha) = \mathrm{coz}(|\alpha|).$$

(2) For all $\alpha, \beta \geq 0$ in $\mathcal{R}L$,

$$\mathrm{coz}(\alpha \wedge \beta) = \mathrm{coz}(\alpha) \wedge \mathrm{coz}(\beta), \quad \mathrm{coz}(\alpha \vee \beta) = \mathrm{coz}(\alpha) \vee \mathrm{coz}(\beta).$$

(3) For all $\alpha \in \mathcal{R}L$ and $p, q \in \mathbb{Q}$,

$$\mathrm{coz}((\alpha - \mathbf{p})^+ \wedge (\mathbf{q} - \alpha)^+) = \alpha(p, q).$$

To put this in perspective it should be added that, in the case of $L = \mathfrak{O}X$ for some space X, if $\alpha \in \mathcal{R}(\mathfrak{O}X)$ corresponds to $a \in C(X)$ then $\mathrm{coz}(\alpha)$, as expected, is just the usual cozero set of a. What is then noteworthy is that the above conditions, totally obvious for cozero *sets*, still hold in the present extended context.

3   The Functional Representation

This section provides a more detailed presentation of results previously outlined in Banaschewski [3]. As in [3], we note that a similar treatment of *archimedean ℓ-groups with weak order unit* is given by Madden [11], motivated by earlier results of Madden-Vermeer [12] which, in turn, are also covered and extended by Ball-Hager [1].

We first consider an arbitrary archimedean bounded f-ring A over $\mathbb{Q}$ and recall from Banaschewski [3] that the uniformly closed ℓ-ideals of A form a compact completely regular frame $\mathfrak{M}A$. This is quite evidently the right frame for our purposes: the spectrum of $\mathfrak{M}A$ is the space of maximal ℓ-ideals of A, and whenever A has bounded inversion – which is the case for the A we are ultimately interested in – this is just $\text{Max}\,A$, making it exactly the space used in the classical proof described in the Introduction.

For the following, let $\langle a \rangle$ be the closed ℓ-ideal generated by $a \in A$ and note that

$$\langle a \rangle \cap \langle b \rangle = \langle a \wedge b \rangle, \quad \langle a \rangle \vee \langle b \rangle = \langle a \vee b \rangle$$

for any $a, b \geq 0$ in A. Further, put

$$\hat{a}(p,q) = \langle (a - p)^+ \wedge (q - a)^+ \rangle$$

for all $a \in A$ and $p, q \in \mathbb{Q}$.

Claim 1. *This defines a frame homomorphism $\hat{a} : \mathcal{L}(\mathbb{R}) \longrightarrow \mathfrak{M}A$, so that we obtain a map $\tau_A : A \longrightarrow \mathcal{R}(\mathfrak{M}A)$ taking $a \in A$ to $\hat{a} \in \mathcal{R}(\mathfrak{M}A)$.*

Claim 2. *τ_A is an ℓ-ring homomorphism.*

Claim 3. *τ_A is an embedding.*

To settle the first claim we have to show that the assignment $(p, q) \mapsto \hat{a}(p, q)$ transforms the defining relations of $\mathcal{L}(\mathbb{R})$ into identities in $\mathfrak{M}A$.

For (R1), we have

$$(a - p)^+ \wedge (q - a)^+ \wedge (a - r)^+ \wedge (s - a)^+ = ((a - p) \wedge (a - r))^+ \wedge ((q - a) \wedge (s - a))^+$$
$$= (a - (p \vee r))^+ \wedge ((q \wedge s) - a)^+,$$

and the result follows from the above rules concerning principal closed ℓ-ideals.

Similarly, for (R2), if $p \leq r < q \leq s$ then

$$((a - p)^+ \wedge (q - a)^+) \vee ((a - r)^+ \wedge (s - a)^+)$$
$$= (a - p)^+ \wedge ((a - p)^+ \vee (s - a)^+) \wedge ((q - a)^+ \vee (a - r)^+) \wedge (s - a)^+.$$

Now

$$(a - p)^+ \vee (s - a)^+ = \left(\left(\left(a - \frac{s+p}{2} \right) \vee \left(\frac{s+p}{2} - a \right) \right) + \frac{s-p}{2} \right)^+$$

$$= \left| a - \frac{s+p}{2} \right| + \frac{s-p}{2} \ge \frac{s-p}{2} > 0,$$

and hence the closed ℓ-ideal generated by this is $\langle 1 \rangle$, and the same argument applies to $(q - a)^+ \vee (a - r)^+$ since $r < q$. It follows that

$$\hat{a}(p, q) \vee \hat{a}(r, s) = \langle (a - p)^+ \wedge (s - a)^+ \rangle = \hat{a}(p, s),$$

as desired.

Next,

$$\langle a \rangle = \bigvee \left\{ \langle (a - r)^+ \rangle \mid 0 < r \le 1 \text{ in } \mathbb{Q} \right\}$$

by the proof of Proposition 2.3 in Banaschewski [3] and hence

$$\langle (a - p)^+ \rangle = \bigvee \left\{ \langle ((a - p)^+ - r)^+ \rangle \mid 0 < r \le 1 \text{ in } \mathbb{Q} \right\}$$
$$= \bigvee \left\{ \langle (a - p')^+ \rangle \mid p < p' \text{ in } \mathbb{Q} \right\}$$

because

$$((a - p)^+ - r)^+ = (a - p - r) \vee (-r) \vee 0$$
$$= (a - p - r) \vee 0 = (a - p') \vee 0,$$

where $p < p' = p + r$. Similarly,

$$\langle (q - a)^+ \rangle = \bigvee \left\{ \langle ((q - a)^+ - r)^+ \rangle \mid 0 < r \le 1 \text{ in } \mathbb{Q} \right\}$$
$$= \bigvee \left\{ \langle (q' - a)^+ \rangle \mid q' < q \text{ in } \mathbb{Q} \right\},$$

showing that

$$(*) \qquad \hat{a}(p, q) = \bigvee \left\{ \hat{a}(p', q') \mid p < p', \ q' < q \text{ in } \mathbb{Q} \right\}.$$

Further, if $q' \le p'$ then $(a - p')^+ \wedge (q' - a)^+ \le ((a - p') \wedge (p' - a))^+ = 0$ and hence the condition in $(*)$ reduces to $p < p' < q' < q$, settling the case of (R3).

Finally, for (R4), note that there exist $p \le a - 1$ and $q \ge a + 1$ since A is bounded, and for these we have

$$(a - p)^+ \wedge (q - a)^+ \ge 1$$

so that already $\hat{a}(p, q) = \langle 1 \rangle$.

For the second claim, taking the easy case of $-$ first, we have

$$(\widehat{-a})(p, q) = \langle (-a - p)^+ \wedge (q + a)^+ \rangle = \hat{a}(-q, -p) = (-\hat{a})(p, q).$$

Next, for $+$, consider any $p, q, r, s, t, u \in \mathbb{Q}$ such that $\langle r, s \rangle + \langle t, u \rangle \subseteq \langle p, q \rangle$, saying that $p \le r + t$ and $s + u \le q$. Then

$$a + b - p \geq (a - r) + (b - t) \geq 2((a - r) \wedge (b - t)),$$
$$q - (a + b) \geq (s - a) + (u - b) \geq 2((s - a) \wedge (u - b)),$$

by the general rule in ℓ-rings that $x + y = (x \vee y) + (x \wedge y) \geq 2(x \wedge y)$, and hence

$$2((a - r)^+ \wedge (s - a)^+ \wedge (b - t)^+ \wedge (u - b)^+) \leq (a + b - p)^+ \wedge (q - (a + b))^+ ,$$

showing that

$$(\hat{a} + \hat{b})(p, q) \leq \widehat{(a + b)}(p, q)$$

for all $p, q \in \mathbb{Q}$. Now, since any element of $\mathcal{L}(\mathbb{R})$ is a join of elements (p, q), it follows that $\hat{a} + \hat{b} \leq \widehat{a + b}$ in the usual (argumentwise) partial order of the homomorphisms between two frames; furthermore, as already pointed out in the previous section, this partial order reduces to equality if the domain is regular, and hence the regularity of $\mathcal{L}(\mathbb{R})$ implies the desired identity.

Regarding $\wedge$, if $\langle r, s \rangle \wedge \langle t, u \rangle \subseteq \langle p, q \rangle$ and hence $p \leq r \wedge t$ and $s \wedge u \leq q$ for p, q, r, s, t, u in $\mathbb{Q}$ then

$$(a - r) \wedge (b - t) \leq (a - p) \wedge (b - p) \leq (a \wedge b) - p.$$

On the other hand,

$$(s - a) \wedge (u - b) \leq q - a, \quad \text{if } s \leq u$$

and

$$(s - a) \wedge (u - b) \leq q - b, \quad \text{if } u \leq s$$

so that

$$(s - a) \wedge (u - b) \leq (q - a) \vee (q - b) = q - (a \wedge b).$$

In all it follows that $(\hat{a} \wedge \hat{b})(p, q) \leq \widehat{(a \wedge b)}(p, q)$ for all $p, q \in \mathbb{Q}$, and consequently $\hat{a} \wedge \hat{b} = \widehat{a \wedge b}$, again by the regularity of $\mathcal{L}(\mathbb{R})$.

As a result we now also have the case of $\vee$ by the familiar fact that $a \vee b = -((-a) \wedge (-b))$ in any ℓ-ring.

Next we note that $\tau_A(1) = 1$, meaning explicity that

$$\langle (1 - p)^+ \wedge (q - 1)^+ \rangle = \begin{cases} \langle 1 \rangle & \text{if } p < 1 < q \\ \langle 0 \rangle & \text{if } 1 \leq p \text{ or } q \leq 1, \end{cases}$$

which is immediately obvious.

Finally, for the most involved case of $\cdot$, consider first $a, b \geq 1$ and let $p, q, r, s, t, u \in \mathbb{Q}$ be such that $\langle r, s \rangle \cdot \langle t, u \rangle \subseteq \langle p, q \rangle$. We have to show that

(†) $\qquad \langle (a - r)^+ \wedge (s - a)^+ \wedge (b - t)^+ \wedge (u - b)^+ \rangle \subseteq \langle (ab - p)^+ \wedge (q - ab)^+ \rangle$

and we first deal with the case $p > 0$. Then none of the r, s, t, u can be 0, and if $s < 0$ then $(s - a)^+ = 0$, so that (†) holds trivially. On the other hand, if $r < 0$ then also $s < 0$, showing that we only have to consider $r > 0$.

Now, since $2b \geq 2$ and $2r > 0$,

$$\langle (a - r)^+ \rangle = \langle (2(a - r)b)^+ \rangle, \quad \langle (b - t)^+ \rangle = \langle (2(b - t)r)^+ \rangle$$

and hence

$$\begin{aligned}
\langle (a - r)^+ \wedge (b - t)^+ \rangle &= \langle 2((a - r)b \wedge (b - t)r)^+ \rangle \\
&\subseteq \langle ((a - r)b + (b - t)r)^+ \rangle = \langle (ab - rt)^+ \rangle \\
&\subseteq \langle (ab - p)^+ \rangle,
\end{aligned}$$

the last step since $p \leq rt$. On the other hand, $u > 0$ since $s > 0$ and therefore, similarly,

$$\langle (s - a)^+ \wedge (u - b)^+ \rangle = \langle 2((s - a)u \wedge (u - b)a)^+ \rangle \subseteq \langle (q - ab)^+ \rangle.$$

proving (†). It follows that $(\hat{a} \cdot \hat{b})(p, q) \leq \widehat{(ab)}(p, q)$ whenever $0 < p$.

Next, note that $\hat{a} \cdot \hat{b}, \widehat{(ab)} \geq 1$ because τ_A preserves 1 and $\leq$. Consider then $\alpha, \beta \geq 1$ in any $\mathcal{R}L$ such that $\alpha(p, q) \leq \beta(p, q)$ for all $p > 0$. Now $1 \leq \gamma$ in $\mathcal{R}L$ means

$$\bigvee \left\{ \gamma(p, r) \mid p < r \text{ in } \mathbb{Q} \right\} \leq \bigvee \left\{ 1(p, r) \mid p < r \text{ in } \mathbb{Q} \right\}$$

for each $r \in \mathbb{Q}$ so that $\alpha, \beta \geq 1$ in particular implies $\alpha(p, q) = 0 = \beta(p, q)$ whenever $q \leq 1$. Further, if $p \leq 0$ and $1 < q$ then by (R2)

$$\alpha(p, q) = \alpha(p, 1) \vee \alpha(\tfrac{1}{2}, q) \leq \beta(p, 1) \vee \beta(\tfrac{1}{2}, q) = \beta(p, q),$$

showing that $\alpha(p, q) \leq \beta(p, q)$ for all $p, q \in \mathbb{Q}$. Now, as already noted, the regularity of $\mathcal{L}(\mathbb{R})$ implies that $\alpha = \beta$, and we conclude that $\tau_A(ab) = \tau_A(a)\tau_A(b)$ for all $a, b \geq 1$ in A.

As the last step, this result is extended to arbitrary $a, b \in A$ by the following general observation: any additive map $\varphi : A \to B$ between bounded f-rings such that $\varphi(1) = 1$ and $\varphi(ab) = \varphi(a)\varphi(b)$ whenever $a, b \geq 1$ preserves all products. Indeed, given any $a, b \in A$, take some natural k such that $a + k, b + k \geq 1$, using that A is bounded, and multiply out the brackets on either side of the resulting equation

$$\varphi((a + k)(b + k)) = (\varphi(a) + k)(\varphi(b) + k).$$

Regarding the final claim, $\hat{a} = 0$ means that $\langle (a - p)^+ \wedge (q - a)^+ \rangle = \langle 1 \rangle$ whenever $p < 0 < q$ in $\mathbb{Q}$. In particular this implies, for any $r > 0$ in $\mathbb{Q}$ that there exist $b \geq 0$ such that

$$1 \leq ((a + r)^+ \wedge (r - a)^+)b = (r - |a|)^+ b$$

and hence $(|a| - r)^+ = 0$, showing that $|a| \leq r$ and therefore $a = 0$ since A is archimedean.

Given that, for any archimedean bounded f-ring A, its ring of fractions inverting all natural $n \in A$ is an extension to an archimedean bounded f-ring over $\mathbb{Q}$ which does not affect the frame $\mathfrak{M}A$ this proves

Proposition 3.1. *Any archimedean bounded f-ring A has an ℓ-ring embedding into* $\mathcal{R}(\mathfrak{M}A)$.

Remark 3.2. The procedure defining $\tau_A : A \longrightarrow \mathcal{R}(\mathfrak{M}A)$ is also employed by Madden [11] to obtain, for any archimedean ℓ-group with weak order unit, an ℓ-group embedding into the additive ℓ-group of a certain $\mathcal{R}L$, taking its unit to $1 \in \mathcal{R}L$. Unlike the present $\mathfrak{M}A$, the frame L is introduced by generators and relations, but some subsequent discussion in [11] shows it is actually isomorphic to $\mathfrak{M}A$ in the case considered here. It should be noted, though, that the *compactness* of the latter can only be derived from [11] by arguments not valid in Zermelo-Fraenkel set theory, and the same applies to the corollary below, showing that the present results cannot be obtained as consequences of [11]. In a slightly different vein, it should be added that an alternative description of $\mathfrak{M}A$ by generators and relations is discussed in Banaschewski [4], but for the purpose of this paper it seemed more straightforward to take it as originally presented in Banaschewski [3].

Remark 3.3. One might compare the above arguments substantiating Claims 1–3 with the somewhat related calculations given in Section 3 of Ball-Hager [1] which employ $\mathfrak{O}\mathbb{R}$ instead of $\mathcal{L}(\mathbb{R})$. It seems clear that this is one of the situations in which the use of the latter offers distinct advantages: it automatically provides a ready made route to follow whereas in [1] the route had to be specifically designed first.

There is an additional refinement concerning the embeddings $\tau_A : A \longrightarrow \mathcal{R}(\mathfrak{M}A)$. For any archimedean bounded f-ring A over $\mathbb{Q}$, the set of all $\hat{a}(p,q)$, $a \in A$ and $p, q \in \mathbb{Q}$, generates $\mathfrak{M}A$: by an earlier observation,

$$\langle a \rangle = \bigvee \left\{ \langle ((a-p)^+) \rangle \mid 0 < p \leq 1 \right\}$$

for any $a \in A$, hence

$$\langle a \rangle = \bigvee \left\{ \hat{a}(p,q) \mid 0 < p \leq 1 \right\}$$

for any $q \geq a + 1$, and the $\langle a \rangle$ trivially generate $\mathfrak{M}A$. This shows that the image of τ_A satisfies the hypothesis of the following pointfree form of the Stone-Weierstrass Theorem (Banaschewski [5]).

For a compact completely regular frame L, any subring S of $\mathcal{R}L$ containing $\mathbb{Q}$ for which $\{\alpha(p,q) \mid \alpha \in S; p, q \in \mathbb{Q}\}$ generates L is uniformly dense in $\mathcal{R}L$.

As a result we now have

Corollary 3.4. *For any uniformly complete archimedean bounded f-ring over $\mathbb{Q}$, the embedding* $\tau_A : A \longrightarrow \mathcal{R}(\mathfrak{M}A)$ *is an isomorphism.*

The obvious question arising at this stage is whether all $\mathcal{R}L$ for compact completely regular L are uniformly complete. It turns out this is indeed the case, but while the corresponding classical result is almost trivial, the present situation seems to require a good deal more work.

To begin with, we need the fact that the correspondence $A \mapsto \mathfrak{M}A$ is functorial for archimedean bounded f-rings over $\mathbb{Q}$ such that the frame homomorphism $\mathfrak{M}\varphi$: $\mathfrak{M}A \longrightarrow \mathfrak{M}B$ corresponding to an ℓ-ring homomorphism $\varphi : A \longrightarrow B$ takes each closed ℓ-ideal J of A to the closed ℓ-ideal of B generated by its image $\varphi[J]$ (Banaschewski [3]). Furthermore, if $\varphi : A \to B$ is a dense embedding then $\mathfrak{M}\varphi : \mathfrak{M}A \longrightarrow \mathfrak{M}B$ is an isomorphism.

Next, for any compact completely regular frame L there is an isomorphism σ_L : $\mathfrak{M}(\mathcal{R}L) \longrightarrow L$ such that $\sigma_L(J) = \bigvee \{ \operatorname{coz}(\alpha) \mid \alpha \in J \}$ (Banaschewski [4], Proposition 12), which in turn implies that $\tau_{\mathcal{R}L} : \mathcal{R}L \longrightarrow \mathcal{R}(\mathfrak{M}(\mathcal{R}L))$ is an isomorphism: for any $\alpha \in \mathcal{R}L$ and $p, q \in \mathbb{Q}$,

$$
\begin{aligned}
(\mathcal{R}\sigma_L)\tau_{\mathcal{R}L}(\alpha)(p,q) &= \sigma_L(\langle (\alpha - \mathbf{p})^+ \wedge (\mathbf{q} - \alpha)^+ \rangle) \\
&= \operatorname{coz}((\alpha - \mathbf{p})^+ \wedge (\mathbf{q} - \alpha)^+) = \alpha(p,q),
\end{aligned}
$$

the last step by the properties of coz listed earlier, hence $(\mathcal{R}\sigma_L)\tau_{\mathcal{R}L} = \operatorname{id}_{\mathcal{R}L}$, and since $\mathcal{R}\sigma_L$ is an isomorphism the same holds for $\tau_{\mathcal{R}L}$.

Finally, for any ℓ-ring homomorphism $\varphi : A \longrightarrow B$ between archimedean bounded f-rings over $\mathbb{Q}$, $\tau_B \varphi = \mathcal{R}(\mathfrak{M}\varphi)\tau_A$: for any $a \in A$,

$$
\begin{aligned}
\mathcal{R}(\mathfrak{M}\varphi)(\hat{a})(p,q) &= ((\mathfrak{M}\varphi)\hat{a})(p,q) \\
&= \mathfrak{M}\varphi(\langle (a - p)^+ \wedge (q - a)^+ \rangle) \\
&= \langle (\varphi(a) - p)^+ \wedge (q - \varphi(a))^+ \rangle = \varphi(a)^\wedge(p,q),
\end{aligned}
$$

hence $\mathcal{R}(\mathfrak{M}\varphi)(\hat{a}) = \varphi(a)^\wedge$ for all $a \in A$, and this proves the assertion.

Now we have

Lemma 3.5. *For any compact completely regular L, $\mathcal{R}L$ is uniformly complete.*

Proof. Since the completion, relative to the uniform topology, of an archimedean bounded f-ring over $\mathbb{Q}$ is a uniformly dense extension it will be enough to show that any uniformly dense embedding $\varphi : \mathcal{R}L \longrightarrow A$ into such an f-ring is an isomorphism. Note that any such φ is an epimorphism in the category of archimedean bounded f-rings over $\mathbb{Q}$ since the homomorphisms between the latter are continuous relative to the uniform topologies which, in turn, are Hausdorff. Now consider the square

$$
\begin{array}{ccc}
\mathcal{R}\mathfrak{M}(\mathcal{R}L) & \xrightarrow{\mathcal{R}\mathfrak{M}\varphi} & \mathcal{R}\mathfrak{M}A \\
{\scriptstyle \tau_{\mathcal{R}L}} \uparrow & & \uparrow {\scriptstyle \tau_A} \\
\mathcal{R}L & \xrightarrow[\varphi]{} & A
\end{array}
$$

which commutes, as just shown. Further, by the other arguments above, $\tau_{\mathcal{R}L}$ and $\mathcal{R}\mathfrak{M}\varphi$ are isomorphisms, hence $\tau_A\varphi$ is an isomorphism, and therefore $\psi\varphi = \mathrm{id}_{\mathcal{R}L}$ for some $\psi : A \longrightarrow \mathcal{R}L$. Finally, since φ is epic this makes it an isomorphism.   ∎

In all this proves the following pointfree version of Stone's result.

Proposition 3.6. *Up to isomorphism, the Stone rings are exactly the $\mathcal{R}L$ for compact completely regular frames L.*

Further, as in the classical case, this readily extends to a category equivalence. The only thing left to show for this is that any ring homomorphism $\varphi : \mathcal{R}L \longrightarrow \mathcal{R}M$ is $\mathcal{R}h$ for some frame homomorphism $h : L \to M$. Actually, this is an immediate consequence of a considerably more general result (Banaschewski [4], Proposition 14); a simple direct proof in the present context is as follows. Note first that φ is an ℓ-ring homomorphism: if $\alpha \geq 0$ in $\mathcal{R}L$ then $\alpha = \beta^2$ for some $\beta \in \mathcal{R}L$ and hence $\varphi(\alpha) = \varphi(\beta)^2 \geq 0$, showing that φ is order preserving, but then also $\varphi(|\alpha|) = |\varphi(\alpha)|$ since $\varphi(|\alpha|) \geq 0$ and $\varphi(|\alpha|)^2 = |\varphi(\alpha)|^2$. It follows that $\mathfrak{M}\varphi$ is defined, and if $h = \sigma_M(\mathfrak{M}\varphi)\sigma_L^{-1}$ then

$$\mathcal{R}h = \mathcal{R}\sigma_M \mathcal{R}\mathfrak{M}\varphi \mathcal{R}\sigma_L^{-1} = \tau_{\mathcal{R}M}^{-1}(\mathcal{R}\mathfrak{M}\varphi)\tau_{\mathcal{R}L} = \varphi$$

by the above argument showing any $\tau_{\mathcal{R}L}$ is an isomorphism and the general identity $\tau_B\varphi = (\mathcal{R}\mathfrak{M}\varphi)\tau_A$ derived subsequently.

This proves

Proposition 3.7. *The functor $\mathcal{R}$ induces an equivalence between the category of compact completely regular frames and the category of Stone rings and ring homomorphisms.*

Finally, since the compact completely regular frames are spatial if the Prime Ideal Theorem is assumed we have

Corollary 3.8. *Given the Prime Ideal Theorem, the functor $X \mapsto C(X)$ induces a dual equivalence between the category of compact Tychonoff spaces and the category of Stone rings and ring homomorphisms.*

Remark 3.9. Note that the duality between *compact Hausdorff* spaces and Stone rings, referred to in the Introduction, seems to require somewhat stronger foundations: the complete regularity of such spaces is obtained by means of Urysohn's Lemma, the usual proof of which involves the Axiom of Countable Dependent Choice. Whether there are weaker choice principles leading to this result appears to be unknown.

Remark 3.10. It may be worth pointing out that nowhere in the above development did one have to appeal to the classical result that every archimedean totally ordered field is isomorphic to a subfield of $\mathbb{R}$ which was crucially used in step 3 of the outline of the proof given in the Introduction. In fact, in the present context that result is a *consequence* of what is proved here: any archimedean totally ordered field A is an archimedean bounded f-ring for which $\mathfrak{M}A \simeq 2$, and Proposition 3.1 then says A has an embedding into $\mathcal{R}2$, but this is isomorphic to $\mathbb{R}$, since $\mathcal{R}2 = \mathcal{R}(\mathfrak{O}1)$ for the one-point space 1, and $\mathcal{R}(\mathfrak{O}1) \cong C(1) \cong \mathbb{R}$.

References

[1] R. N. Ball and A. W. Hager, *On the localic Yosida representation of an archimedean lattice ordered group with weak order unit.* J. Pure & Appl. Alg. **70** (1991), 17-43.

[2] B. Banaschewski, *The power of the ultrafilter theorem.* J. London Math. Soc. **27** (1983), 193-202.

[3] B. Banaschewski, *Pointfree topology and the spectra of f-rings.* in *Ord. Alg. Struct. Proc. Curaçao Conf. on Part. Ord. Alg. Systems* (1995); W. C. Holland & J. Martinez, Eds. (1997), Kluwer Acad. Publ.,Dordrecht, 123–148.

[4] B. Banaschewski, *The Real Numbers in Pointfree Topology.* Textos de Matemática Série B, **12** (1997), Departamento de Matemática da Universidade Coimbra.

[5] B. Banaschewski, *f-Rings and the Stone-Weierstrass Theorem.* Order; to appear.

[6] B. Banaschewski, *Ring theory and pointfree topology.* (2001), McMaster University preprint.

[7] E. Bishop, *Foundations of Constructive Analysis.* McGraw-Hill (1967), New York.

[8] M. P. Fourman and J. M. E. Hyland, *Sheaf Models for Analysis.* Applications of Sheaves, Springer LNM **753** (1979), 280-301.

[9] P. T. Johnstone, *Stone spaces.* Cambridge Studies Adv. Math. **3** (1982), Cambridge University Press, Cambridge.

[10] A. Joyal, *Theorie des topos et le théorème de Barr.* Tagungsbericht Category Theory Meeting (1977), Oberwolfach.

[11] J. J. Madden, *Frames associated with an abelian group.* Trans. AMS **331** (1992), 265-279.

[12] J. J. Madden and J. Vermeer, *Epicomplete archimedean ℓ-groups via a localic Yosida theorem.* J. Pure & Appl. **68** (1990), 243-252.

[13] M. H. Stone, *A general theory of spectra I.* Proc. Nat. Acad. Sci. USA **26** (1940), 280-283.

[14] J. J. C. Vermeulen, *The constructive expansion of the binomial* $(1-X)^{1/2}$. (1999), University of Cape Town preprint.

[15] S. Vickers, *Topology via Logic.* Cambridge Tracts Theor. Comp. Sci. **5** (1985), Cambridge University Press, Cambridge.

[16] E. Witt, *Lectures on the infinitesimal calculus.* Summer Semester 1948, Hamburg University.

Department of Mathematics and Statistics, McMaster University,
Hamilton, ON L8S 4K1, Canada

The Range of Lattice Homomorphisms on f-Algebras[1]

Karim Boulabiar

I would like to dedicate this paper to Professor Paul Conrad,
whose papers influenced my research.

ABSTRACT. Let A be an archimedean f-algebra with unit element e, B be an archimedean semiprime f-algebra and $T : A \to B$ be a lattice (or Riesz) homomorphism. The main purpose of this paper is to show, in straightforward and elementary manner, that the range $R(T)$ of T is an f-subalgebra of B if and only if Te is idempotent in B.

1 Introduction

In the course of time much attention has been paid to the behavior of lattice (or Riesz) homomorphisms defined on an f-algebra. Perhaps the most striking theorem in this direction, due to A. W. Hager and L. C. Robertson [4], is that a lattice homomorphism between two archimedean unital f-algebras that preserves identity is an algebra (or multiplicative) homomorphism. There are many results of this kind in the literature. We mention other results in that direction. It is proven by B. van Putten in his thesis [10] that the set of all lattice homomorphisms between two archimedean unital f-algebras that preserve identity coincides with the set of extreme points of the convex set of all Markov linear operators (i.e., positive linear operators that preserve identity). Highlight is the important paper of C. B. Huijsmans and B. de Pagter [6], in which the connection between lattice homomorphisms and algebra (or multiplicative) homomorphisms on f-algebras is considered in great detail. One of the major results presented in [6] is that a lattice homomorphism T from an archimedean f-algebra A with unit element e into an archimedean semiprime f-algebra B is an algebra homomorphism if and only if Te is idempotent (i.e., $(Te)^2 = Te$).

In the present paper we intend to make some contributions to this area by investigating the range of a lattice homomorphism defined on an f-algebra. To be more

---

[1] The author is grateful to both referees for pointing out some obscurities in the original version of this paper.

J. Martínez (ed.), Ordered Algebraic Structures, 179–188.
© 2002 Kluwer Academic Publishers.

precise, we consider a lattice homomorphism T from an archimedean f-algebra A into an archimedean vector lattice B and we prove that the formula

$$Tf * Tg = T(fg); \quad f, g \in A$$

defines a multiplication in the range $R(T)$ of T, under which, $R(T)$ is an archimedean f-algebra. On the other hand, we assume that A has in addition a unit element and B is a semiprime f-algebra and we give necessary and sufficient conditions on T in order that $R(T)$ be an f-subalgebra of B. We point out that, in the literature, the above formula played a key role in various directions. See, for instance, the works [7] and [6] of C. B. Huijsmans and B. de Pagter.

We assume that the reader is familiar with the notion of vector lattices (or Riesz spaces). For terminology, notations and concepts not explained in this paper we refer to the standard monographs [1] and [8].

2   Some preliminaries

In order to avoid unnecessary repetition, we shall assume throughout this paper that all (real) vector lattices under consideration are archimedean.

Throughout this section, A and B stand for (archimedean) vector lattices. First, we recall some notions about (linear) operators on vector lattices. Let T be an operator from A into B. We say that T is *order bounded* if T maps order intervals of A into order intervals of B. The operator T is said to be *positive* if $f \geq 0$ in A implies $Tf \geq 0$ in B. We call T a *lattice homomorphism* whenever $f \wedge g = 0$ in A implies $Tf \wedge Tg = 0$ in B (equivalently, $|Tf| = T|f|$ for all $f \in A$). Obviously, every lattice homomorphism is automatically positive. Our main reference about linear operators on vector lattices is [1].

In the next lines, we list the definition and some elementary properties of f-algebras and orthomorphisms. We call a vector lattice A a *lattice-ordered algebra* (briefly, ℓ-*algebra*) if there exists an associative multiplication in A with the usual algebra properties such that $0 \leq fg$ for all $0 \leq f, g \in A$. Such an ℓ-algebra A is called an f-*algebra* if A has the additional property that $f \wedge g = 0$ implies $fh \wedge g = hf \wedge g = 0$ for all $0 \leq h \in A$. If A is an f-algebra and $f, g \in A$ such that $f \wedge g = 0$ then $fg = 0$. Every (archimedean) f-algebra is commutative and has positive squares. The f-algebra A is said to be *semiprime* if 0 is the only nilpotent element in A. The f-algebra A is semiprime if and only if $f^2 = 0$ implies $f = 0$. Every f-algebra with unit element is semiprime. More about f-algebras can be found in [9].

The following paragraph deals with basic facts on orthomorphisms. The order bounded operator T of the vector lattice A is said to be an *orthomorphism* if $|f| \wedge |g| = 0$ implies $|Tf| \wedge |g| = 0$. The collection Orth (A) of all orthomorphisms of A is an (archimedean) unital f-algebra with respect to the usual vector space operations and composition as multiplication. Of course, the identity mapping on A is the unit element of Orth (A). Consider an f-algebra A, take an arbitrary $f \in A$ and denote the multiplication by f with π_f. Since A is an f-algebra, $\pi_f \in$ Orth (A). Consequently, an operator ρ can be defined as follows

$$\rho: \quad A \quad \to \quad \text{Orth}\,(A)$$
$$f \quad \mapsto \quad \pi_f$$

For later reference, we state the following proposition, in which, we collect some properties of ρ.

Proposition 2.1. *The operator ρ is a lattice and algebra homomorphism. Moreover, ρ is injective if and only if A is semiprime and ρ is bijective if and only if A has a unit element.*

For the proof of the previous proposition, as well as for more informations about orthomorphisms, we refer the reader to [9]. Finally, the following property will be useful for later purposes; it was proved by de Pagter in his thesis [9] and for the sake of completeness we reproduce the simple proof.

Proposition 2.2. *Let A be an archimedean f-algebra and $0 \le f, g \in A$. The following statement holds.*

$$0 \le fg - fg \wedge ng \le \frac{1}{n}f^2 g \quad (n = 1, 2, ...)$$

Proof. From

$$(fg - fg \wedge ng) \wedge (ng - fg \wedge ng) = 0$$

it follows that

$$\begin{aligned} 0 &= (fg - fg \wedge ng) \wedge \left[\frac{1}{n}f\,(ng - fg \wedge ng)\right] \\ &= (fg - fg \wedge ng) \wedge \left(fg - \frac{1}{n}f^2 g \wedge fg\right). \end{aligned}$$

Consequently

$$fg = (fg \wedge ng) \vee \left(\frac{1}{n}f^2 g \wedge fg\right).$$

Finally

$$0 \le fg - fg \wedge ng \le \frac{1}{n}f^2 g \wedge fg \le \frac{1}{n}f^2 g$$

and the proof is finished.   ∎

3   The main results

Throughout the sequel, T stands for a lattice homomorphism from an f-algebra A into a vector lattice B. Our first result in this section (Proposition 3.2) states that the range $R(T)$ of T is an archimedean f-algebra in its own right. In order to assist in the exposition, we give the following lemma, which follows immediately, as in [5, Proposition 3.1], from Proposition 2.2. Recall that an *order ideal* in a vector lattice A is a linear subspace I of A such that $0 \leq |f| \leq |g|$ and $g \in I$ imply $f \in I$.

Lemma 3.1. *Let A be an archimedean f-algebra, B be an archimedean vector lattice and $T : A \to B$ be a lattice homomorphism. Then $\ker(T)$ is an ℓ-ideal (ring and order ideal) in A.*

Consider now the range $R(T)$ of T. Clearly, $R(T)$ is a vector sublattice of B. The lattice operations in $R(T)$ are

$$Tf \wedge Tg = T(f \wedge g) \quad \text{and} \quad Tf \vee Tg = T(f \vee g)$$

for all $f, g \in A$. Moreover, a multiplication $*$ can be introduced in $R(T)$ by putting

$$Tf * Tg = T(fg)$$

for all $f, g \in A$. Indeed, let $f, g, u, v \in A$ such that $Tf = Tu$ and $Tg = Tv$. It ensues from Lemma 3.1 that

$$fg - uv = (f - u)g + u(g - v) \in \ker(T).$$

Hence

$$T(fg) - T(uv) = T(fg - uv) = 0.$$

Finally

$$T(fg) = T(uv)$$

and thus the above multiplication $*$ is well-defined. Obviously, the multiplication $*$ is associative. With respect to this multiplication, $R(T)$ is an archimedean f-algebra as it is shown in the next proposition.

Proposition 3.2. *Let A be an archimedean f-algebra, B be an archimedean vector lattice and $T : A \to B$ be a lattice homomorphism. The range $R(T)$ of T, which is a vector sublattice of B, is an f-algebra with respect to the multiplication $*$ defined in $R(T)$ by*

$$Tf * Tg = T(fg)$$

for all $f, g \in A$. Moreover, if A has a unit element e then $R(T)$ has Te as unit element.

Proof. Let $u, v, w \in R(T)$ such that $u \wedge v = 0$ and $w \geq 0$. Since T is a lattice homomorphism, there exist $0 \leq f, g, h \in A$ such that $u = Tf$, $v = Tg$ and $w = Th$. From $u \wedge v = 0$ it follows that $T(f \wedge g) = 0$ and, in view of the fact that $\ker(T)$ is an ℓ-ideal (see Lemma 3.1), $T(h(f \wedge g)) = 0$. So, we can write

$$Tf = Tf - T(f \wedge g) = T(f - f \wedge g) \tag{1}$$

and

$$T(hg) = T(hg) - T(h(f \wedge g)) = T(hg - h(f \wedge g)). \tag{2}$$

On the other hand,

$$(f - f \wedge g) \wedge (g - f \wedge g) = 0.$$

Thus

$$(f - f \wedge g) \wedge (hg - h(f \wedge g)) = 0.$$

and therefore

$$T(f - f \wedge g) \wedge T(hg - h(f \wedge g)) = 0.$$

This equality together with (1) and (2) implies

$$
\begin{aligned}
u \wedge (w * v) &= Tf \wedge (Th * Tg) = Tf \wedge T(hg) \\
&= T(f - f \wedge g) \wedge T(hg - h(f \wedge g)) = 0.
\end{aligned}
$$

We deduce that $R(T)$ is an f-algebra.

Suppose now that A has, in addition, a multiplicative identity e. So

$$Tf * Te = T(fe) = Tf$$

for all $f \in A$. It follows that Te is the unit element of the f-algebra $R(T)$ and the proof of the proposition is complete. ∎

As an illustration of Proposition 3.2, we make the next remark.

Remark 3.3. Let X be a locally compact Hausdorff space and $C_K(X)$ be the vector lattice of all real continuous functions on X with compact support. Since $C_K(X)$ can not be furnished with a structure of unital f-algebra, there is not a surjective lattice homomorphism from $C(X)$ onto $C_K(X)$, where $C(X)$ is the unital f-algebra of all real continuous functions on X, as the latter can be identified with $\mathrm{Orth}(C_K(X))$. ([11])

Now, let B be a vector lattice such that there exists a surjective lattice homomorphism T from $\mathrm{Orth}(B)$ onto B. In this case, and in view of Proposition 3.2, B is a unital f-algebra with respect to the multiplication induced by T. Under this multiplication, $B \cong \mathrm{Orth}(B)$ (where we use Proposition 2.1). However, T itself need not be an isomorphism as it is shown in the next example.

Example 3.4. Let $C_b\left(\left[1,+\infty\right[\right)$ denote the f-algebra of all bounded real continuous functions on $\left[1,+\infty\right[$ and i the function defined on $\left[1,+\infty\right[$ by $i\left(x\right)=x$ for all $x\in\left[1,+\infty\right[$. Consider

$$B=\left\{g\in C_b\left(\left[1,+\infty\right[\right):ig\in C_b\left(\left[1,+\infty\right[\right)\right\}$$

Evidently, B is a semiprime f-subalgebra of $C_b\left(\left[1,+\infty\right[\right)$.

First, we claim that Orth (B) can be identified with $C_b\left(\left[1,+\infty\right[\right)$. To this end, fix $f\in C_b\left(\left[1,+\infty\right[\right)$. Obviously, the operator τ_f defined by $\tau_f\left(g\right)=fg$ for all $g\in B$ is an orthomorphism of B. Therefore, a mapping τ can be defined from $C_b\left(\left[1,+\infty\right[\right)$ into Orth (B) by putting $\tau\left(f\right)=\tau_f$ for all $f\in C_b\left(\left[1,+\infty\right[\right)$. Actually, τ is a lattice and algebra isomorphism. The only part that needs some details is the surjectivity of τ. To do this, let $\pi\in$ Orth (B) and set $f=i\pi\left(i^{-1}\right)$, where $i^{-1}\left(x\right)=1/x$ for all $x\in\left[1,+\infty\right[$. Observe that $f\in C_b\left(\left[1,+\infty\right[\right)$ and $\tau_f\left(i^{-1}\right)=\pi\left(i^{-1}\right)$. Now, i^{-1} is a week order unit in B (i.e., $i^{-1}\wedge g=0$ and $g\in B$ imply $g=0$). This implies that $\tau_f=\pi$ (see, [9, Corollary 9.7]).

At this point, let's define the operator T from Orth $(B)\cong C_b\left(\left[1,+\infty\right[\right)$ into B by putting $\left(Tf\right)\left(x\right)=f\left(2x\right)/x$ for all $f\in C_b\left(\left[1,+\infty\right[\right)$ and $x\in\left[1,+\infty\right[$. It is not hard to show that T is a lattice homomorphism. Furthermore, T is surjective. Indeed, for a given $g\in B$, put $f\in C_b\left(\left[1,+\infty\right[\right)$ defined by

$$f\left(x\right)=\begin{cases}xg\left(x/2\right)/2 & if \quad x\in\left[2,+\infty\right[\\ g\left(1\right) & if \quad x\in\left[1,2\right[\end{cases}$$

Clearly, $Tf=g$. However, T is evidently not injective.

Finally, note that the multiplication $*$ in $B=R\left(T\right)$ is given by

$$g_1*g_2=ig_1g_2;\qquad g_1,g_2\in B.$$

and i^{-1} is the unit element of B with respect to the multiplication $*$. The relationship between the multiplication $*$ and the pointwise multiplication in B is also given by the following formula

$$g_1g_2=i^{-2}*g_1*g_2;\qquad g_1,g_2\in B$$

which is not surprising in view of [3, Theorem 2.5].

According to Proposition 3.2, the fact that A is a *unital* f-algebra implies that $R\left(T\right)$ is also a *unital* f-algebra. Consequently, it seems to be natural to ask whether $R\left(T\right)$ is a *semiprime* f-algebra when A is a *semiprime* f-algebra. The next example shows that in general this is not true.

Example 3.5. Let $A=C\left(\mathbb{R}\right)$ with the usual operations and order and define a multiplication $\bullet$ in A by putting

$$\left(f\bullet g\right)\left(x\right)=xf\left(x\right)g\left(x\right)$$

for all $f, g \in A$ and $x \in \mathbb{R}$. Clearly, A is a semiprime f-algebra with respect to the multiplication $\bullet$. Now, let T be the lattice homomorphism of A defined by

$$T(f)(x) = f(0)$$

for all $f \in A$ and $x \in \mathbb{R}$. So $R(T)$ is the subspace of constants. Observe that

$$Tf * Tg = T(f \bullet g) = 0.$$

for all $f, g \in A$. This implies that, under $*$, $R(T)$ is not semiprime.

At this point, suppose that A has a unit element e and B is a semiprime f-algebra (both multiplications, in A as well as B, are denoted by juxtaposition). The range $R(T)$ of the lattice homomorphism $T : A \to B$ is a vector sublattice of B and an f-algebra with Te as unit element under the multiplication defined in $R(T)$ by

$$Tf * Tg = T(fg)$$

for all $f, g \in A$. In general, $R(T)$, equipped with $*$, is not a subalgebra of B. The following example illustrates this.

Example 3.6. Consider $A = B = \mathbb{R}^{\mathbb{N}}$, the unital f-algebra of real sequences with the usual operations and order, and $T : A \to B$ defined par $T\left(\{u_n\}_{n \geq 0}\right) = \{u_1 n\}_{n \geq 0}$. It is not hard to show that T is a lattice homomorphism and its range is given by

$$R(T) = \left\{\{\lambda n\}_{n \geq 0} : \lambda \in \mathbb{R}\right\}.$$

Equipped with the multiplication $*$ defined by

$$\{\lambda n\}_{n \geq 0} * \{\mu n\}_{n \geq 0} = \{\lambda \mu n\}_{n \geq 0}$$

$R(T)$ is not an f-subalgebra of B.

Next, we will establish necessarily and sufficient conditions on T in order that $R(T)$ be a subalgebra of B. To this end, we need the following proposition, which is of some independent interest.

Proposition 3.7. *Let L be a vector sublattice of an archimedean semiprime f-algebra B and assume that L is furnished with a multiplication $*$ in such a manner that L is an f-algebra with unit element ω. Then the following conditions are equivalent:*

(i) *Furnished with $*$, L is an f-subalgebra of B (i.e., $*$ coincides with the multiplication in B).*

(ii) *ω is idempotent in B (i.e., $\omega^2 = \omega$).*

Proof. (i)⇒(ii) is trivial. (ii)⇒(i) is much less evident. Embedding, if necessary, B in the unital f-algebra Orth (B) of all orthomorphisms of B, we can assume that B has a unit element u. Let $f \in L$ such that $0 \leq f \leq \omega$. We have

$$0 \leq |f\omega - f| = |\omega - u| f \leq |\omega - u|\omega = |\omega^2 - \omega| = 0.$$

Hence

$$f\omega = f. \tag{3}$$

Furthermore, it follows from [2, Proposition 4.1] that

$$0 \leq \bigwedge_{k=0}^{n} \left| f - \frac{k}{n}\omega \right| \leq \frac{1}{n}\omega \quad (n = 1, 2, ...)$$

This gives, on the one hand,

$$0 \leq f * f - \bigvee_{k=0}^{n} \left(\frac{2k}{n}f - \frac{k^2}{n^2}\omega \right) = \bigwedge_{k=0}^{n} \left(f - \frac{k}{n}\omega \right) * \left(f - \frac{k}{n}\omega \right) \leq \frac{1}{n^2}\omega, \tag{4}$$

for all $n \in \{1, 2, ...\}$, and on the other hand,

$$0 \leq f^2 - \bigvee_{k=0}^{n} \left(\frac{2k}{n}f - \frac{k^2}{n^2}\omega \right) = \bigwedge_{k=0}^{n} \left(f - \frac{k}{n}\omega \right)^2 \leq \frac{1}{n^2}\omega, \tag{5}$$

for all $n \in \{1, 2, ...\}$ (where we use (3)). Combining (4) and (5), we obtain

$$f * f = f^2. \tag{6}$$

This identity can be straightforwardly generalized to an arbitrary $f \in L^+$ via the inequalities

$$0 \leq f - f \wedge n\omega \leq \frac{1}{n}f * f \quad (n = 1, 2, ...)$$

which are obtained replacing g by ω in Proposition 2.2. Now, since $f = f^+ - f^-$ and $f^+ * f^- = f^+ f^- = 0$, the equality (6) holds for an arbitrary $f \in L$.

Finally, let $f, g \in L$ and observe that

$$\begin{aligned}
f * g &= \left(\frac{f+g}{2} \right) * \left(\frac{f+g}{2} \right) - \left(\frac{f-g}{2} \right) * \left(\frac{f-g}{2} \right) \\
&= \left(\frac{f+g}{2} \right)^2 - \left(\frac{f-g}{2} \right)^2 = fg.
\end{aligned}$$

The proof of the proposition is now complete.   ∎

Combining Proposition 3.2 and Proposition 3.7, we get immediately the main result of this paper.

Theorem 3.8. *Let A be an archimedean f-algebra with unit element e, B be an archimedean semiprime f-algebra and $T : A \to B$ be a lattice homomorphism. Consider the range $R(T)$ of T, which is a vector sublattice of B and an f-algebra with Te as unit element under the multiplication $*$ defined in $R(T)$ by*

$$Tf * Tg = T(fg)$$

for all $f, g \in A$. The following statements are equivalent:

(i) *Equipped with $*$, $R(T)$ is an f-subalgebra of B.*

(ii) *Te is idempotent in f (i.e., $Te = (Te)^2$).*

(iii) *T is an algebra homomorphism from A into B.*

At last, we recall that the above equivalence (ii) $\Leftrightarrow$ (iii) was already proven by C. B. Huijsmans and B. de Pagter in their paper [6]. Here we presented an alternative proof of this equivalence.

References

[1] C. D. Aliprantis and O. Burkinshaw, *Positive operators*. Academic Press, (1985) Orlando.

[2] F. Beukers, C. B. Huijsmans and B. de Pagter, *Unital embedding and complexification of f-algebras*. Math. Z. **183** (1983), 131-144.

[3] K. Boulabiar, *A relationship between two almost f-algebra products*. Algebra Univ. **43** (2000), 347-367.

[4] A. W. Hager and L. C. Robertson, *Representing and ringifying a Riesz space*. Symposia Math. **21** (1977), 411-431.

[5] C. B. Huijsmans and B. de Pagter, *Ideal theory in f-algebras*. Trans. Amer. Math. Soc. **269** (1982), 225-245.

[6] C. B. Huijsmans and B. de Pagter, *Subalgebras and Riesz subspaces of an f-algebra*. Proc. London Math. Soc. **48** (1984), 161-174.

[7] C. B. Huijsmans and B. de Pagter, *Averaging operators and positive projections*. J. Math. Anal. Appl. **113** (1986), 163-184.

[8] W. A. J. Luxemburg and A. C. Zaanen, *Riesz spaces, I*. North-Holland (1971), Amsterdam-London.

[9] B. de Pagter, *f-Algebras and Orthomorphisms*. Thesis (1981), Leiden.

[10] B. van Putten, *Disjunctive linear operators and partial multiplication in Riesz spaces*. Thesis (1980), Wageningen.

[11] A. C. Zaanen, *Examples of orthomorphisms*. J. Approx. Theory **13** (1975), 192-204.

Département des Classes Préparatoires, Institut Préparatoire
aux Etudes Scientifiques et Techniques, BP 51, 2070-La Marsa, Tunisia
karim.boulabiar@ipest.rnu.tn

The Bornological Tensor Product
of two Riesz spaces:
Proof and Background Material

G. Buskes[1] and A. van Rooij

ABSTRACT.   This paper is a companion to [2]. The main result in this paper is Theorem 7.1.

All Riesz spaces that occur here are understood to be Archimedean. Most undefined terms concerning Riesz spaces are to be found in [7] or [1].

1   Fremlin's Results (as far as relevant to us)

Fremlin in [3] (respectively [4]) proved the existence of a Riesz tensor product $E\overline{\otimes}F$ for two Archimedean Riesz space E and F (respectively, some consequences for that tensor products if E and F are Banach lattices). In [2], Theorem 2.6, we indicated how to unify and extend each of the two universal mapping properties that Fremlin proved, by studying bornological Riesz spaces and maps of bounded variation. We now provide all the details of a rather similar result in Theorem 7.1. We first repeat from [3] and [4] all the information that we will need.

Remark 1.1. $E\overline{\otimes}F$ is a Riesz space containing the vector space tensor product $E \otimes F$ as a linear subspace. If $a \in E^+$ and $b \in F^+$, then $a \otimes b \in (E\overline{\otimes}F)^+$.

Remark 1.2. $E \otimes F$ is uniformly dense in $E\overline{\otimes}F$ in the sense that for all $u \in E\overline{\otimes}F$ there exist $a \in E^+$, $b \in F^+$ and $u_1, u_2, \ldots \in E \otimes F$ such that $|u-u_n| \leq n^{-1}a \otimes b$ for all n. ([3], Th. 4.2(iii))

Remark 1.3. If G is a uniformly complete Riesz space and $T : E \times F \to G$ is bilinear and positive (i.e., $T(E^+, F^+) \subseteq G^+$), then there exists a unique positive linear $T^{\otimes} : E\overline{\otimes}F \to G$ with $T = T^{\otimes} \circ \otimes$. ([3], Th. 5.3)

[1]The first author acknowledges support from Office of Naval Research grant ONR N00014-01-1-0322, in the summer of 2001. Both authors acknowledge support of NATO CRG grant 940605.

J. Martínez (ed.), Ordered Algebraic Structures, 189–203.
© 2002 Kluwer Academic Publishers.

Remark 1.4. If $E_0 \subseteq E$ and $F_0 \subseteq F$ are Riesz subspaces, then the Riesz subspace of $E \overline{\otimes} F$ generated by $\{x \otimes y : x \in E_0, y \in F_0\}$ is (naturally isomorphic to) the Riesz tensor product of E_0 and F_0. ([3], Cor. 4.5)

In a second paper ([4]) E and F are Banach lattices. Fremlin shows that the formula

$$\|u\|_{|\pi|} := \inf \left\{ \sum_n \|x_n\| \, \|y_n\| \; : \; x_1, \ldots, x_N \in E^+, \; y_1, \ldots, y_N \in F^+, \; |u| \leq \sum x_n \otimes y_n \right\}$$

defines a Riesz norm on $E \overline{\otimes} F$, the *projective product norm*, and that:

Remark 1.5. $E \otimes F$ is norm dense in $E \overline{\otimes} F$. ([4], 1B(a))

Remark 1.6. If G is a Banach lattice and $T : E \times F \longrightarrow G$ is bilinear and positive (hence continuous) then $T^\otimes$ is continuous; actually,

$$\|T^\otimes\| = \|T\| := \sup \{ \|T(x,y)\| \; : \; x \in E, \|x\| \leq 1, \; y \in F, \|y\| \leq 1 \}. \quad ([4], 1E(iii))$$

Remark 1.7. The cone generated by $\{x \otimes y : x \in E^+, y \in F^+\}$ in $E \overline{\otimes} F$ is norm dense in $(E \overline{\otimes} F)^+$. ([4], 1B(b))

In the proofs of these statements (except for the automatic continuity in 1.6) the norm completeness of E and F is not used. We will apply them to general normed Riesz spaces.

2 A Notation

Definition & Remarks 2.1. Let E be a Riesz space, $a \in E^+$. A *partition* of a is a finite sequence $x = (x_1, \ldots, x_N)$ of elements of E^+ whose sum is a. The partitions of a form a set Πa.

If $f \in E^\sim$, then the element $|f|$ of $E^\sim$ is determined by:

$$(1) \qquad |f|(a) = \sup_{x \in \Pi a} \sum_n |f(x_n)| \quad (a \in E^+).$$

Definition & Remarks 2.2. If $x = (x_1, \ldots, x_N)$ and $y = (y_1, \ldots, y_M)$ are partitions of a we call x a *refinement* of y if the set $\{1, \ldots, N\}$ can be written as a disjoint union of sets $I(1), \ldots, I(M)$ in such a way that

$$y_m = \sum_{n \in I(m)} x_n \quad (m = 1, \ldots, M).$$

Any two partitions have a common refinement. (This follows from [7], 15.6(i).) Thus, in a natural way Πa is a directed set.

If $f \in E^\sim$, then for every $a \in E^+$ the net

$$\left(\sum_n |f(x_n)| \right)_{x \in \Pi a}$$

is increasing, so that (1) and (2), above, may also be written as:

(1) $$|f|(a) = \lim_{x \in \Pi a} \sum_n |f(x_n)| \qquad (a \in E^+).$$

Remark 2.3. Let $a, b \in E^+$, $c = a + b$. Obviously, if $(x_1, \ldots, x_N) \in \Pi a$ and $(y_1, \ldots, y_M) \in \Pi b$, then $(x_1, \ldots, x_N, y_1, \ldots, y_M) \in \Pi c$. On the other hand, if $(z_1, \ldots, z_K) \in \Pi c$, then for each $k \in \{1, \ldots, K\}$ one can, by the Riesz interpolation property, write z_k as a sum $x_k + y_k$ such that $(x_1, \ldots, x_K) \in \Pi a$, $(y_1, \ldots, y_K) \in \Pi b$ (and $(x_1, \ldots, x_K, y_1, \ldots, y_K)$ is a refinement of $(z_1, \ldots, z_K)$.)

3   Bilinear Functions of Order Bounded Variation

Definition & Remarks 3.1. Let E, F be Archimedean Riesz spaces. We say that a bilinear function $g : E \times F \to \mathbb{R}$ is *of order bounded variation* if for all $a \in E^+$ and $b \in F^+$ the set

$$\left\{ \sum_{n,m} |g(x_n, y_m)| \; : \; x \in \Pi a, \, y \in \Pi b \right\}$$

is bounded. The bilinear functions $E \times F \to \mathbb{R}$ that are of order bounded variation form a vector space $\mathrm{Bil}_{\mathrm{bv}}(E, F; \mathbb{R})$ containing all positive bilinear functions. We render $\mathrm{Bil}_{\mathrm{bv}}(E, F; \mathbb{R})$ an ordered vector space by taking the set of all positive bilinear functions as its positive cone, $\mathrm{Bil}_{\mathrm{bv}}(E, F; \mathbb{R})^+$.

Theorem 3.2. *Let E, F be Archimedean Riesz spaces. Then $\mathrm{Bil}_{\mathrm{bv}}(E, F; \mathbb{R})$ is a Dedekind complete Riesz space. For $g \in \mathrm{Bil}_{\mathrm{bv}}(E, F; \mathbb{R})$, $|g|$ is determined by*

(1) $$|g|(a, b) = \lim_{(x,y) \in \Pi a \times \Pi b} \sum_{n,m} |g(x_n, y_m)| \qquad (a \in E^+, b \in F^+).$$

The map $h \mapsto h \circ \otimes$ $(h \in (E \overline{\otimes} F)^\sim)$ is a Riesz isomorphism of $(E \overline{\otimes} F)^\sim$ onto $\mathrm{Bil}_{\mathrm{bv}}(E, F; \mathbb{R})$.

(Regarding the interpretation of (1), note that Πa and Πb, and thereby $\Pi a \times \Pi b$, are directed sets.)

Proof. If $h_0 \in (E \overline{\otimes} F)^{\sim +}$, then $h_0 \circ \otimes$ is a positive bilinear function on $E \times F$. Hence, $h \mapsto h \circ \otimes$ is a positive linear map of $(E \overline{\otimes} F)^\sim$ into $\mathrm{Bil}_{\mathrm{bv}}(E, F; \mathbb{R})$. If $h_1 \in (E \overline{\otimes} F)^\sim$ and $h_1 \circ \otimes = 0$, then h_1 vanishes on the vector space tensor product

$E \otimes F$ and, by 1.2, $h_1 = 0$. It follows that $h \mapsto h \circ \otimes$ is injective. Furthermore, it maps $(E \overline{\otimes} F)^{\sim +}$ *onto* $\mathrm{Bil}_{\mathrm{bv}}(E, F; \mathbb{R})^+$ because of 1.3.

So far, our inference is that the map $h \mapsto h \circ \otimes$ is a linear bijection and an order isomorphism of $(E \overline{\otimes} F)^{\sim}$ onto a linear subspace of $\mathrm{Bil}_{\mathrm{bv}}(E, F; \mathbb{R})$ that contains the positive cone.

Take $g \in \mathrm{Bil}_{\mathrm{bv}}(E, F; \mathbb{R})$. Define $\overline{g}_+ : E^+ \times F^+ \to [0, \infty)$ by

$$(2) \qquad \overline{g}_+(a, b) := \sup_{x \in \Pi a, y \in \Pi b} \sum_{n,m} |g(x_n, y_m)| \qquad (a \in E^+, b \in F^+).$$

A little exercise using 2.3 shows that $\overline{g}_+$ extends to a positive bilinear function $\overline{g} : E \times F \to \mathbb{R}$. As $\overline{g} - g$ is also bilinear and positive, we see that g is a difference of two elements of $\mathrm{Bil}_{\mathrm{bv}}(E, F; \mathbb{R})^+$.

Consequently, $h \mapsto h \circ \otimes$ maps $(E \overline{\otimes} F)^{\sim}$ onto $\mathrm{Bil}_{\mathrm{bv}}(E, F; \mathbb{R})$. Then $\mathrm{Bil}_{\mathrm{bv}}(E, F; \mathbb{R})$, as an ordered vector space, is isomorphic to $(E \overline{\otimes} F)^{\sim}$ and is a Dedekind complete Riesz space. (Corollary 8.3.5 in [7])

Take $g \in \mathrm{Bil}_{\mathrm{bv}}(E, F; \mathbb{R})$; it remains only to prove (1). Let $\overline{g}_+$ and $\overline{g}$ be as above. Clearly, $\overline{g} \geq g$ and $\overline{g} \geq -g$, so $\overline{g} \geq |g|$. On the other hand, if $g_0 \in \mathrm{Bil}_{\mathrm{bv}}(E, F; \mathbb{R})$, $g_0 \geq g$ and $g_0 \geq -g$, then for all $a \in E^+$, $b \in F^+$, $x \in \Pi a$, $y \in \Pi b$ we have

$$g_0(a, b) = \sum_{n,m} g_0(x_n, y_m) \geq \sum_{n,m} |g(x_n, y_m)|;$$

then $g_0(a, b) \geq \overline{g}_+(a, b) = \overline{g}(a, b)$ $(a \in E^+, b \in F^+)$, i.e., $g_0 \geq \overline{g}$. Thus, $\overline{g} = |g|$. Now (1) follows from (2) and the fact that the net

$$\left(\sum_{n,m} |g(x_n, y_m)| \right)_{(x,y) \in \Pi a \times \Pi b}$$

is increasing.   ∎

Remark 3.3. Let E, F be Archimedean Riesz spaces. By the above, for every $g \in \mathrm{Bil}_{\mathrm{bv}}(E, F; \mathbb{R})$ there exists a unique $g^{\otimes} \in (E \overline{\otimes} F)^{\sim}$ for which

$$g = g^{\otimes} \circ \otimes.$$

The map $g \mapsto g^{\otimes}$ is a Riesz isomorphism of $\mathrm{Bil}_{\mathrm{bv}}(E, F; \mathbb{R})$ onto $(E \overline{\otimes} F)^{\sim}$.

4   The Normed Riesz Space Case

Definition & Remarks 4.1. Let E, F be normed Riesz spaces. If $S : E \longrightarrow F$ is linear we define the *norm variation* of S to be

$$\mathrm{Var}\, S := \sup \left\{ \left\| \sum |Sx_n| \right\| : x_1, \ldots, x_N \in E^+, \left\| \sum x_n \right\| \leq 1 \right\}$$

and we say that S is *of norm bounded variation* if $\operatorname{Var} S$ is finite.

If $S : E \longrightarrow F$ is linear, then $\|S\| \leq \operatorname{Var} S$; equality holds as soon as S is positive. Moreover, for all $a \in E^+$:

$$(1) \qquad \left\| \sum |Sx_n| \right\| \leq \|a\| \operatorname{Var} S \qquad (x \in \Pi a).$$

The linear maps $E \longrightarrow F$ that are of norm bounded variation form a vector space containing all positive continuous linear maps.

Definition & Remarks 4.2. Let E, F, G be normed Riesz spaces. The *norm variation* of a bilinear $T : E \times F \longrightarrow G$ is

$$\operatorname{Var} T := \sup \left\| \sum_{n,m} |T(x_n, y_m)| \right\|,$$

over all $x_1, \ldots, x_N \in E^+$, with $\|\sum x_n\| \leq 1$, and $y_1, \ldots, y_M \in F^+$, with $\|\sum y_m\| \leq 1$. T is *of norm bounded variation* if $\operatorname{Var} T < \infty$.

If $T : E \times F \to G$ is bilinear, then $\|T\| \leq \operatorname{Var} T$; equality holds if T is positive. For all $a \in E^+$ and $b \in F^+$:

$$(1) \qquad \left\| \sum_{n,m} |T(x_n, y_m)| \right\| \leq \|a\| \, \|b\| \operatorname{Var} T \qquad (x \in \Pi a, \, y \in \Pi b).$$

Theorem 4.3. *Let E and F be normed Riesz spaces, G a Banach lattice; let $T : E \times F \longrightarrow G$ be bilinear and of norm bounded variation. Then there exists a unique continuous linear map $T^\otimes : E \overline{\otimes} F \to G$ with*

$$T(x, y) = T^\otimes(x \otimes y) \qquad (x \in E, y \in F).$$

$T^\otimes$ *is of norm bounded variation; in fact,* $\operatorname{Var} T = \operatorname{Var} T^\otimes$.

(Here $E \overline{\otimes} F$ is provided with the projective product norm; see Section 1.)

Proof. **(I)** The uniqueness is a simple consequence of 1.5.

(II) Take $f \in G'$. If $a \in E^+$ and $b \in F^+$, then for all $x \in \Pi a$ and $y \in \Pi b$ we have (with 4.2(1)):

$$\sum_{n,m} |f \, T(x_n, y_m)| \; \leq \; \sum_{n,m} |f| \, |T(x_n, y_m)| = |f| \Big(\sum_{n,m} |T(x_n, y_m)| \Big)$$

$$\leq \; \|f\| \, \Big\| \sum_{n,m} |T(x_n, y_m)| \Big\| \leq \|f\| \, \|a\| \, \|b\| \operatorname{Var} T.$$

It follows that $f \circ T \in \operatorname{Bil}_{\mathrm{bv}}(E, F; \mathbb{R})$ and (using 3.2(1)) that

$$|f \circ T|(a, b) \leq \|f\| \, \|a\| \, \|b\| \operatorname{Var} T \qquad (a \in E^+, \, b \in F^+).$$

By 3.3, $f \circ T$ and $|f \circ T|$ induce elements $(f \circ T)^{\otimes}$ and $|f \circ T|^{\otimes}$ of $(E \overline{\otimes} F)^{\sim}$, and $|(f \circ T)^{\otimes}| = |f \circ T|^{\otimes}$.

Take $u \in E \overline{\otimes} F$. For all $x_1, \ldots, x_K \in E^+$ and $y_1, \ldots, y_K \in F^+$ with $|u| \leq \sum x_k \otimes y_k$ we have

$$|(f \circ T)^{\otimes}(u)| \leq |(f \circ T)^{\otimes}|(|u|) = |f \circ T|^{\otimes}(|u|)$$

$$\leq \sum |f \circ T|^{\otimes}(x_k \otimes y_k) = \sum |f \circ T|(x_k, y_k)$$

$$\leq \sum \|f\| \|x_k\| \|y_k\| \operatorname{Var} T,$$

so that, by the definition of the norm on $E \overline{\otimes} F$,

$$|(f \circ T)^{\otimes}(u)| \leq \|f\| \|u\|_{|\pi|} \operatorname{Var} T.$$

Consequently, $(f \circ T)^{\otimes}$ is norm continuous and

$$\|(f \circ T)^{\otimes}\| \leq \|f\| \operatorname{Var} T.$$

(III) From the above we obtain a continuous linear map $f \mapsto (f \circ T)^{\otimes}$ of G' into $(E \overline{\otimes} F)'$, inducing a continuous linear $T^{\circ} : E \overline{\otimes} F \longrightarrow G''$, by

$$(T^{\circ} u)(f) := (f \circ T)^{\otimes}(u) \quad (u \in E \overline{\otimes} F, f \in G').$$

Observe that

$$\|T^{\circ}\| \leq \operatorname{Var} T.$$

If $x \in E$, $y \in F$, then for all $f \in G'$, $T^{\circ}(x \otimes y)(f) = f\big(T(x, y)\big)$, so T° maps the vector space tensor product $E \otimes F$ into the canonical image $\widehat{G}$ of G in G''. Then, by 1.5 and the completeness of G, T° maps all of $E \overline{\otimes} F$ into $\widehat{G}$ and we can define $T^{\otimes} : E \overline{\otimes} F \to G$ by

$$f(T^{\otimes} u) = (T^{\otimes} u)(f) \quad (f \in G', u \in E \overline{\otimes} F).$$

$T^{\otimes}$ is linear and continuous, and $T^{\otimes} \circ \otimes = T$. Note that

$$f \circ T^{\otimes} = (f \circ T)^{\otimes} \quad (f \in G').$$

(IV) It remains to prove that $\operatorname{Var} T^{\otimes} = \operatorname{Var} T$.

First, if $a \in E^+$, $b \in F^+$, then for all $x \in \Pi a$ and $y \in \Pi b$:

$$\left\| \sum_{n,m} |T(x_n, y_m)| \right\| = \left\| \sum_{n,m} |T^{\otimes}(x_n \otimes y_m)| \right\| \leq \|a \otimes b\|_{|\pi|} \operatorname{Var} T^{\otimes} \leq \|a\| \|b\| \operatorname{Var} T^{\otimes},$$

according to 4.1(1). It follows that $\operatorname{Var} T \leq \operatorname{Var} T^{\otimes}$.

For the reverse inequality, take $u_1, \ldots, u_K \in (E \otimes F)^+$, $\| \sum u_k \|_{|\pi|} \leq 1$; we prove that $\| \sum |T^\otimes u_k| \| \leq \mathrm{Var}\, T$. Take $f \in (G')^+$. In view of [8], Formula (2) on Page 87, it suffices to prove that

$$f\left(\sum |T^\otimes u_k|\right) \leq \|f\| \mathrm{Var}\, T.$$

Now using the Banach lattice isomorphism $G \longrightarrow \widehat{G}$ we see that

$$
\begin{aligned}
f\left(\sum |T^\otimes u_k|\right) &= \sum f(|T^\otimes u_k|) = \sum |T^\circ u_k|(f) \quad (2.2(1)) \\
&= \sum_k \lim_{g \in \Pi f} \sum_i |(T^\circ u_k)(g_i)| \\
&= \sum_k \lim_{g \in \Pi f} \sum_i |(g_i \circ T)^\otimes (u_k)| \\
&= \lim_{g \in \Pi f} \sum_i \sum_k |(g_i \circ T)^\otimes (u_k)| \\
&\leq \lim_{g \in \Pi f} \sum_i |(g_i \circ T)^\otimes| \left(\sum_k u_k\right) \quad (3.3) \\
&= \lim_{g \in \Pi f} \sum_i |g_i \circ T|^\otimes \left(\sum_k u_k\right) \\
&= \sup_{g \in \Pi f} \left(\sum_i |g_i \circ T|\right)^\otimes \left(\sum_k u_k\right).
\end{aligned}
$$

Thus, we are done if we can prove that

$$\left\| \left(\sum_i |g_i \circ T|\right)^\otimes \right\| \leq \|f\| \mathrm{Var}\, T \quad (g \in \Pi f)$$

i.e., because of 1.6,

(∗)
$$\left\| \sum_i |g_i \circ T| \right\| \leq \|f\| \mathrm{Var}\, T \quad (g \in \Pi f).$$

Take $a \in E^+$, $b \in F^+$. Applying 3.2(i) we obtain for every $g \in \Pi f$:

$$
\begin{aligned}
\sum_i |g_i \circ T|(a, b) &= \sum_i \lim_{(x,y) \in \Pi a \times \Pi b} \sum_{n,m} |(g_i \circ T)(x_n, y_n)| \\
&= \lim_{(x,y) \in \Pi a \times \Pi b} \sum_{n,m} \sum_i |g_i\left(T(x_n, y_m)\right)| \\
&\leq \sup_{(x,y) \in \Pi a \times \Pi b} \sum_{n,m} \sum_i g_i |T(x_n, y_m)| \\
&= \sup_{(x,y) \in \Pi a \times \Pi b} \sum_{n,m} f\left(|T(x_n, y_m)|\right)
\end{aligned}
$$

$$\leq \sup_{(x,y)\in\Pi a\times\Pi b} \|f\| \|\sum_{n,m} |T(x_n,y_m)|\|$$

$$\leq \|f\| \|a\| \|b\| \operatorname{Var} T.$$

$(*)$ follows. ∎

Remark 4.4. $T^\otimes$ is positive if and only if T is. (One implication is trivial, the other follows from 1.3.)

5   Bornological Riesz Spaces

The following is a straighforward adaptation to Riesz space theory of the basic notions of bornological vector spaces. Throughout, E is an Archimedean Riesz space.

Definition & Remarks 5.1. A *Riesz disk* in E is a nonempty subset A that is convex and solid (i.e., if $x \in A$, $y \in E$, $|y| \leq |x|$, then $y \in A$.) If A is a Riez disk in E we denote by E_A the linear hull of A. This E_A is a Riesz ideal in E containing A as an absolutely convex absorbent subset; (refer to [6] for an account of these terms.) A determines a gauge function $p_A : E_A \longrightarrow \mathbb{R}$:

$$p_A(x) := \inf\{\lambda \in [0,\infty) : x \in \lambda A\} \quad (x \in E_A).$$

p_A is a Riesz seminorm. This follows from combining [1], p. 132, with the solidity of A, though the reader is advised that our terminology for convex sets from [6] differs slightly from [1].

Definition & Remarks 5.2. A *Riesz bornology* on E is a collection $\mathcal{A}$ of subsets of E with the following properties.

(i) If $X \in \mathcal{A}$, $Y \subseteq X$, then $Y \in \mathcal{A}$.

(ii) If $X, Y \in \mathcal{A}$, then $X + Y \in \mathcal{A}$.

(iii) $\mathcal{A}$ covers E.

(iv) For every $X \in \mathcal{A}$ there is a Riesz disk A with $X \subseteq A \in \mathcal{A}$.

If $\mathcal{A}$ is such a Riesz bornology on E the pair $(E, \mathcal{A})$ (mostly just called E) is a *bornological Riesz space*; the sets belonging to $\mathcal{A}$ are $\mathcal{A}$-*bounded* or, simply, *bounded*.

The Riesz bornology (or: the bornological Riesz space) is called *separated* if no 1-dimensional linear subspace is bounded. This is the case if and only if for every bounded Riesz disk A the gauge p_A is a norm.

Definition & Remarks 5.3. A Riesz disk A in E is *completant* if p_A is a norm and renders E_A a Banach space (hence a Banach lattice.) A Riesz bornology (or: a bornological Riesz space) is said to be *(bornologically) complete* if every bounded set is contained in a completant bounded Riesz disk.

Definition & Remarks 5.4. The order bounded subsets of E form a separated Riesz bornology, the *order bornology*. It is complete if and only if E is uniformly complete. ([7], Section 4.2.)

Definition & Remarks 5.5. Suppose E is a normed Riesz space. The norm bounded sets form a separating Riesz bornology, the *von Neumann bornology*. It is complete if and only if E is norm complete. (3.5, Proposition 1, [5].)

Definition & Remarks 5.6. Let $\mathcal{A}$ be a separated Riesz bornology on E. If $a, x_1, x_2, \ldots \in E$ we write

$$x_n \to a \ (\mathcal{A})$$

if there exists an $\mathcal{A}$-bounded Riesz disk A such that $a - x_n \in E_A$ for sufficiently large n and $p_A(a - x_n) \to 0$. (In the situations of 5.4 and 5.5 this convergence is relatively uniform convergence and norm convergence, respectively.) The *bornological pseudoclosure* of a set $X \subseteq E$ is

$$\{ a \in E : \text{there exist } x_1, x_2, \ldots \in X \text{ with } x_n \to a \ (\mathcal{A}) \}.$$

Definition & Remarks 5.7. Let E and F be bornological Riesz spaces. A linear map $S : E \longrightarrow F$ is said to be *of bounded variation* (relative to the given bornologies) if for every bounded Riesz disk A in E the set

$$\left\{ \sum |S x_n| : x_1, \ldots, x_N \in E^+, \sum x_n \in A \right\}$$

is bounded.

If E, F, G are bornological Riesz spaces and $T : E \times F \longrightarrow G$ is bilinear, T is called *of bounded variation* if for all bounded Riesz disks $A \subseteq E$ and $B \subseteq F$ the set

$$\left\{ \sum |T(x_n, y_m)| : x_1, \ldots, x_N \in E^+, \sum x_n \in A; \ y_1, \ldots, y_M \in F^+, \sum y_m \in B \right\}$$

is bounded.

It is clear how these definitions tie in with those presented in 3.1, 4.1 and 4.2.

6   The Bornological Tensor Product

In this section, E and F are Archimedean Riesz spaces; $E\overline{\otimes}F$ is a Riesz tensor product of E and F.

Remark 6.1. Let $A \subseteq E$ and $B \subseteq F$ be Riesz disks. By $[A \otimes B]$ we indicate the smallest Riesz disk in $E\overline{\otimes}F$ that contains $\{ x \otimes y : x \in A, y \in B \}$.

It is not hard to see that an element u of $E\overline{\otimes}F$ lies in $[A \otimes B]$ if and only if

$$(1) \qquad\qquad |u| \leq \sum \mu_n a_n \otimes b_n,$$

for suitable $a_1, \ldots, a_N \in A \cap E^+$, $b_1, \ldots, b_N \in B \cap F^+$ and $\mu_1, \ldots, \mu_N \in [0, \infty)$ with $\sum \mu_n = 1$.

According to 1.4, the Riesz subspace of $E\overline{\otimes}F$ generated by $\{x \otimes y : x \in E_A, y \in F_B\}$ is a Riesz tensor product of E_A and F_B; we denote it $E_A\overline{\otimes}F_B$. By the above

$$(2) \qquad\qquad E_A\overline{\otimes}F_B \subseteq (E\overline{\otimes}F)_{[A\otimes B]};$$

in fact, $(E\overline{\otimes}F)_{[A\otimes B]}$ is the Riesz ideal of $E\overline{\otimes}F$ generated by $E_A\overline{\otimes}F_B$.

We have Riesz seminorms p_A, p_B and $p_{[A\otimes B]}$ on E_A, F_B and $(E\overline{\otimes}F)_{[A\otimes B]}$, respectively. They are connected by:

Lemma 6.2. Let $A \subseteq B$, $B \subseteq F$ be Riesz disks. Then for all $u \in (E\overline{\otimes}F)_{[A\otimes B]}$,

$$(1) \qquad\qquad p_{[A\otimes B]}(u) = \inf \sum p_A(x_n)p_B(y_n),$$

over all $x_1, \ldots, x_N \in E_A^+$ and $y_1, \ldots, y_N \in F_B^+$, with $|u| \leq \sum x_n \otimes y_n$.

Proof. First, suppose $x_1, \ldots, x_N \in E_A^+$, $y_1, \ldots, y_N \in F_B^+$, $|u| \leq \sum x_n \otimes y_n$. For all choices of real numbers $\alpha_n > p_A(x_n)$ and $\beta_n > p_B(y_n)$, $(n = 1, \ldots, N)$, we have

$$p_{[A\otimes B]}(u) \leq \sum p_{[A\otimes B]}(x_n \otimes y_n) = \sum \alpha_n \beta_n p_{[A\otimes B]}(\alpha_n^{-1}x_n \otimes \beta_n^{-1}y_n) \leq \sum \alpha_n \beta_n.$$

It follows that $p_{[A\otimes B]}(u)$ is at most equal to the right hand member of (1). On the other hand, if $\lambda > p_{[A\otimes B]}(u)$, then (see 6.1(1)) there exist $a_1, \ldots, a_N \in A \cap E^+$, $b_1, \ldots, b_N \in B \cap F^+$ and $\mu_1, \ldots, \mu_N \in [0, \infty)$ with $\sum \mu_n = 1$ and

$$|\lambda^{-1}u| \leq \sum \mu_n a_n \otimes b_n.$$

Then $|u| \leq \sum \lambda\mu_n a_n \otimes b_n$ and

$$\sum p_A(\lambda\mu_n a_n)p_B(b_n) = \lambda \sum \mu_n p_A(a_n)p_B(b_n) \leq \lambda \sum \mu_n = \lambda.$$

The remaining part of (1) follows.   ∎

Corollary 6.3. *Let $A \subseteq E$, $B \subseteq F$ be Riesz disks such that p_A and p_B are norms. Then the restriction of $p_{[A \otimes B]}$ to $E_A \overline{\otimes} F_B$ is precisely Fremlin's projective product norm.* (See Section 1)

Then, in particular, $p_{[A \otimes B]}$ is a norm on $E_A \overline{\otimes} F_B$. In order to see that $p_{[A \otimes B]}$ is a norm on $(E \overline{\otimes} F)_{[A \otimes B]}$ we need the following lemma, saying in effect that $E_A \overline{\otimes} F_B$ is order dense in $(E \overline{\otimes} F)_{[A \otimes B]}$.

Lemma 6.4. *Suppose E and F have strong units, e_E and e_F. Let $a \in E^+$, $b \in F^+$, $u \in (E \overline{\otimes} F)^+$, $u \leq a \otimes b$ and $\varepsilon \in (0, \infty)$. Then there exists a $v \in (E_a \overline{\otimes} F_b)^+$ with $0 \leq u - v \leq \varepsilon e_E \otimes e_F$.* (Here E_a and F_b are the principal ideals generated by a and b.)

Proof. By the Yosida Representation Theorem ([7], 45.4) we may assume the existence of compact Hausdorff spaces X and Y such that E and F are norm dense Riesz subspaces of $C(X)$ and $C(Y)$, respectively, and $e_E = \mathbf{1}_X$, $e_F = \mathbf{1}_Y$. Moreover, by [3], Corollary 4.4 we can realize $E \overline{\otimes} F$ as the Riesz subspace of $C(X \times Y)$ generated by the functions $f \otimes g$ ($f \in E$, $g \in F$), where

$$(f \otimes g)(x, y) := f(x)g(y) \qquad (x \in X, \ y \in Y).$$

(I) Let $(x_0, y_0) \in X \times Y$ and $0 < \delta < 1$; we first make a $w \in (E_a \overline{\otimes} F_b)^+$ with $w \leq u$, $w(x_0, y_0) \geq (1 - \delta)^3 u(x_0, y_0)$.

If $u(x_0, y_0) = 0$ we take $w := 0$. Now assume $u(x_0, y_0) > 0$; then $a(x_0) > 0$, $b(y_0) > 0$. Choose open sets $X' \subseteq X$ and $Y' \subseteq Y$ such that

$x_0 \in X'$, $y_0 \in Y'$;

$u(x, y) \geq (1 - \delta)u(x_0, y_0)$ for $x \in X'$, $y \in Y'$;

$2a(x) \geq a(x_0)$ if $x \in X'$; $2b(y) \geq b(y_0)$ if $y \in Y'$.

Next, choose $f \in C(X)^+$ and $g \in C(Y)^+$ with

$$f(x_0) = 1, \quad g(y_0) = 1, \quad f \leq \mathbf{1}_{X'}, \quad g \leq \mathbf{1}_{Y'}.$$

Finally, choose $f_1 \in E^+$ and $g_1 \in F^+$ with

$$f - \delta \mathbf{1}_X \leq f_1 \leq f \quad \text{and} \quad g - \delta \mathbf{1}_Y \leq g_1 \leq g,$$

and set

$$w := (1 - \delta)u(x_0, y_0)f_1 \otimes g_1.$$

We have $0 \leq f_1 \leq f \leq \mathbf{1}_{X'} \leq 2a(x_0)^{-1}a$, whence $f_1 \in E_a^+$. Similarly, $g_1 \in F_b^+$, so $w \in (E_a \overline{\otimes} F_b)^+$. Furthermore,

$$w \leq (1 - \delta)u(x_0, y_0)\mathbf{1}_{X' \times Y'} \leq u\mathbf{1}_{X' \times Y'} \leq u$$

and $w(x_0, y_0) = (1 - \delta)u(x_0, y_0)f_1(x_0)g_1(y_0) \geq (1 - \delta)^3 u(x_0, y_0)$.

(II) It follows that the pointwise supremum of the set

$$W := \{\, w \in (E_a \overline{\otimes} F_b)^+ \,:\, w \leq u \,\}$$

is u. Then for given $\varepsilon > 0$ we can choose $w_1, \ldots, w_M \in W$ such that

$$\bigcup_m \Big\{ (x, y) \in X \times Y \,:\, w_m(x, y) > u(x, y) - \varepsilon \Big\}$$

is $X \times Y$. Take $v := w_1 \vee \ldots \vee w_M$.   ∎

Remarks 6.5. The lemma has various consequences. Let $A \subseteq E$ and $B \subseteq F$ be Riesz disks.

(i) $E_A \overline{\otimes} F_B$ *is order dense in* $(E \overline{\otimes} F)_{[A \otimes B]}$.

Proof. Let $u \in (E \overline{\otimes} F)_{[A \otimes B]}$, $u > 0$; we show that there exists a $v \in E_A \overline{\otimes} F_B$ with $0 < v \leq u$. There are $a \in A$, $b \in B$ such that $a \geq 0$, $b \geq 0$, $u \leq a \otimes b$. As $u \in E \overline{\otimes} F$ there exist $a' \in E^+$, $b' \in F^+$ with $u \in E_{a'} \overline{\otimes} F_{b'}$ and $a \leq a'$, $b \leq b'$. As $u > 0$ there is an $\varepsilon > 0$ for which $u \not\leq \varepsilon a' \otimes b'$. Applying the lemma to the Riesz spaces $E_{a'}$ and $F_{b'}$ we obtain a $v \in (E_a \overline{\otimes} F_b)^+$ with $0 \leq u - v \leq \varepsilon a' \otimes b'$. Then $0 < v \leq u$.   ∎

(ii) *For every* $u \in (E \overline{\otimes} F)_{[A \otimes B]}$ *there exist* $c \in (E \overline{\otimes} F)^+$ *and* $v_1, v_2, \ldots \in E_A \overline{\otimes} F_B$ *such that* $|u - v_n| \leq n^{-1} c$ *($n \in \mathbb{N}$).*

Proof. We may assume $u \geq 0$. Take a', b' as above, $c := a' \otimes b'$ and apply the reasoning of the proof of (i).   ∎

(iii) *If the gauges* p_A *and* p_B *are norms, then so is* $p_{[A \times B]}$. This follows from (i) and the observation preceding Lemma 6.4.

Remark 6.6. Suppose $\mathcal{A}$ and $\mathcal{B}$ are Riesz bornologies on E and F, respectively. The sets $[A \otimes B]$ where $A \subseteq E$ and $B \subseteq F$ are bounded Riesz disks generate a Riesz bornology $\mathcal{A} \overline{\otimes} \mathcal{B}$ on $E \overline{\otimes} F$. A subset of $E \overline{\otimes} F$ is $\mathcal{A} \overline{\otimes} \mathcal{B}$-bounded if and only if it is contained in $[A \otimes B]$ for certain Riesz disks $A \in \mathcal{A}$, $B \in \mathcal{B}$.

From 6.5(iii) we see: if $\mathcal{A}$ and $\mathcal{B}$ are separated, so is $\mathcal{A} \overline{\otimes} \mathcal{B}$.

Two special cases: if $\mathcal{A}$ and $\mathcal{B}$ are the order bornologies of E and F, then $\mathcal{A} \overline{\otimes} \mathcal{B}$ is the order bornology of $E \overline{\otimes} F$; if E and F carry Riesz norms and $\mathcal{A}$ and $\mathcal{B}$ are the von Neumann bornologies, then $\mathcal{A} \overline{\otimes} \mathcal{B}$ is the von Neumann bornology induced by the projective-product norm on $E \overline{\otimes} F$.

7 The Main Result

Theorem 7.1. *Let E, F, G be separated bornological Riesz spaces. Assume that G is bornologically complete and that in G the bornological pseudoclosure of any bounded set is bounded. Let $T : E \times F \to G$ be a bilinear map that is of bounded variation. Then there exists a unique linear map $T^\otimes : E \overline{\otimes} F \to G$ that is of bounded variation and for which*

$$T(x, y) = T^\otimes(x \otimes y) \quad (x \in E, y \in F).$$

Proof. (I) As earlier, the uniqueness follows from 1.2.

(II) A "useful triple" is a triple (A, B, C) where A, B, C are bounded Riesz disks in E, F, G respectively, C is completant and

$$\left[x_1, \dots, x_N \in E^+, \ \sum x_n \in A, \ y_1, \dots, y_M \in F^+, \ \sum y_m \in B \right] \Rightarrow \sum |T(x_n, y_m)| \in C.$$

Whenever $A \subseteq E$ and $B \subseteq F$ are bounded Riesz disks, there exists a C such that (A, B, C) is a useful triple.

If (A, B, C) is a useful triple, the restriction of T to $E_A \times F_B$ is a bilinear map $T_{ABC} : E_A \times F_B \to G_C$ that is of bounded variation relative to the von Neumann bornologies. (Indeed, its variation is at most 1.) Then by Theorem 4.3 there is a unique norm continuous linear $T^\otimes_{ABC} : E_A \overline{\otimes} F_B \to G_C$ with

$$T(x, y) = T^\otimes_{ABC}(x \otimes y) \quad (x \in E_A, y \in F_B),$$

and (as $\operatorname{Var} T^\otimes_{ABC} = \operatorname{Var} T_{ABC} \le 1$):

$$v_1, \dots, v_K \in (E_A \overline{\otimes} F_B)^+, \ \sum v_k \in [A \otimes B] \implies \sum |T^\otimes_{ABC} v_k| \in C.$$

(III) Let (A, B, C) and (A', B', C') be useful triples. There is a useful triple (A'', B'', C'') with $A \cup A' \subseteq A''$, $B \cup B' \subseteq B''$, $C \cup C' \subseteq C''$. Then (A, B, C'') is a useful triple. By the uniqueness of $T^\otimes_{ABC''}$ and the norm continuity of the natural maps $G_C \to G_{C''}$ and $E_A \overline{\otimes} F_B \to E_{A''} \overline{\otimes} F_{B''}$ we have $T^\otimes_{ABC} = T^\otimes_{ABC''}$ on $E_A \overline{\otimes} F_B$ and $T^\otimes_{A''B''C''} = T^\otimes_{ABC''}$ on $E_A \overline{\otimes} F_B$, so that $T^\otimes_{ABC} = T^\otimes_{A''B''C''}$ on $E_A \overline{\otimes} F_B$. Similarly, $T^\otimes_{A'B'C'} = T^\otimes_{A''B''C''}$ on $E_{A'} \overline{\otimes} F_{B'}$. Hence,

$$T^\otimes_{ABC} = T^\otimes_{A'B'C'} \quad \text{on} \ (E_A \overline{\otimes} F_B) \cap (E_{A'} \overline{\otimes} F_{B'}).$$

Consequently, there exists a unique $S : E \overline{\otimes} F \to G$ such that for all useful triples (A, B, C)

$$S = T^\otimes_{ABC} \quad \text{on} \ E_A \overline{\otimes} F_B$$

and in particular

(1) $\qquad v_1, \dots, v_K \in (E_A \overline{\otimes} F_B)^+, \ \sum v_k \in [A \otimes B] \implies \sum |S v_k| \in C.$

It is clear that S is linear and $S \circ \otimes = T$. In view of 1.5 it only remains to show that S is of bounded variation.

(IV) Let $A \subseteq E$ and $B \subseteq F$ be bounded Riesz disks. Choose a bounded Riesz disk C in G such that (A, B, C) is useful and let $\overline{C}$ be the bornological pseudoclosure of C. We are done if we can prove:

(*) $u_1, \ldots, u_K \in (E \overline{\otimes} F)^+, \quad \sum u_k \in [A \otimes B] \implies \sum |Su_k| \in \overline{C}.$

Take $u_1, \ldots, u_K$ as in the premise of (*). There exist $a \in A$, $a \geq 0$ and $b \in B$, $b \geq 0$ with

(2) $$\sum u_k \leq a \otimes b.$$

Every u_k is a finite combination of elementary tensors $x \otimes y$ under the operations $+$, $\vee$, $\wedge$. Hence, there exist $a' \in E^+$ and $b' \in F^+$ such that

(3) $$u_1, \ldots, u_K \in E_{a'} \overline{\otimes} F_{b'}$$

where $E_{a'}$ and $F_{b'}$ are the principal Riesz ideals of E and F generated by a' and b'. Without loss of generality, let

(4) $$a \leq a', \quad b \leq b'.$$

Choose a useful triple (A', B', C') with $A \cup \{a'\} \subseteq A'$ and $B \cup \{b'\} \subseteq B'$.

In order to obtain the conclusion of (*) it suffices to show that $\sum |Su_k|$ is a $p_{C'}$-limit of a sequence in C. To this end we prove for every positive number ε:

(**)   There is a $w \in C$ with $p_{C'}\left(\sum |Su_k| - w\right) \leq \varepsilon,$

Take $\varepsilon > 0$. It follows from (3), (2) and Lemma 6.4 (applied to the Riesz spaces $E_{a'}$ and $F_{b'}$) that for each $k \in \{1, \ldots, K\}$ we can find a $v_k \in (E_a \overline{\otimes} F_b)^+$ with

$$0 \leq u_k - v_k \leq \frac{\varepsilon}{K}\left(a' \otimes b'\right).$$

Take $w := \sum |Sv_k|$.

As $v_k \in (E_A \overline{\otimes} F_B)^+$ for every k and $\sum v_k \leq \sum u_k \in [A \otimes B]$, from (1) we get $w \in C$. Furthermore, $u_k - v_k \in (E_{A'} \overline{\otimes} F_{B'})^+$ for each k, and $\sum (u_k - v_k) \leq \varepsilon a' \otimes b'$. Hence, again by (1), $\sum |S(u_k - v_k)| \in \varepsilon C'$. But

$$\left| \sum |Su_k| - w \right| = \left| \sum (|Su_k| - |Sv_k|) \right| \leq \sum |S(u_k - v_k)|.$$

Therefore, $\sum |Su_k| - w \in \varepsilon C'$ and $p_{C'}(\sum |Su_k| - w) \leq \varepsilon$.   ∎

Remark 7.2. In the situation of the previous theorem, $T^{\otimes}$ is positive if and only if T is. Indeed, if $T^{\otimes}$ is positive, then so is $T^{\otimes} \circ \otimes$, which is T. Conversely, if T is positive, then so is every $T^{\otimes}_{ABC}$ (4.4) and so is $T^{\otimes}$.

Remark 7.3. The condition, mentioned in the theorem, that in G all bounded sets have bounded bornological pseudoclosures, is often satisfied. It is, e.g., in the two cases covered by Fremlin's theorems, viz. if G is a uniformly complete Riesz space with the order bornology or a Banach lattice with the von Neumann bornology.

References

[1] C. D. Aliprantis and O. Burkinshaw, *Positive Operators*. (1985) Acad. Press, New York-London.

[2] G. Buskes and A. van Rooij, *The tensor product of two bornological Riesz spaces*. In these Proceedings, 3-9.

[3] D. H. Fremlin, *Tensor products of Archimedean vector lattices*. Amer. J. Math. **94** (1972), 777-798.

[4] D. H. Fremlin, *Tensor products of Banach lattices*. Math. Ann. **211** (1974), 87-106.

[5] H. Hogbe-Nlend, *Bornologies and Functional Analysis*. Math. Studies **26** (1977), North Holland, Amsterdam-New York-Oxford.

[6] G. Köthe, *Topological Vector Spaces I*. Springer Verlag (1969), Berlin-Heidelberg-New York.

[7] W. A. J. Luxemburg and A. C. Zaanen, *Riesz Spaces, I*. (1971) North-Holland, Amsterdam-New York-Oxford.

[8] A. C. Zaanen, *Riesz Spaces, II*. (1983) North-Holland, Amsterdam-New York-Oxford.

Department of Mathematics, University of Mississippi, University, MI 38677, USA
mmbuskes@hilbert.math.olemiss.edu

Catholic University, Department of Mathematics, Toernooiveld, 6525 ED Nijmegen, the Netherlands

Non-distributive Cancellative Residuated Lattices

James A. Cole

ABSTRACT.   Cancellative residuated lattices are a natural generalization of lattice-ordered groups (ℓ-groups). In studying this variety, several questions have occurred about residuated lattice orders on free monoids and commutative free monoids. One of these questions is whether every residuated lattice order on a (commutative) free monoid is distributive, a fact known about ℓ-groups. We will construct two examples that shows that this is not necessarily the case.

1   Introduction

A *residuated lattice-ordered monoid*, or *residuated lattice* for short, is an algebra $\mathbf{L} = \langle L, \wedge, \vee, \cdot, e, \backslash, / \rangle$ such that $\langle L, \wedge, \vee \rangle$ is a lattice, $\langle L, \cdot, e \rangle$ is a monoid, and multiplication is both left and right residuated, with $\backslash$ and $/$ as residuals, i.e., for all $a, b, c \in L$

$$a \cdot b \leq c \quad \Leftrightarrow \quad a \leq c/b \quad \Leftrightarrow \quad b \leq a \backslash c.$$

In [2], the class of residuated lattices is shown to be a variety. A more general discussion of this variety can be found in [3].

There are three subvarieties of residuated lattices that we will be concerned about in this paper. The variety of commutative residuated lattices, denoted $\mathcal{CRL}$, is the collection of residuated lattices which have a commutative monoid reduct. The variety of distributive residuated lattices, $\mathcal{DRL}$, is the collection of residuated lattices which have a distributive lattice reduct. Finally, the variety of cancellative residuated lattices, $\mathcal{CanRL}$, is the collection of residuated lattices whose monoid reduct is cancellative.

While $\mathcal{CRL}$ and $\mathcal{DRL}$ are clearly varieties, it is not as clear that $\mathcal{CanRL}$ is a variety, since cancellativity is a quasiequation in monoids. However, the following shows that cancellativity is equationally defined in residuated lattices.

Lemma 1.1. *A residuated lattice is right cancellative as a monoid if and only if the identity $xy/y \approx x$ holds.*

J. Martínez (ed.), Ordered Algebraic Structures, 205–212.
© 2002 *Kluwer Academic Publishers*.

Proof. The identity $(xy/y)y = xy$ holds in any residuated lattice since $xy/y \leq xy/y$ implies $(xy/y)y \leq xy$, and $xy \leq xy$ implies $x \leq xy/y$, hence $xy \leq (xy/y)y$. By right cancellativity, we have $xy/y = x$.

Conversely, suppose $xy/y = x$ holds, and consider elements a, b, c such that $ac = bc$. Then $a = ac/c = bc/c = b$, so right cancellativity is satisfied.   ∎

Hence a residuated lattice is cancellative iff it satisfies $xy/y \approx x$ and $y \backslash yx \approx x$.

The most widely studied subvariety of cancellative residuated lattices are lattice-ordered groups, or ℓ-groups. One interesting property of lattice-ordered groups is that their lattice reducts are distributive (see, for example, [1]). This leads to the question of whether this holds for any residuated lattice ordering of a cancellative monoid. More specifically, is any residuated lattice ordering of a free monoid distributive? Of a commutative free monoid?

2   A non-distributive order on a free monoid

We will now construct an example of a non-distributive residuated lattice order on the free monoid X^* generated by $X = \{a, b, c\}$.

For $\alpha \in X^*$ and $x \in X$, let $|\alpha|_x$ be the number of times x appears in α. For example, $|aabcb|_a = 2, |aabcb|_b = 2$, and $|aabcb|_c = 1$. For each triplet (l, m, n) of non-negative integers, let

$$B(l, m, n) = \{\alpha \in X^* \,||\alpha|_a = l, |\alpha|_b = m, |\alpha|_c = n\}.$$

We will order elements of $B(l, m, n)$ in the following way: $\alpha \leq_p \beta$ if in $B(l, m, n)$ there is a sequence of words $\alpha = \alpha_0 \leq \alpha_1 \leq \cdots \leq \alpha_p = \beta$, where $\alpha_i = \alpha_i' x_1 x_2 \alpha_i''$ and $\alpha_{i+1} = \alpha_i' x_1 x_2 \alpha_i''$, where $x_1 = c$, $x_2 = a$; $x_1 = b$, $x_2 = a$; or $x_1 = c$, $x_2 = b$. We will show that $\leq_p$ is a lattice order on each $B(l, m, n)$.

Lemma 2.1. Let $\alpha, \beta \in B(l, m, n)$ such that $\alpha \leq_p \beta$. Let α', β' be prefixes of the same length for α and β, respectively. Then we have the following:

(i) $|\alpha'|_c \geq |\beta'|_c$

(ii) $|\alpha'|_a \leq |\beta'|_a$

Proof. Notice that in order to move from a smaller word to a larger word, we cannot place c's into a prefix nor remove a's from a prefix.   ∎

Lemma 2.2. Let $\alpha\beta \in B(l, m, n)$ such that $\alpha \leq_p \beta$. Let α', β' be prefixes of α and β such that $|\alpha|_b = |\beta|_b$ and $\alpha = \alpha'b\alpha''$ and $\beta = \beta'b\beta''$. Then we have the following:

(i) $|\alpha'|_a \leq |\beta'|_a$

(ii) $|\beta'|_c \leq |\alpha'|_c$

Proof. For (i), if $|\alpha'|_a > |\beta'|_a$, then $|\alpha'|_c < |\beta'|_c$, which contradicts 2.1(i) above. A similar argument can be made for (ii) using 2.1(ii) above. ∎

Lemma 2.3. *Let $\alpha \in B(l, m, n)$.*

(i) *Let $\alpha = \alpha' a \alpha''$ such that $|\alpha'|_a = 0$. Then*

$$\min\{\beta \in aB(l-1, m, n) | \beta \geq_p \alpha\} = a\alpha'\alpha''$$

(ii) *Let $\alpha = \alpha' c \alpha''$ such that $|\alpha'|_c = 0$. Then*

$$\max\{\beta \in cB(l, m, n-1) | \beta \leq_p \alpha\} = c\alpha'\alpha''$$

(iii) *If there is a $\beta \in bB(l, m-1, n)$ such that $\beta \geq_p \alpha$, then $\alpha = c^n b \alpha''$ and*

$$\min\{\beta \in bB(l, m-1, n) | \beta \geq_p \alpha\} = bc^n \alpha''$$

(iv) *If there is a $\beta \in bB(l, m-1, n)$ such that $\beta \leq_p \alpha$, then $\alpha = a^n b \alpha''$ and*

$$\max\{\beta \in bB(l, m-1, n) | \beta \leq_p \alpha\} = ba^n \alpha''$$

Proof. The first two statements are applications of the statements in the first lemma, and the second two statements are applications of the second lemma. Since the arguments are similar for all four statements, we will only prove the first one.

For this proof, we will also place an ordering on the generators a, b, and c by $a > b > c$. (This ordering will not necessarily hold later on.) It should be clear that $a\alpha'\alpha'' \geq_p \alpha$. Let $\beta \in aB(l-1, m, n)$ such that $\beta \geq_p \alpha$. We can write $\beta = x_1 x_2 \cdots x_k$, where each $x_i \in \{a, b, c\}$. We have that $x_1 = a$. Let $\alpha_1 = \alpha'\alpha''$. Now, we want to write $\alpha_1 = \alpha_1' x_2 \alpha_1''$ such that $|\alpha_1'|_q = 0$ for $q \geq x_2$. If $x_2 = a$, then we can always find such an α_1'. If $x_2 = b$ and the prefix before the first b in α_1 contains an a, then $\alpha \not\leq_p \beta$. So, the condition holds for $x_2 = b$. Finally, assume that $x_2 = c$. Since the first two letters of β contain a c, and $\alpha \leq_p \beta$, we have that one of the first two letters of α is c. If the first letter of α is c, then we are done ($\alpha_1 = c\alpha_1''$). If the first letter of α is not c, then the second letter must be. The first letter, then is either a or b. The second letter cannot be b, since that would imply that no word starting with ac is above α. In particular, it would contradict $\alpha \leq_p \beta$. Thus, the second letter must be a. Therefore, we can write $\alpha_1 = \alpha_1' x_2 \alpha_1''$.

Continuing, we have that $\alpha \leq_p a\alpha_1 = a\alpha_1' x_2 \alpha_1'' \leq_p ax_2\alpha_1'\alpha_1'' = ax_2\alpha_2$. Let $\alpha_{i+1} = \alpha_i'\alpha_i''$. Then using the arguments above, we can show that $\alpha_{i+1} = \alpha_{i+1}' x_{i+2} \alpha_{i+1}''$, with $|\alpha_{i+1}|_q = 0$ for $q \geq x_{i+1}$. From here we have that

$$\begin{aligned}
a\alpha_1 &= x_1\alpha_1 = x_1\alpha_1' x_2\alpha_1'' \leq x_1 x_2\alpha_2 = x_1 x_2\alpha_2' x_3\alpha_2'' \\
&\leq x_1 x_2 x_3\alpha_3 = \cdots = x_1 x_2 x_3 \cdots x_k = \beta.
\end{aligned}$$

Thus, we have that $\min\{\beta \in aB(l-1,m,n)|\beta \geq_p \alpha\} = a\alpha'\alpha''$. ■

Before we continue, we want to make a few observations about multiplication. For any $\gamma \in X^*$, it should be clear that the canonical bijection from

$$\gamma B(l,m,n) = \{\gamma\alpha|\alpha \in B(l,m,n)\}$$

to $B(l,m,n)$ is a order isomorphism. (Note that

$$\gamma B(l,m,n) \subseteq B(l+|\gamma|_a, m+|\gamma|_b, n+|\gamma|_c).$$

So, $\gamma\alpha \leq_p \gamma\beta \Leftrightarrow \alpha \leq_p \beta$. Further, if $\alpha, \beta \in B(l,m,n)$ and $B(l,m,n)$ is a lattice under $\leq_p$, then $\gamma(\alpha \vee \beta)$ is the smallest word above both $\gamma\alpha$ and $\gamma\beta$ with a prefix of γ.

Lemma 2.4. $B(l,m,n)$ *is a lattice under* $\leq_p$ *for every triplet* (l,m,n).

Proof. We will prove this by induction. Clearly $B(0,0,0) = \{e\}$ is a lattice under $\leq_p$. Let (l,m,n) be a triplet. Assume that if $r \leq l, s \leq m, t \leq n$ and $r+s+t < l+m+n$, then $B(r,s,t)$ is a lattice under $\leq_p$. Let $\alpha, \beta \in B(l,m,n)$. We will show that $\alpha \vee \beta$ exists in $B(l,m,n)$. Since each $B(l,m,n)$ is finite, we will be done.

Case 1: Assume $\alpha = a\alpha_1$. We can write $\beta = \beta'a\beta''$. By Lemma 2.3(i), the least word above β that begins with an a is $a\beta'\beta'' = a\beta_1$. The least word that begins with an a above both α and β then is $a(\alpha_1 \vee \beta_1)$. If $\alpha\beta \leq_p \delta$, then $\delta = a\delta_1$, since any word above α must also start with an a. Since $\alpha, \beta \leq_p \delta$, we have that $\alpha_1, \beta_1 \leq_p \delta_1$ in $B(l,m,n)$. So, $\alpha_1 \vee \beta_1 \leq_p \delta_1$ in $B(l,m,n)$, which implies that $a(\alpha_1 \vee \beta_1) \leq_p \delta$. Thus, $\alpha \vee \beta$ exists.

Case 2: Assume $\alpha = b\alpha_1$. We only need to consider the cases when $\beta = b\beta_1$ and $\beta = c\beta_1$. Assume that $\beta = b\beta_1$. Consider $b(\alpha_1 \vee \beta_1)$. If $\alpha, \beta \leq_p \delta$, then either $\delta = a\delta_1$ or $\delta = b\delta_1$. If $\delta = b\delta_1$, then it should be clear that $\delta \leq_p b(\alpha_1 \vee \beta_1)$. Assume $\delta = a\delta_1$. Since $b\alpha_1 \leq_p \delta$, we have that $\delta = a^n b\delta_2$, and, by Lemma 2.3(iv), $ba^n\delta_2$ is the greatest element below δ that begins with a b. So, $\alpha, \beta \leq_p ba^n\delta_2$, which implies that $b(\alpha_1 \vee \beta_1) \leq_p ba^n\delta_2 \leq_p \delta$. Thus, $\alpha \vee \beta$ exists in this case.

Now, assume that β starts with c. Then, either $\beta = c^r a\beta_1$ or $\beta = c^r b\beta_1$. If $\beta = c^r b\beta$, then the smallest word above β that starts with b is $bc^r\beta$, by Lemma 2.3(iii). Using similar arguments from above, we can show that $\alpha \vee \beta = b(\alpha_1 \vee c^r\beta)$. Assume that $\beta = c^r a\beta$. Then, there is not a word above β that starts with b, and the smallest word above β that starts with a is $ac^r\beta$. We can also rewrite $\alpha = \alpha_2 a\alpha_3, |\alpha_2|_a = 0$, and we have that $a\alpha_2\alpha_3$ is the smallest word above α that starts with a. Again, using arguments similar to those above, we have that $\alpha \vee \beta = a(\alpha_2\alpha_3 \vee c^r\beta_1)$.

Case 3: Now assume that $\alpha = c\alpha_1$. All we need to consider is $\beta = c\beta_1$. If $\alpha, \beta \leq_p \delta$, then $\delta = \delta_1 c\delta_2$, where $|\delta_1|_c = 0$, and, by Lemma 2.3(ii), the largest word below δ that starts with c is $c\delta_1\delta_2$. Since $\alpha, \beta \leq_p \delta, \alpha, \beta \leq_p c\delta_1\delta_2$. So, $\alpha_1, \beta_1 \leq_p \delta_1\delta_2$ and $\alpha_1 \vee \beta_1 \leq_p \delta_1\delta_2$. Thus, $c(\alpha_1 \vee \beta_1) \leq_p c\delta_1\delta_2 \leq_p \delta$ and $\alpha \vee \beta = c(\alpha_1 \vee \beta_1)$. ■

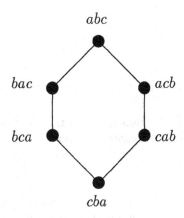

Figure 1: The ordering of $B(1,1,1)$ under $\leq_p$

Example 2.5. Let $X = \{a, b, c\}$. Let $f : X^* \to \mathbb{R}$ defined by $f(a) = -1, f(b) = -\sqrt{2}, f(c) = -\sqrt{3}$, where the operation on $\mathbb{R}$ is addition. Define an ordering $\leq$ by $\alpha \leq \beta$ is $f(\alpha) \leq f(\beta)$; or $f(\alpha) = f(\beta)$ and $\alpha \leq_p \beta$. Note that $f(\alpha) = f(\beta$ if and only if $\alpha, \beta \in B(l, m, n)$ for some l, m, n. If $f(\alpha) \neq f(\beta)$, then either $\alpha \leq \beta$ or $\beta \leq \alpha$. In either case $\alpha \vee \beta$ and $\alpha \wedge \beta$ exist. If $f(\alpha) = f(\beta)$, then it should be clear that $\alpha \vee \beta$ and $\alpha \wedge \beta$ are the same as under $\leq_p$. Thus, this ordering forms a lattice.

We now want to show that this lattice is residuated by showing that $\alpha\backslash\beta$ and β/α exist. If $\alpha \leq \beta$, then we have that $\alpha\backslash\beta = \beta/\alpha = e$. Assume that $\alpha \not\leq \beta$. The argument for $\alpha\backslash\beta$ and β/α are similar, and so we will only show that β/α exists. Let $r = f(\alpha) - f(\beta)$. If $\gamma\alpha \leq \beta$, then $f(\gamma\alpha) = f(\gamma)f(\alpha) \leq f(\beta)$, or $f(\gamma) \leq r$. Also note that if $f(\gamma) < r$, then $\gamma\alpha \leq \beta$. Either there exists $\gamma \in f^{-1}(r)$ such that $\gamma\alpha \leq \beta$, or there does not. If there is such a $\gamma \in f^{-1}(r)$ then it should be clear that $\vee f^{-1}(r) = \beta/\alpha$.

Assume no such $\gamma \in f^{-1}(r)$ exists. Let $t = \max\{s \in f(X^*)|s < r\}$. Note that t exists and $t \neq r$. So, $\gamma\alpha \leq \beta$ if and only if $f(\gamma) < t$. That means that $\gamma \leq \vee f^{-1}(t) = \beta/\alpha$. Thus, β/α exists. A similar argument shows that $\alpha\backslash\beta$ exists.

To show that this forms a non-distributive lattice order, consider $B(1,1,1)$ (shown in Figure 1). We have that

$$(bca \vee cab) \wedge bac = abc \wedge bac = bac \quad \text{and} \quad (bca \wedge bac) \vee (cab \wedge bac) = bca \vee cba = bca.$$

Thus, this gives an example of a non-distributive lattice order on the free monoid generated by 3 elements.

The methods and ideas used in this section can be extended to produce examples on a larger number of variables. Also, the function f that is chosen in the above example is rather arbitrary and only serves to place a residuated total order on the free commuative monoid generated by $\{a, b, c\}$, which could be replaced by any other residuated total order to produce different examples.

3 Non-distributive order on a commutative free monoid

The example in the previous section relied on the fact that the monoid was not commutative. However, we can also construct an example to show that commutivity and cancellativity do not imply distributivity. A complete explanation of this example can be found in [4].

Example 3.1. Let $\mathbf{F}$ be the free commuative monoid generated by $\{a, b, c\}$. We can define an order on $\mathbf{F}$ by $\alpha \leq \beta$ if and only if $|\alpha| > |\beta|$ or $|\alpha| = |\beta|$ and $|\alpha|_b \geq |\beta|_b$ and $|\alpha|_c \geq |\beta|_c$. Let $\mathcal{W}_r = \{\alpha \in \mathcal{F} : |\alpha| = r\}$. It can be shown that each $\mathcal{W}_r$ is a finite join-semilattice, and hence every join in $\mathbf{F}$ can be found by a finite subjoin. To show that this order is residuated, it suffices to show that the multiplication distributes over the join, which is relatively straightforward. We get cancellativity for free, since the underlying monoid is free. To show that this order is not distributive, we can see that $bb \vee (ab \wedge cc) = bb$ and $(bb \vee ab) \wedge (bb \vee cc) = ab$.

The above example, while not distributive, is join-semidistributive. A lattice is *join-semidistributive* if whenever $\alpha \vee \beta = \alpha \vee \gamma = \delta$, then $\alpha \vee (\beta \wedge \gamma) = \delta$.

Example 3.2. Let $\mathbf{F} = \mathbf{F}_{CM}$ be the commutative free monoid generated by $\{a, b, c, d\}$. Order $\mathbf{F}$ by $\alpha \leq \beta$ if $|\alpha| > |\beta|$ or $|\alpha| = |\beta|$ and $|\alpha|_x \geq |\beta|_x$ for $x \in \{b, c, d\}$. Using the same arguments as above, we have that $\mathbf{F}$ is residuated under this order, and that this is a non-distributive lattice ordering. However, note that now $b \vee c = b \vee d = a$ while $b \vee (c \wedge d) = b \vee (aa) = b$. So, this ordering is not join-semidistributive either.

4 Conclusion

Cancellative residuated lattices are one of the more obvious generalizations of ℓ-groups. However, there are many differences between these two varieties. As we can see from above, cancellative residuated lattices do not have to be distributive. In fact, one result in [3] shows that any lattice can be embedded into a cancellative residuated lattice. We are currently investigating whether the same can be said for commutative cancellative residuated lattices. That is, are any lattice equations forced on the variety of commutative cancellative residuated lattices? Alternatively, what is the variety generated by the lattice reducts of commutative cancellative residuated lattices?

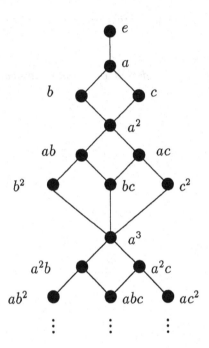

Figure 2: A non-distributive order on **F**

References

[1] M. Anderson and T. Feil, *Lattice-Ordered Groups: an introduction.*(1988) D. Reidel Publishing Company.

[2] K. Blount and C. Tsinakis, *The structure of Residuated Lattices.* Preprint.

[3] P. Jipsen and C. Tsinakis, *A Survey of Residuated Lattices.* (2002) In these Proceedings, 19-56.

[4] P. Bahls, J. Cole, P. Jipsen, N. Galatos, and C. Tsinakis, *Cancellative Residuated Lattices.* Preprint.

Department of Mathematics, Vanderbilt University, Nashville, TN 37250, USA
coleja@math.vanderbilt.edu

Reconstructing $\mathcal{C}(G)$ from a Plenary Subset

Michael R. Darnel

To Paul Conrad,
in honor of his 80th birthday.

ABSTRACT.   The well known method of using ultrafilters on an index set to identify all minimal prime subgroups of a cardinal product of o-groups is generalized to a method of using prime ideal filters on the special values of a special-valued ℓ-group to identify prime subgroups. Among those that can be identified this way are all minimal prime subgroups, all closed prime subgroups, and all branch prime subgroups. The method is then used to give necessary and sufficient conditions for a value to be such a prime subgroup.

With the exception of a few representations of representable ℓ-groups based upon polars, all other representations of ℓ-groups depend upon prime subgroups or, to go a step further, upon plenary subsets of values. Naturally enough, plenary subsets of values also suffice to determine the order of an ℓ-group and are used to verify the satisfaction of identities. Specifically, for an ℓ-group G and plenary subset Δ of values, $g \geq h$ if and only if $Vg \geq Vh$ for every $V \in \Delta$; G satisfies an identity "$w(\vec{x}) = e$" if and only if for every $V \in \Delta$, $V\vec{g} = V$ for every substitution $\vec{x} \to \vec{g}$ in G.

Let us first remind the reader of some concepts and definitions of lattice-ordered groups, all of which can be found in [D95]. $\Gamma(G)$ will denote the set of all values of G. A *plenary subset* Δ of values is a subset of $\Gamma(G)$ such that if $V_1 \subseteq V_2$ in $\Gamma(G)$ and $V_1 \in \Delta$, then $V_2 \in \Delta$ and such that $\cap \{V \in \Delta\} = \{e\}$. $\Gamma(G)$ is contained in $\mathrm{Spec}(G)$, the set of *all* prime subgroups of G, including G itself, which in turn is contained in $\mathcal{C}(G)$, the set of all convex ℓ-subgroups of G. Under inclusion, $\mathcal{C}(G)$ is a complete *Brouwerian lattice*: a distributive lattice such that $A \wedge \bigvee_{\lambda \in \Lambda} B_\lambda = \bigvee_{\lambda \in \Lambda} (A \wedge B_\lambda)$. On the other hand, under inclusion, $\mathrm{Spec}(G)$ and $\Gamma(G)$ are root systems: no two incomparable elements have a common lower bound. Any plenary subset Δ, being a dual ideal of $\Gamma(G)$ whose intersection is the group identity, is then necessarily a root system.

Now both $\mathrm{Spec}(G)$ and $\Gamma(G)$ are determined within $\mathcal{C}(G)$. Recall that within a lattice, an element d is *meet-irreducible* if $d = \wedge_{\lambda \in \Lambda} a_\lambda$ implies that $d = a_\lambda$ for some $\lambda \in \Lambda$, while d is *finitely meet-irreducible* if $d = \wedge_{i=1}^{n} a_i$ implies $d = a_i$ for some

213

J. Martínez (ed.), Ordered Algebraic Structures, 213–230.
© 2002 *Kluwer Academic Publishers.*

$1 \leq i \leq n$. $\mathrm{Spec}(G)$ is then the set of finitely meet-irreducible elements of $\mathcal{C}(G)$ and $\Gamma(G)$ is the set of meet-irreducible elements of $\mathcal{C}(G)$. Conversely, $\mathcal{C}(G)$ can be constructed from $\Gamma(G)$ as for any $C \in \mathcal{C}(G)$, C is the intersection of all values V that contain C, and so $\mathcal{C}(G)$ can be formed from $\Gamma(G)$ by taking intersections of dual ideals (though different dual ideals may produce the same convex ℓ-subgroup).

Thus any one of $\mathcal{C}(G)$, $\mathrm{Spec}(G)$, or $\Gamma(G)$ can be used to produce the other two.

However, in general, $\mathcal{C}(G)$ cannot be reconstructed in full from an arbitrary plenary subset Δ of $\Gamma(G)$. The difficulty is best shown by an example.

Let G be $\prod_{i=1}^{\infty} \mathbb{Z}$, the ℓ-group of all integer sequences, with pointwise order and addition. The (unique) minimal plenary subset of $\Gamma(G)$ is $\Delta = \{M_i : 1 \leq i < \infty\}$, where $M_i = \{g \in G : g(i) = 0\}$. Each M_i is a minimal prime subgroup. G, though, has other minimal prime subgroups, which are readily constructed from Δ by using ultrafilters on Δ. If $\mathcal{U}$ is an ultrafilter on Δ, let $M_{\mathcal{U}} = \{g \in G : Z(g) \in \mathcal{U}\}$, where $Z(g) = \{i : g(i) = 0\}$. Then $M_{\mathcal{U}}$ is a minimal prime subgroup of G and the mapping $\mathcal{U} \leftrightarrow M_{\mathcal{U}}$ is a bijection between the ultrafilters on Δ and the minimal prime space $\mathrm{Min}(G)$ consisting of all minimal prime subgroups of G and with the topology induced by taking as a basis all sets $M(g) = \{M \in \mathrm{Min}(G) : g \notin M\}$. Thus $\mathrm{Min}(G) \cong \beta \mathbb{N}$. Now Γ is stranded (every prime subgroup contains a unique minimal prime subgroup), and it is known that if $\mathcal{U}$ is a nonprincipal ultrafilter on Δ, then the set of values in $\Gamma(G)$ properly containing $M_{\mathcal{U}}$ is an η_1-set [P61]. So without introducing specific elements of G, there is no way to separately construct different values in such η_1-sets.

The difficulty in reconstructing $\mathcal{C}(G)$ from a plenary subset Δ is obvious once we recall that two ℓ-groups may have isomorphic plenary subsets but radically different lattices of convex ℓ-subgroups. For example, let $G = \prod_{i=1}^{\infty} \mathbb{Z}$ and $H = \sum_{i=1}^{\infty} \mathbb{Z}$. As before for G, for $1 \leq i < \infty$, let $M_i = \{g \in G : g(i) = 0\}$, and for H, let $H_i = \{h \in H : h(i) = 0\}$. Let $\Delta_G = \{M_i : 1 \leq i < \infty\}$ and $\Delta_H = \{H_i : 1 \leq i < \infty\}$. Then Δ_G and Δ_H are, respectively, the unique minimal plenary subsets of $\Gamma(G)$ and $\Gamma(H)$, and are clearly isomorphic to one another via $M_i \leftrightarrow H_i$. Now $\Gamma(H) = \Delta_H$ [C67], and so $|\Gamma(H)| = \omega$, while $|\Gamma(G)| = 2^{2^{\omega}}$.

In the more general setting, an ℓ-group G may have *branch primes*: P is a branch prime subgroup if P is a proper prime subgroup and is the join of a set of at least two minimal prime subgroups. Closely linked to the idea of branch prime subgroups is that of the *lex kernel* of an ℓ-group. Recall [D95] that the lex kernel of an ℓ-group G, written $\mathrm{lex}(G)$, is the join of all minimal prime subgroups of G. For a convex ℓ-subgroup C of G, $\mathrm{lex}(C) = \bigvee \{C \cap M : M \in \mathrm{Min}(G) \text{ and } C \not\subseteq M\}$. For a proper prime subgroup P, $\mathrm{lex}(P)$ is then clearly the join of all minimal prime subgroups contained in P. Note then that $P = \mathrm{lex}(P)$ implies either P is a minimal prime subgroup or P is a branch prime subgroup. A useful characterization of $\mathrm{lex}(G)$ is the following: $e < g \in G$ is a *nonunit* [D95] if there exists $e < h \in G$ such that $g \wedge h = e$. If N is the set of nonunits of G, then $\mathrm{lex}(G)$ is merely the subgroup of G generated by N.

So for an arbitrary ℓ-group G, what prime convex ℓ-subgroups should we expect to be able to construct from a plenary subset Δ of $\Gamma(G)$? The following seem to be

modest goals for any process.
  i)   All values in Δ;
 ii)   Any prime subgroup which is the intersection or union of chains in Δ;
iii)   Any minimal prime subgroup;
 iv)   Branch prime subgroups.

We now outline a process which meets at least these goals. We begin by defining filters and ultrafilters of ideals on a partially ordered set, and giving five basic propositions about such filters and ultrafilters. Though nothing profound, the proofs of most are included. Recall that an *ideal* of a partially ordered set is any subset I such that if $x \leq y \in I$, then $x \in I$.

Definition A. Let Δ be a poset. $\mathcal{F}$ is an *ideal filter on* Δ if:
  i) $\mathcal{F}$ is a set of ideals of Δ;
 ii)   if I_1, I_2 are ideals of Δ in $\mathcal{F}$, then $I_1 \cap I_2 \in \mathcal{F}$;
iii) $\mathcal{F}$ does not contain the empty ideal; and
 iv)   if J is an ideal of Δ containing $I \in \mathcal{F}$, then $J \in \mathcal{F}$.
A maximal ideal filter on Δ is called an *ideal ultrafilter*.

Proposition 1. *For any poset* Δ, *ideal ultrafilters exist on* Δ.

Proposition 2. *Let* Δ *be a poset and* $\mathcal{F}$ *be an ideal filter on* Δ. $\mathcal{F}$ *is an ideal ultrafilter on* Δ *if and only if for any ideal* I *of* Δ *not in* $\mathcal{F}$, *there is* $J \in \mathcal{F}$ *such that* $J \cap I = \emptyset$.

  Proof. Suppose that $\mathcal{F}$ is an ideal ultrafilter on Δ and I is an ideal of Δ not in $\mathcal{F}$. Suppose by way of contradiction that for any $J \in \mathcal{F}$, $I \cap J \neq \emptyset$.
  Let $\mathcal{U}$ be the set of ideals V of Δ such that there exists $J \in \mathcal{F}$ so that $V \supseteq I \cap J$. If $V_1, V_2 \in \mathcal{U}$ with corresponding ideals $J_1, J_2 \in \mathcal{F}$ such that $I \cap J_n \subseteq V_n$, then $V_1 \cap V_2 \supseteq I \cap (J_1 \cap J_2)$, and so $\mathcal{U}$ is closed with respect to intersections. Conditions (iii) and (iv) of Definition A are then trivially true, and so $\mathcal{U}$ is an ideal filter of Δ properly containing $\mathcal{F}$ since $I \in \mathcal{U}$. Since this contradicts $\mathcal{F}$ being an ideal ultrafilter, there must be $J \in \mathcal{F}$ such that $I \cap J = \emptyset$.
  Conversely, clearly $\mathcal{F}$ cannot be extended to a larger ideal filter. ∎

Proposition 3. *Let* I *be an ideal of a poset* Δ. *Then there exists a unique maximal ideal* J *(possibly empty) of* Δ *such that* $J \cap I = \emptyset$.

  Proof. Let J be the union of all ideals K of Δ such that $K \cap I = \emptyset$. ∎

Definition B. Let I be an ideal of a poset Δ. I^* will denote the (possibly empty) largest ideal of Δ such that $I \cap I^* = \emptyset$.

Proposition 4. *An ideal filter* $\mathcal{F}$ *on a poset* Δ *is an ultrafilter if and only if for any ideal* I *of* Δ, *either* I *or* I^* *is in* $\mathcal{F}$.

Proof. Suppose that $\mathcal{F}$ is an ideal ultrafilter on Δ and I is an ideal of Δ not in $\mathcal{F}$. Then there exists $J \in \mathcal{F}$ such that $J \cap I = \emptyset$. Since then $J \subseteq I^*$, $I^* \in \mathcal{F}$.

The converse is obvious from Proposition 2. ∎

Definition C. An ideal filter $\mathcal{F}$ on a poset Δ is *prime* if $I_1 \cup I_2 \in \mathcal{F}$ implies $I_1 \in \mathcal{F}$ or $I_2 \in \mathcal{F}$.

Proposition 5. *An ideal ultrafilter $\mathcal{U}$ on a poset Δ is prime.*

Proof. Let I_1 and I_2 be ideals of Δ such that $I_1 \cup I_2 \in \mathcal{U}$. Suppose by way of contradiction that neither I_1 nor I_2 is in $\mathcal{U}$. Then $I_1^*, I_2^* \in \mathcal{U}$, and so $I_1^* \cap I_2^* \in \mathcal{U}$. But then $\emptyset = (I_1^* \cap I_2^*) \cap (I_1 \cup I_2) \in \mathcal{U}$, which is clearly impossible. ∎

We are now ready to define a process by which we can construct convex ℓ-subgroups by means of the filters on a plenary subset Δ.

Definition D. Let G be an ℓ-group and Δ be a plenary subset of $\Gamma(G)$. For $g \in G$, let $S(g) = \{V \in \Delta : g \notin V\}$. For $C \in \mathcal{C}(G)$, let $\psi(C)$ be the set of ideals I of Δ such that there exists $g \notin C$ with $S(g) \subseteq I$.

Note that $S(g)$ is an ideal $\Delta(G)$.

Proposition 6. *Let G be an ℓ-group and Δ be a plenary subset of $\Gamma(G)$. Let $C_1 \subseteq C_2$ be convex ℓ-subgroups of G. Then $\psi(C_2) \subseteq \psi(C_1)$.*

Proof. For $I \in \psi(C_2)$, there exists $e < g \notin C_2$ such that $S(g) \subseteq I$. But then $g \notin C_1$ implies $I \in \psi(C_1)$. ∎

Proposition 7. *Let G be an ℓ-group and Δ be a plenary subset of $\Gamma(G)$. Then for any prime ℓ-subgroup P of G, $\psi(P)$ is an ideal filter on Δ.*

Proof. Clearly $\psi(P)$ does not contain the empty set, and is a set of ideals of Δ such that if $I \in \psi(P)$ and J is an ideal of Δ such that $I \subseteq J$, then $J \in \psi(P)$.

So suppose that $I_1, I_2 \in \psi(P)$. Then there exist $e < g_1, g_2 \notin P$ such that $S(g_1) \subseteq I_1$ and $S(g_2) \subseteq I_2$. But as P is prime, $g_1 \wedge g_2 \notin P$ and thus

$$S(g_1 \wedge g_2) = S(g_1) \cap S(g_2) \subseteq I_1 \cap I_2,$$

showing that $I_1 \cap I_2 \in \psi(P)$. ∎

Definition E. Let G be an ℓ-group and Δ be a plenary subset of $\Gamma(G)$. For a *prime* ideal filter $\mathcal{F}$ of Δ, let $\alpha(\mathcal{F}) = \{g \in G : S(g) \notin \mathcal{F}\}$.

Proposition 8. *Let G be an ℓ-group and Δ be a plenary subset of $\Gamma(G)$. For a prime ideal filter $\mathcal{F}$ of Δ, $\alpha(\mathcal{F})$ is a prime subgroup of G.*

Proof. Let $g, h \in \alpha(\mathcal{F})$. Then $S(gh^{-1}) \subseteq S(g) \cup S(h)$. So if $gh^{-1} \notin \alpha(\mathcal{F})$, then $S(gh^{-1}) \in \mathcal{F}$, and so $S(g) \cup S(h) \in \mathcal{F}$. But then, since $\mathcal{F}$ is prime, either $S(g) \in \mathcal{F}$ or $S(h) \in \mathcal{F}$, contradicting that $g, h \in \alpha(\mathcal{F})$. Thus $\alpha(\mathcal{F})$ is a subgroup of G.

If $g \in \alpha(\mathcal{F})$, then $S(g \vee e) \subseteq S(g) \notin \mathcal{F}$ implies $S(g \vee e) \notin \mathcal{F}$, and hence $g \vee e \in \alpha(\mathcal{F})$. So $\alpha(\mathcal{F})$ is an ℓ-subgroup.

If $e \leq h \leq g \in \alpha(\mathcal{F})$, then $S(h) \subseteq S(g)$ implies $S(h) \notin \mathcal{F}$ and so $h \in \alpha(\mathcal{F})$. So $\alpha(\mathcal{F})$ is convex.

Finally, suppose that $a \wedge b = e$ in G, and that $a \notin \alpha(\mathcal{F})$. Then $S(a) \in \mathcal{F}$. Now if $b \notin \alpha(\mathcal{F})$, then $S(b) \in \mathcal{F}$ and so $\emptyset = S(a) \cap S(b) \in \mathcal{F}$, which is false. So $a \notin \alpha(\mathcal{F})$ implies $b \in \alpha(\mathcal{F})$. ∎

Note that if $\mathcal{F}$ is a prime ideal filter, then $\psi(\alpha(\mathcal{F})) \subseteq \mathcal{F}$.

Now suppose that for some prime subgroup P, $\psi(P)$ is a prime ideal filter. Then $\alpha(\psi(P))$ would be defined and be a prime subgroup. If for *every* prime subgroup P, $\psi(P)$ is a prime ideal filter, then we could examine $\psi(\alpha(\psi(P)))$, etc.

Unfortunately, it is not the case that for every prime subgroup P and plenary subset Δ, $\psi(P)$ is a prime ideal filter.

Example. Let A be the ℓ-group of eventually constant integer sequences, with pointwise order and addition, and $\Delta = \{A_i : 1 \leq i < \infty\}$, where $A_i = \{a \in A : a(i) = 0\}$. Then $P = \sum_{i=1}^{\infty} \mathbb{Z}$ is a prime subgroup and $\psi(P)$ is the filter of cofinite subsets of Δ. But $\psi(P)$ is not a prime ideal filter, for let $I_1 = \{A_{2i} : 1 \leq i < \infty\}$ and $I_2 = \{A_{2i-1} : 1 \leq i < \infty\}$. Then $I_1 \cup I_2 = \Delta \in \psi(P)$, but neither I_1 nor I_2 is in $\psi(P)$.

There is, however, an instance in which for every prime subgroup P, $\psi(P)$ is a prime ideal filter.

Recall [D95] that a *component* of an element $e < g$ is an element $e < x$ such that $x \wedge gx^{-1} = e$. If the only component of g is itself, g is called *indecomposable*.

Also recall [D95] that an element g is *special* if g has only one value, which is then called a *special value*. An ℓ-group G is called *special-valued* if the special values form a plenary subset of $\Gamma(G)$; this is equivalent to each positive element g being the join of a set of mutually disjoint special elements, which are called the *special components* of g. Special components are indeed components, and in a special-valued ℓ-group, an element is indecomposable if and only if it is special. For G a special-valued ℓ-group and $g \in G$, $\Delta(G)$ will denote the plenary subset of special values and $\Delta(g)$ will denote the special values of g. If $G_\delta \in \Delta(g)$, g_δ will denote the unique special component of g with value G_δ. If for $S \subseteq \Delta(g)$, $a = \vee_S g_\delta$ exists, then a is called the *projection* of g onto S, and it is necessarily a component of g.

A convex ℓ-subgroup C of an ℓ-group G is *closed* if whenever $\{a_\lambda\}_{\lambda \in \Lambda} \subseteq C^+$ such that $\vee_\Lambda a_\lambda$ exists in G, then $\vee_\Lambda a_\lambda \in C$. If G is special-valued, a value V is closed if and only if V is special [D95].

Also recall [D95] that an ℓ-group G is *laterally complete* if every set of mutually disjoint positive elements has a least upper bound.

Combining the two conditions, we will now see that if G is special-valued and laterally complete, then $\psi(P)$ is a prime ideal filter for every prime subgroup P.

Proposition 9. *Let G be a laterally complete, special-valued ℓ-group and $\Delta(G)$ be the minimal plenary subset of special values. If P is a prime subgroup of G, then $\psi(P)$ is a prime ideal filter of $\Delta(G)$.*

Proof. Suppose that I_1, I_2 are ideals of $\Delta(G)$ such that $I_1 \cup I_2 \in \psi(P)$. Then there exists $e < g \notin P$ such that $S(g) \subseteq I_1 \cup I_2$. Let $A_1 = \{G_\delta \in \Delta(g) : G_\delta \in I_1 \setminus I_2\}$, $A_2 = \{G_\delta \in \Delta(g) : G_\delta \in I_2 \setminus I_1\}$, and $A_3 = \{G_\delta \in \Delta(g) : G_\delta \in I_1 \cap I_2\}$. Since G is laterally complete and special-valued, the projections g_1, g_2, and g_3 of g onto A_1, A_2, and A_3, respectively, exist in G, where g_i is the join of $\{g_\delta : G_\delta \in J_i\}$, and are pairwise disjoint. Since $g = g_1 \vee g_2 \vee g_3 \notin P$, exactly one of g_1, g_2, and g_3 is not in P. If $g_1 \notin P$, then $S(g_1) = J_1 \subseteq I_1$ implies $I_1 \in \psi(P)$. Likewise, if $g_2 \notin P$, then $I_2 \in \psi(P)$. Finally, if $g_3 \notin P$, then both $I_1, I_2 \in \psi(P)$.   ∎

We will hereafter assume that G is a laterally complete special-valued lattice-ordered group, and that Δ is the plenary subset of special values in $\Gamma(G)$. This assumption, while strong, actually provides useful information for most normal-valued ℓ-groups. For if H is normal-valued and Δ is a plenary subset closed with respect to conjugation ($V \in \Delta$ implies that for any $h \in H$, $h^{-1}Vh \in \Delta$), then H can be embedded into a laterally complete special-valued ℓ-group K whose plenary subset of special values is isomorphic to Δ, where if $H_\delta \in \Delta$ is a value of $h \in H$, then the corresponding $K_\delta \in \Delta(K)$ is a special value of h in K, and $H_\delta = H \cap K_\delta$. Since then H is an ℓ-subgroup of K, any convex ℓ-subgroup of H arises by intersection of a convex ℓ-group of K with H.

Proposition 10. *Let P be a prime subgroup of G. Then $\alpha(\psi(P)) \subseteq P$.*

Proof. Let $e \leq g \in \alpha(\psi(P))$; then $S(g) \notin \psi(P)$ implies that $g \in P$.   ∎

Proposition 11. *Let $\mathcal{F}_1 \subseteq \mathcal{F}_2$ be prime ideal filters of G. Then $\alpha(\mathcal{F}_2) \subseteq \alpha(\mathcal{F}_1)$.*

Proof. $g \in \alpha(\mathcal{F}_2)$ implies that $S(g) \notin \mathcal{F}_2$, and so $S(g) \notin \mathcal{F}_1$. Thus $g \in \alpha(\mathcal{F}_1)$.   ∎

Propositions 6 and 11 show that, under reverse inclusion, the maps ψ and α establish a Galois connection between primes subgroups of G and prime ideal filters of $\Delta(G)$.

Proposition 12. *Let P be a prime subgroup of G. Then $\psi(\alpha(\psi(P))) = \psi(P)$.*

Proof. The proof is obvious when we couple the observation after Definition D with Proposition 11.   ∎

Proposition 13. *Let P be a prime subgroup of G. Then $\alpha(\psi(\alpha(\psi(P)))) = \alpha(\psi(P))$.*

In the following, we will examine those prime subgroups P such that $P = \alpha(\psi(P))$, which are constructable from Δ, and show that this class of prime subgroups contains the four types of primes subgroups mentioned earlier.

It is easy now to see that any minimal prime subgroup M is in this class, as $\alpha(\psi(M))$ is a prime subgroup which by Proposition 10 is contained in M, and thus equals $\alpha(\psi(M))$. In Propositions 15 and 16, more will be said about how minimal primes arise in this way.

Proposition 14. *Let $\mathcal{C}$ be a nonempty chain in $\Delta(G)$; let $P = \cap\mathcal{C}$ and $Q = \cup\mathcal{C}$. Then $P = \alpha(\psi(P))$ and $Q = \alpha(\psi(Q))$.*

Proof. Suppose $e < g \notin P$. Then g has a value V containing P, and so there exists $G_\delta \in \mathcal{C}$ such that $G_\delta \subseteq V$. Since $\Delta(G)$ is a dual ideal of $\Gamma(G)$, $V \in \Delta(G)$, and $\psi(P) = \{\text{ideals } I \text{ of } \Delta(G) : \mathcal{C} \cap I \neq \emptyset\}$. But then if $e < x \notin \alpha(\psi(P))$, $S(x) \in \psi(P)$ implies $x \notin P$, and so $P \subseteq \alpha(\psi(P))$, which shows equality.

If $Q \neq G$, then $e < h \notin Q$ means that h has a value $V \supseteq Q$, and so for any $G_\delta \in \mathcal{C}$, $G_\delta \subseteq V$, and thus $V \in \Delta$. So $\psi(Q) = \{\text{ideals } I \text{ of } \Delta : \mathcal{C} \subseteq I\}$. But then if $e < y \notin \alpha(\psi(Q))$, $S(y) \in \psi(Q)$, and so y has a value exceeding each $G_\delta \in \mathcal{C}$, which means $y \notin Q$. So again $Q = \alpha(\psi(Q))$. ∎

Note that for any $V \in \Delta$, $\{V\}$ is a chain in Δ, and so $V = \alpha(\psi(V))$.

For the next proposition, we remind the reader ([D95]) that a prime subgroup P is minimal if and only if for any $e \leq g \in P$, there exists $e < h \notin P$ such that $g \wedge h = e$.

Proposition 15. *$\mathcal{U}$ is an ideal ultrafilter on Δ if and only if $\alpha(\mathcal{U})$ is a minimal prime subgroup of G.*

Proof. ($\Rightarrow$) Since $\mathcal{U}$ is a prime filter, $\alpha(\mathcal{U})$ is a prime subgroup of G. So let $e < g \in \alpha(\mathcal{U})$. Then $S(g) \notin \mathcal{U}$ implies $S(g)^* \in \mathcal{U}$, and thus $S(g)^* \neq \emptyset$. Let Ω be a maximal antichain in $S(g)^*$. Since we are assuming that G is laterally complete and special-valued, there exists $e < h \in G$ such that $\Delta(h) = \Omega$. Now $h \in \alpha(\mathcal{U})$ would imply that $S(h) \notin \mathcal{U}$, and so $S(h)^* \in \mathcal{U}$. But then $S(g)^* \cap S(h)^* \in \mathcal{U}$. But let $G_\delta \in S(g)^* \cap S(h)^*$ and s be special with value G_δ. Then $s \wedge g = s \wedge h = e$, which contradicts that Ω was maximal. So $h \notin \alpha(\mathcal{U})$, implying that $g \in h^\perp$. Thus $\alpha(\mathcal{U})$ is minimal.

($\Leftarrow$) Suppose that I is an ideal of Δ such that $I \notin \mathcal{U}$. Let Ω be a maximal antichain in I and let $e < g \in G$ such that $\Delta(g) = \Omega$. Then $S(g) \subseteq I$ implies $S(g) \notin \mathcal{U}$ and so $g \in \alpha(\mathcal{U})$. Since $\alpha(\mathcal{U})$ is a minimal prime subgroup, there exists $e < h \notin \alpha(\mathcal{U})$ such that $g \wedge h = e$. As $S(h) \in \mathcal{U}$, then $I^* \supseteq S(h)$ is also in $\mathcal{U}$. ∎

Proposition 16. *If P is a minimal prime subgroup of G, then $\psi(P)$ is an ideal ultrafilter on $\Delta(G)$.*

Proof. Suppose that I is an ideal of $\Delta(G)$ such that $I \notin \psi(P)$. Let Λ be a maximal antichain in I and $e < g \in G$ such that $\Delta(g) = \Lambda$. Then as $S(g) \subseteq I$, $S(g) \notin \psi(P)$, and so $g \in P$. Since P is a minimal prime, there exists $e < h \notin P$ such that $g \wedge h = e$. So $S(h) \in \psi(P)$ and so $S(h) \subseteq I^*$ implies $I^* \in \psi(P)$. Hence, $\psi(P)$ is an ideal ultrafilter.   ∎

We now start a series of propositions that will show that branch primes are of the form $\alpha(\psi(P))$.

Proposition 17. *Let P_1, P_2 be prime subgroups of G. Then $\psi(P_1) = \psi(P_2)$ implies P_1 is comparable to P_2.*

The proof follows easily from the definition of ψ.

The following example shows that the converse to Proposition 16 is not generally true and also that if $\psi(P_1) = \psi(P_2)$, we cannot conclude that $P_1 = P_2$.

Example. Let $G = \prod_{i=1}^{\infty} \mathbb{R}$. As mentioned before, $\Gamma(G)$ is stranded, and so any value V contains a unique minimal prime subgroup, M. It is easily checked that $\psi(V) = \psi(M)$, and so for any value V, $\psi(V)$ is an ultrafilter on the trivially ordered set $\Delta(G)$ of special values. Now $V = M$ if and only if V is a special value. So for any value V which is not special, we have a counterexample to the converse of Proposition 16.

Also, if we consider any two comparable prime subgroups $P_1 \subset P_2$ in G, we then have $\psi(P_1) = \psi(P_2)$.

Proposition 18. *If $P = \alpha(\psi(P))$, then $g \in P$ if and only if $S(g) \notin \psi(P)$.*

Proof. If $S(g) \notin \psi(P)$, then $g \in \alpha(\psi(P)) = P$. Conversely, if $g \in P$, then $g \in \alpha(\psi(P))$ implies $S(g) \notin \psi(P)$.   ∎

Proposition 19. *Let $P_1 \subseteq P_2$ be prime subgroups of G. Then $\alpha(\psi(P_1)) \subseteq \alpha(\psi(P_2))$.*

Proof. $g \in \alpha(\psi(P_1))$ implies $S(g) \notin \psi(P_1) \supseteq \psi(P_2)$, and thus $S(g) \notin \psi(P_2)$. So $g \in \alpha(\psi(P_2))$.   ∎

Proposition 20. *For any branch prime subgroup P of G, $P = \alpha(\psi(P))$.*

Proof. Since P is a branch prime subgroup, there is a set of at least two minimal primes $\{M_\lambda\}_{\lambda \in \Lambda} \subseteq \mathrm{Min}(G)$ such that $P = \vee_{\lambda \in \Lambda} M_\lambda$. Since for each $\lambda \in \Lambda$, $M_\lambda = \alpha(\psi(M_\lambda)) \subseteq \alpha(\psi(P))$, $P = \vee_\lambda M_\lambda \subseteq \alpha(\psi(P)) \subseteq P$.   ∎

As a corollary, we have:

Corollary 21. *If P_1 and P_2 are incomparable prime subgroups of G, then $P_1 \vee P_2 = \alpha(\psi(P_1 \vee P_2))$.*

In light of Proposition 20, it would seem likely that if P_1, P_2 are incomparable primes, then $\psi(P_1 \vee P_2)$ should equal $\psi(P_1) \cap \psi(P_2)$. This, though, is not the case. Returning to the example before Proposition 18, let $\mathcal{U}_1$ and $\mathcal{U}_2$ be distinct nonprincipal ultrafilters on $\Delta(G)$. Then $M_{\mathcal{U}_1} \neq M_{\mathcal{U}_2}$ and since $\Gamma(G)$ is stranded, $M_{\mathcal{U}_1} \vee M_{\mathcal{U}_2} = G$. So $\psi(M_{\mathcal{U}_1} \vee M_{\mathcal{U}_2}) = \emptyset$. But $\mathcal{U}_1 = \psi(M_{\mathcal{U}_1})$ and $\mathcal{U}_2 = \psi(M_{\mathcal{U}_2})$. Since both $\mathcal{U}_1$ and $\mathcal{U}_2$ are nonprincipal, any cofinite subset is contained in their intersection.

However, we do have the following result.

Proposition 22. *If P_1 and P_2 are incomparable prime subgroups of G, then $P_1 \vee P_2 = \alpha(\psi(P_1) \cap \psi(P_2))$.*

Proof. Since $P_1, P_2 \subset P_1 \vee P_2$, $\psi(P_1 \vee P_2) \subseteq \psi(P_1) \cap \psi(P_2)$, and thus

$$\alpha(\psi(P_1) \cap \psi(P_2)) \subseteq \alpha(\psi(P_1 \vee P_2)) = P_1 \vee P_2.$$

On the other hand, $\alpha(\psi(P_1)), \alpha(\psi(P_2)) \subseteq \alpha(\psi(P_1) \cap \psi(P_2))$ implies

$$P_1 \vee P_2 = \alpha(\psi(P_1)) \vee \alpha(\psi(P_2)) \subseteq \alpha(\psi(P_1) \cap \psi(P_2)).$$

∎

We now look at prime subgroups of the form $\alpha(\psi(V))$, where $V \in \Gamma(G)$. In the following, $\widehat{V}$ will denote the cover of V: the least convex ℓ-group of G properly containing V.

Proposition 23. *Let $V \in \Gamma(G)$ and $v \in \widehat{V} \setminus V$. Then $\mathcal{U}_v = \{I \cap \Delta(v) : I \in \psi(V)\}$ is an ultrafilter on $\Delta(v)$.*

Proof. Let $I \in \psi(V)$; then there exists $e < g \notin V$ such that $S(g) \subseteq I$. But then $g \wedge |v| \notin V$, and so $\emptyset \neq S(g \wedge |v|) \cap \Delta(v)) \subseteq I \cap \Delta(v)$. Thus $\mathcal{U}_v$ does not contain the empty set. It is also clear, now, that $\mathcal{U}_v$ has the finite intersection property.

Suppose $J \subseteq \Delta(v)$ and $I \cap \Delta(v) \subseteq J$ for some $I \in \psi(V)$. Let t be the projection of v onto J. Since $t \wedge vt^{-1} = e$ while $v = (vt^{-1})t \notin V$, either $t \in V$ or $vt^{-1} \in V$, but not both. But $S(vt^{-1}) \subseteq I^*$ shows that $vt^{-1} \in V$. So $t \notin V$, and $J = S(t) \cap \Delta(v) \in \mathcal{U}_v$.

Finally, suppose $J \subseteq \Delta(v)$, and $J \notin \mathcal{U}_v$. Then $K = \Delta(v) \setminus J \neq \emptyset$. Let s be the projection of v onto K and $t = vt^{-1}$. Again, exactly one of s or t is in V. If $t \notin V$, then $J = S(t) \cap \Delta(v) \in \mathcal{U}_v$. So $s \notin V$ and so $K \in \mathcal{U}_v$. ∎

Proposition 24. *Let $V \in \Gamma(G)$ and $e < v \in \widehat{V} \setminus V$. Then for any $w \in \widehat{V} \setminus V$, $\Delta(w) \cap \Delta(v) \in \mathcal{U}_v$.*

Proof. $S(w) \in \psi(V)$, and so $S(w) \cap \Delta(v) \in \mathcal{U}_v$. Let t be the projection of v onto $S(w) \cap \Delta(v)$. Then if $vt^{-1} \notin V$, $S(vt^{-1}) \in \psi(V)$. But since $t \wedge vt^{-1} = e$, $S(vt^{-1}) \subseteq S(t)^*$. So $(S(w) \cap \Delta(v)) \cap \Delta(vt^{-1}) = \emptyset$, which cannot be if $vt^{-1} \notin V$. So $t \notin V$. But then $\Delta(w) \cap \Delta(v) = \Delta(t) = S(t) \cap \Delta(v) \in \mathcal{U}_v$. ∎

Note that a corollary to Proposition 24 is that if $e < v, w \in \widehat{V} \setminus V$, then for any $W \in \mathcal{U}_w$, $W \cap \Delta(v) \in \mathcal{U}_v$.

Proposition 25. *Let Ω be an antichain in $\Delta(G)$ and $\mathcal{U}$ be an ultrafilter on Ω. Let $\mathcal{F}_\mathcal{U}$ be the set of ideals I of $\Delta(G)$ such that $I \cap \Omega \in \mathcal{U}$. Then $\mathcal{F}_\mathcal{U}$ is a prime ideal filter on $\Delta(G)$.*

Proof. Clearly $\mathcal{F}_\mathcal{U}$ has the finite intersection property and if $I \in \mathcal{F}_\mathcal{U}$ and $I \subset J$ for some ideal J of $\Delta(G)$, then $J \in \mathcal{F}_\mathcal{U}$. So $\mathcal{F}_\mathcal{U}$ is an ideal filter on $\Delta(G)$.

Now let I_1, I_2 be ideals of $\Delta(G)$ such that $I_1 \cup I_2 \in \mathcal{F}_\mathcal{U}$. Then $(I_1 \cap \Omega) \cup (I_2 \cap \Omega) \in \mathcal{U}$, and so at least one of $I_1 \cap \Omega, I_2 \cap \Omega \in \mathcal{U}$. But then at least one is in $\mathcal{F}_\mathcal{U}$. ∎

Recall [D95] that if G is a normal-valued ℓ-group and $g \in G$, then

$$I(g) = \{x \in G : |x| \ll |g|\}$$

is a convex ℓ-group of G, where $e \leq a \ll b$ means that for any integer n, $a^n < b$. In [BCD86], it was shown that $\cap\{G_\delta : G_\delta \in \Delta(v)\} = I(v) \oplus v^\perp$.

Proposition 26. *Let $V \in \Gamma(G)$. Then*

$$\alpha(\psi(V)) = \{g \in G : \text{there exists } v \in \widehat{V} \setminus V \text{ such that } g \in I(v) \oplus v^\perp\}$$

and for any $e < v \in \widehat{V}$, $\cap\{G_\delta : G_\delta \in \Delta(v)\} \subseteq \alpha(\psi(V))$.

Proof. Let $g \in \alpha(\psi(V))$; then $S(g) \notin \psi(V)$. Now if $S(g) \cap \Delta(v) = \emptyset$, then $g \in G_\delta$ for every $G_\delta \in \Delta(v)$, and so $g \in I(v) \oplus v^\perp$.

So suppose $S(g) \cap \Delta(v) \neq \emptyset$; let t be the projection of v onto $S(g) \cap \Delta(v)$. Then $S(t) \subseteq S(g)$. Thus if $t \notin V$, then $S(g) \in \psi(V)$, which is not true. So $t \in V$ and thus $vt^{-1} \in \widehat{V} \setminus V$. But then $g \wedge vt^{-1} = e$ implies $g \in I(vt^{-1}) \oplus (vt^{-1})^\perp$.

Conversely, suppose there exists $v \in \widehat{V} \setminus V$ such that $g \in I(v) \oplus v^\perp$. Then $S(g) \in \psi(V)$ would imply that there exists $e < w \notin V$ such that $S(w) \subseteq S(g)$. But then $\emptyset = S(g) \cap \Delta(v) \supseteq S(w) \cap \Delta(v) \neq \emptyset$. So $S(g) \notin \psi(V)$, and thus $g \in \alpha(\psi(V))$.

That $\cap\{G_\delta : G_\delta \in \Delta(v)\} \subseteq \alpha(\psi(V))$ now follows easily from [BCD86]. ∎

Now if $V \in \Delta(G)$ and $e < v$ is special with value V, then $V = I(v) \oplus v^\perp = \alpha(\psi(V))$. Thus we have shown again that any element of $\Delta(G)$ is of the form $\alpha(\psi(P))$ for some prime subgroup P.

Proposition 27. *Let $V \in \Gamma(G)$. Then*

$$\text{lex}(G) = \{g \in G : \exists e < v \in \widehat{V} \setminus V \text{ such that } \forall G_\delta \in \Delta(v), |g| \wedge v_\delta \in \text{lex}(G_\delta)\}.$$

Proof. Let

$$P = \{g \in G : \exists\, e < v \in \widehat{V} \setminus V \text{ such that } \forall\, G_\delta \in \Delta(v), |g| \wedge v_\delta \in \text{lex}(G_\delta)\}.$$

P is a subgroup of G, for if $g, h \in P$, then there exist $e < v, w \in \widehat{V} \setminus V$ such that for all $G_\delta \in \Delta(v)$, $|g| \wedge v_\delta \in \text{lex}(G_\delta)$, and for all $G_\delta \in \Delta(w)$, $|g| \wedge w_\delta \in \text{lex}(G_\delta)$. Since $\Delta(v) \cap \Delta(w) \subseteq \mathcal{U}_v \cap \mathcal{U}_w$, we can assume by Proposition 24 that $v = w$ is the projection of v onto $\Delta(v) \cap \Delta(w)$. So

$$|gh^{-1}| \wedge v_\delta \leq |g||h||g| \wedge v_\delta \leq (|g| \wedge v_\delta)(|h| \wedge v_\delta)(|g| \wedge v_\delta) \in \text{lex}(G_\delta)$$

for all $G_\delta \in \Delta(v)$. Hence $gh^{-1} \in P$ and P is a subgroup of G.

Let $g \in P$ and $e < v \in \widehat{V} \setminus V$ such that for all $G_\delta \in \Delta(v)$, $|g| \wedge v_\delta \in \text{lex}(G_\delta)$. Then $e \leq (g \vee e) \wedge v_\delta = (g \wedge v_\delta) \vee e \leq |g| \wedge v_\delta \in \text{lex}(G_\delta)$. So P is an ℓ-subgroup and is clearly convex.

Now suppose that $a \wedge b = e$ in G and $a \notin P$. Let $e < v \in \widehat{V} \setminus V$. Since $a \notin P$, $\{G_\delta \in \Delta(v) : a \wedge v_\delta \in \text{lex}(G_\delta)\} \notin \mathcal{U}_v$, else if w is the projection of v onto this set (if nonempty), then $w \in \widehat{V} \setminus V$, and we would have the situation that

$$\{G_\delta \in \Delta(v) : a \wedge v_\delta \in \text{lex}(G_\delta)\} = \{G_\delta \in \Delta(v) : a \wedge w_\delta \in \text{lex}(G_\delta)\},$$

and so $a \in P$. So $X = \{G_\delta \in \Delta(v) : a \wedge v_\delta \notin \text{lex}(G_\delta)\} \in \mathcal{U}_v$. Since $\text{lex}(G_\delta)$ is prime and $b \wedge a \wedge v_\delta = e$, $b \in \text{lex}(G_\delta)$ for all $G_\delta \in X$. Letting x be the projection of v onto X, then for all $G_\delta \in \Delta(x)$, $b \wedge v_\delta \in \text{lex}(G_\delta)$. So $b \in P$ shows that P is prime.

Let M be a minimal prime of G such that $M \subseteq V$. Suppose $M \nsubseteq P$; then M is incomparable to P. So there exists $y \in M \setminus P$ and $x \in P \setminus M$ such that $y \wedge x = e$. Again, let $e < v \in \widehat{V} \setminus V$ such that $\{G_\delta \in \Delta(v) : y \wedge v_\delta \in \text{lex}(G_\delta)\} \notin \mathcal{U}_v$. Then $T = \{G_\delta \in \Delta(v) : y \wedge v_\delta \notin \text{lex}(G_\delta)\} \in \mathcal{U}_v$. Let t be the projection of v onto T and $s = y \wedge t$. $t \notin P$ by definition and so $s \in M \setminus P$. Since $y \wedge x = e$, $s \wedge x = e$. Also, since $s_\delta = y \wedge v_\delta \notin \text{lex}(G_\delta)$, s_δ is special and for any $g \in G$, $g \wedge s_\delta = e$ if and only if $g \wedge v_\delta = e$. Thus $t \wedge x = e$. But then, since $x \notin M$, $t \in M$, which is false, since $t \notin V$. So $M \subseteq P$, and hence $\text{lex}(V) \subseteq P$.

Now let $e \leq g \in P$ and $e < v \in \widehat{V} \setminus V$ such that for all $G_\delta \in \Delta(v)$, $g \wedge v_\delta \in \text{lex}(G_\delta)$. Note that if $g \wedge v = e$, then as $v \notin V$, $g \in V$. Assume $g \wedge v > e$. Let A be the set of $\delta \in \Delta(v)$ such that $g_\delta = g \wedge v_\delta > e$. Then for any $\delta \in A$, $g \wedge v_\delta \ll v_\delta$. Thus if $x = \bigvee_A g_\delta$, $x \ll v$, and so $x \in V$. As $gx^{-1} \wedge v = e$, $gx^{-1} \in V$, and so $g \in V$. Hence $P \subseteq V$.

Now let $e < g \in P$. If g is not special, let s be a special component of g. Then $gs^{-1} > e$ and $s \wedge gs^{-1} = e$. Thus both s and gs^{-1} are nonunits of P and so are in $\text{lex}(P)$, showing that $g \in \text{lex}(P)$. If g is special, let $e < v \in \widehat{V} \setminus V$ such that for all $G_\delta \in \Delta(v)$, $g \wedge v_\delta \in \text{lex}(G_\delta)$. But then as g is special, either $g \wedge v = e$ or there exists a unique v_δ such that $g \ll v_\delta$. In either of these cases, g is once again in $\text{lex}(V)$. So $P \subseteq \text{lex}(V)$, and hence $P = \text{lex}(V)$. ∎

Lemma 28. *Let $V \in \Gamma(G)$. Then $\mathrm{lex}(V) \subseteq \alpha(\psi(V))$ and if $\alpha(\psi(V))$ is a branch prime subgroup, then $\alpha(\psi(V)) = \mathrm{lex}(V)$.*

Proof. Let M be a minimal prime subgroup such that $M \subseteq V$. Then $M = \alpha(\psi(M)) \subseteq \alpha(\psi(V))$ shows that $\mathrm{lex}(V) \subseteq \alpha(\psi(V))$.

Now if $\alpha(\psi(V))$ is a branch prime subgroup, then $\alpha(\psi(V))$ is the join of all minimal prime subgroups contained in $\alpha(\psi(V))$, which are precisely those contained in V. So $\mathrm{lex}(V) = \alpha(\psi(V))$. ∎

Proposition 29. *Let $V \in \Gamma(G)$. $\alpha(\psi(V))$ is a branch prime subgroup if and only if there exists $e < v \in \widehat{V} \setminus V$ such that for all $G_\delta \in \Delta(v)$, G_δ is a branch prime subgroup.*

Proof. Suppose $\alpha(\psi(V))$ is a branch prime subgroup. We first show that there cannot be a $e < v \in \widehat{V} \setminus V$ such that for all $G_\delta \in \Delta(v)$, $G_\delta \neq \mathrm{lex}(G_\delta)$. So by way of contradiction, suppose that such a v does exist. Then for any $G_\delta \in \Delta(v)$, $\mathrm{lex}(G_\delta) \subset G_\delta$. Let $e < g \in G_\delta \setminus \mathrm{lex}(G_\delta)$. Since then $e < g \wedge v \in G_\delta \setminus \mathrm{lex}(G_\delta)$, we can assume that $g \ll v$ and that every special value of g is contained in G_δ. Thus there exists $G_{\beta_\delta} \in \Delta(G)$ such that $\mathrm{lex}(G_\delta) \subseteq G_{\beta_\delta} \subset G_\delta$.

The set $\{G_{\beta_\delta} : G_\delta \in \Delta(v)\}$ is an antichain in $\Delta(G)$ and as G is laterally complete, there exists $e < x \in G$ such that $\Delta(x) = \{G_{\beta_\delta} : G_\delta \in \Delta(v)\}$. We must have that $x \ll v$ and so $x \in I(v) \oplus v^\perp$, which by Proposition 26 gives us that $x \in \alpha(\psi(V))$. But then by Lemma 28, $x \in \mathrm{lex}(V)$. So by Proposition 27, there exists $e < w \in \widehat{V} \setminus V$ such that for any $G_\gamma \in \Delta(w)$, $x \wedge w_\gamma \in \mathrm{lex}(G_\gamma)$. But as $e < w \in \widehat{V} \setminus V$, $\Delta(v) \cap \Delta(w)$ is not empty, and so for any $G_\delta \in \Delta(v) \cap \Delta(w)$, $x \wedge w_\delta \in \mathrm{lex}(G_\delta)$. But $x \wedge w_\delta = x_{\beta_\delta} \notin \mathrm{lex}(G_\delta)$.

Thus there exists $e < v \in \widehat{V} \setminus V$ such that the set $\{G_\delta \in \Delta(v) : G_\delta = \mathrm{lex}(G_\delta)\} \neq \emptyset$. Let $C = \{G_\delta \in \Delta(v) : G_\delta = \mathrm{lex}(G_\delta)\}$ and $D = \{G_\delta \in \Delta(v) : G_\delta \neq \mathrm{lex}(G_\delta)\}$. As $C \cup D = \Delta(v)$ and $C \cap D = \emptyset$, exactly one of C or D is in $\mathcal{U}_v$. If $D \in \mathcal{U}_v$ and w is the projection of v onto D, we then have that $w \in \widehat{V} \setminus V$ and for all $G_\delta \in \Delta(w)$, $\mathrm{lex}(G_\delta) \subset G_\delta$, which we have just shown above cannot happen. So $C \in \mathcal{U}_v$, and if we let v now be the projection of itself onto C, we see that for all $G_\delta \in \Delta(v)$, $G_\delta = \mathrm{lex}(G_\delta)$.

But then for any $G_\delta \in \Delta(v)$, G_δ is either a minimal prime subgroup or a branch prime subgroup. So let

$$A = \{G_\delta \in \Delta(v) : G_\delta \text{ is a minimal prime subgroup }\}$$

and

$$B = \{G_\delta \in \Delta(v) : G_\delta \text{ is a branch prime subgroup }\}.$$

Then $\Delta(v) = A \cup B$ and $A \cap B = \emptyset$. Now if $A \in \mathcal{U}_v$, then $\alpha(\psi(V))$ is a minimal prime subgroup, which is false. So $B \in \mathcal{U}_v$. Thus if we now take v to be the projection of itself onto B, then $v \in \widehat{V} \setminus V$ and any $G_\delta \in \Delta(v)$ is a branch prime subgroup.

Conversely, suppose there exists $e < v \in \widehat{V} \setminus V$ such that for all $G_\delta \in \Delta(v)$, G_δ is a branch prime subgroup.

Let us first explore what it means for special value G_δ in G to be a branch prime subgroup. Let $e < s \in G$ be special with value G_δ. Obviously, G_δ contains at least two incomparable minimal prime subgroups, M_1 and M_2. Thus there exist disjoint x, y such that $x \in M_1 \setminus M_2$ and $y \in M_2 \setminus M_1$. Since $s \notin M_1$ and $s \notin M_2$, without loss of generality, we can assume that $x = x \wedge s$, $y = y \wedge s$, and $x, y \ll s$. But then any special value G_ρ of x is incomparable to any special G_σ of y, and both $G_\rho, G_\sigma \subset G_\delta$. It will also be useful to note that if G_β is a special value properly contained in G_δ, then as $G_\beta \subset G_\delta = \text{lex}(G_\delta)$, there exists a minimal prime subgroup M such that $M \subset G_\delta$ and M is incomparable to G_β. Using the same argument as above (that showed that G_δ contained two incomparable special values), we now see that for any special $G_\beta \subset G_\delta$, there exists another special value G_γ also contained in G_δ such that G_γ is incomparable to G_β.

Now let $e < g \in \alpha(\psi(V))$. As shown before, if g is not special, then $g \in \text{lex}(\alpha(\psi(V)))$. So suppose that g is special. If $g \wedge v = e$, let G_β be a special value properly contained in some $G_\delta \in \Delta(v)$ and $e < t$ be special with value G_β. Then $t \in \alpha(\psi(V))$ by Proposition 26, and $g \wedge t = e$. So g is a nonunit in $\alpha(\psi(V))$ and hence is in $\text{lex}(\alpha(\psi(V)))$. If $g \wedge v \neq e$, then $g \ll v$ and hence the value G_β of g is strictly contained in a special value G_δ of v. But then there exists another special value $G_\gamma \subset G_\delta$ such that G_γ is incomparable to G_β. Let $e < s$ be special with value G_γ. Then $s \ll v$ shows that $s \in \alpha(\psi(V))$ and $s \wedge g = e$ shows that g is a nonunit in $\alpha(\psi(V))$.

So $\alpha(\psi(V)) = \text{lex}(\alpha(\psi(V)))$. Hence the proposition will be proved once we show that $\alpha(\psi(V))$ contains at least two incomparable prime subgroups.

As shown above, each $G_\delta \in \Delta(v)$ properly contains two incomparable special values, G_{ρ_δ} and G_{σ_δ}. Let $\Omega_\rho = \{G_{\rho_\delta} : G_\delta \in \Delta(v)\}$ and $\Omega_\sigma = \{G_{\sigma_\delta} : G_\delta \in \Delta(v)\}$.

For each $U \in \mathcal{U}_v$, define U_ρ to equal $\{G_{\rho_\delta} : G_\delta \in U\}$. Then $W_\rho = \{U_\rho : U \in \mathcal{U}_v\}$ is an ultrafilter on Ω_ρ since the mapping $G_\delta \mapsto G_{\rho_\delta}$ is a one-to-one correspondence between $\Delta(v)$ and Ω_ρ.

Let $\mathcal{F}_{W_\rho}$ be the set of ideals I of $\Delta(G)$ such that $I \cap \Omega_\rho \in W_\rho$. As shown in Proposition 25, $\mathcal{F}_{W_\rho}$ is a prime ideal filter on $\Delta(G)$.

Now let $J \in \psi(V)$; there exists $e < g \notin V$ such that $S(g) \subseteq J$, and so $J \cap \Delta(v) \in \mathcal{U}_v$. Let $A = (J \cap \Delta(v))_\rho$ and $I = \{G_\alpha \in \Delta(G) : G_\alpha \subseteq G_\delta \in A\}$. Then $I \in \mathcal{F}_{W_\rho}$. Now for any $G_\alpha \in I$, $g \notin G_\alpha$; hence $I \subseteq S(g) \subseteq J$, and so $J \in \mathcal{F}_{W_\rho}$. So $\psi(V) \subseteq \mathcal{F}_{W_\rho}$, and thus $P = \alpha(\mathcal{F}_{W_\rho}) \subseteq \alpha(\psi(V))$.

Repeating the process of the last three paragraphs with Ω_ρ rather than Ω_σ produces another prime Q also contained in $\alpha(\psi(V))$. Thus for each $G_\delta \in \Delta(v)$, let $e < x_\delta, y_\delta$ be special elements with values G_{ρ_δ} and G_{σ_δ}, respectively, and then let $x = \vee x_\delta$ and $y = \vee y_\delta$. By Proposition 26, both x and y are in $\alpha(\psi(V))$. Since $\Delta(x) = \Omega_\rho$, $x \notin P$, and likewise $y \notin Q$. Since $x \wedge y = e$, $x \in Q$ and $y \in P$. So P and Q are incomparable prime subgroups contained in $\alpha(\psi(V))$. Thus, $\alpha(\psi(V))$ is a branch prime subgroup. ∎

For $V \in \Gamma(G)$, $\widehat{V}$ is also a prime subgroup of G, and obviously $\alpha(\psi(V)) \subseteq \alpha(\psi(\widehat{V}))$. Whether $\alpha(\psi(V)) = \alpha(\psi(\widehat{V}))$ or not depends upon characteristics of the ultrafilters $\mathcal{U}_v$ for $e < v \in \widehat{V} \setminus V$.

Recall that an ultrafilter has the *countable intersection property* if the intersection of any countable set of elements of the ultrafilter is also in the ultrafilter. Obviously, any principal ultrafilter $\mathcal{U}$ on a set Λ has the countable intersection property; the existence of nonprincipal ultrafilters with the countable intersection property is equivalent to the existence of measurable cardinals [GJ60].

Suppose, though, that $\mathcal{U}$ is an ultrafilter on a set Λ that does not have the countable intersection property; this is equivalent to there being $U_1 \supset U_2 \supset \cdots$ in $\mathcal{U}$ such that $\cap_{i=1}^{\infty} U_i = \emptyset$.

Lemma 30. *Let $V \in \Gamma(G)$ and $e < v, w \in \widehat{V} \setminus V$. Then $\mathcal{U}_v$ has the countable intersection property if and only if $\mathcal{U}_w$ has the countable intersection property.*

Proof. Suppose by way of contradiction that $\mathcal{U}_v$ has the countable intersection property while $\mathcal{U}_w$ does not. Let $W_1 \supset W_2 \supset \ldots$ be in $\mathcal{U}_w$ such that $\cap_{i=1}^{\infty} W_i = \emptyset$. Then $U_i = \Delta(v) \cap W_i \in \mathcal{U}_v$, and clearly $U_1 \supseteq U_2 \supseteq \ldots$ and $\cap_{i=1}^{\infty} U_i = \emptyset$, which is impossible. ∎

Theorem 31. *Let $V \in \Gamma(G)$. Then $\alpha(\psi(V)) \subset \alpha(\psi(\widehat{V}))$ if and only if for any $e < v \in \widehat{V} \setminus V$, $\mathcal{U}_v$ has the countable intersection property.*

Proof. Assume $\alpha(\psi(V)) \subset \alpha(\psi(\widehat{V}))$ and, by way of contradiction, that there exists $e < v \in \widehat{V} \setminus V$ such that $\mathcal{U}_v$ does not have the countable intersection property.

Let $I \in \psi(V)$ and $e < g \notin V$ such that $S(g) \subseteq I$. Now if $g \notin \widehat{V}$, then $I \in \psi(\widehat{V}) \subseteq \psi(V)$ since $V \subset \widehat{V}$. If $g \in \widehat{V}$, then $e < g \in \widehat{V} \setminus V$ and so, as $\mathcal{U}_v$ does not have the countable intersection property, then by Lemma 30, $\mathcal{U}_g$ also does not have the countable intersection property. Thus there exist $U_1 \supset U_2 \supset \cdots$ in $\mathcal{U}_g$ with empty intersection."

For $G_\delta \in U_n \setminus U_{n+1}$, let $w_\delta = g_\delta^n$, and let $w = \vee_{\Delta(g)} w_\delta$. w exists since G is laterally complete and $\{ w_\delta : G_\delta \in \Delta(g) \}$ is a pairwise disjoint set of positive elements. If $w \in \widehat{V}$, there exists an integer k such that $Vg^k > Vw$, and so $g^k w^{-1} \vee e \in \widehat{V} \setminus V$. So

$$W = \Delta(g^k w^{-1} \vee e) \cap \Delta(g) \in \mathcal{U}_g,$$

and for all $G_\delta \in W$, $g_\delta^k > w_{gd}$. But then for $n > k$, $W \cap U_n \in \mathcal{U}_g$ and for all $G_\delta \in W \cap U_n$, $g_\delta^k > w_\delta > g_\delta^n$, which is absurd. So $w \notin \widehat{V}$.

But since $\Delta(w) = \Delta(g)$, $S(w) = S(g) \subseteq I$, and so $I \in \psi(\widehat{V})$. Hence $\psi(V) = \psi(\widehat{V})$, which contradicts $\alpha(\psi(V)) \subset \alpha(\psi(\widehat{V}))$.

Conversely, suppose that for any $e < v \in \widehat{V} \setminus V$, $\mathcal{U}_v$ has the countable intersection property but $\alpha(\psi(V)) = \alpha(\psi(\widehat{V}))$. Then $\psi(V) = \psi(\widehat{V})$, and so for any $e < v \notin V$, there exists $e < w \notin \widehat{V}$ such that $S(w) \subseteq S(v)$.

So let $e < v \in \widehat{V} \setminus V$ and $e < w \notin \widehat{V}$ such that $S(w) \subseteq S(v)$. Let

$$A = \{G_\delta \in \Delta(v) : \text{ there exists } G_\alpha \in \Delta(w) \text{ such that } G_\alpha \subseteq G_\delta\};$$

let $B = \Delta(v) \setminus A$. If $A \notin \mathcal{U}_v$, then $B \in \mathcal{U}_v$, and so the projection, z, of v onto B is not in V while $w \wedge z = e$ (since $S(w) \subseteq S(v)$). But then $w \in V \subset \widehat{V}$, contradicting $w \notin \widehat{V}$. So $A \in \mathcal{U}_v$.

Without loss of generality, let v be its projection onto A and $w = w \vee v$; we then have that $e < v \in \widehat{V} \setminus V$, $e < w \notin \widehat{V}$, but also that $w > v$ and $S(w) = S(v)$. This, though, must mean that $\Delta(w) = \Delta(v)$.

But now for $n = 1, 2, \ldots$, let $W_n = \{G_\delta \in \Delta(v) : v_\delta^n \leq w_\delta < v_\delta^{n+1}\}$. Then $\{W_n : n = 1, 2, \ldots\}$ partition $\Delta(v)$. Since $\mathcal{U}_v$ has the countable intersection property, exactly one $W_n \in \mathcal{U}_v$. But then $Vv^{n+1} \geq Vw$, and so $w \in \widehat{V}$, which again contradicts our choice of w.

Thus $\alpha(\psi(V)) \subset \alpha(\psi(\widehat{V}))$. ∎

Proposition 32. *Let $V \in \Gamma(G)$ such that for any $e < v \in \widehat{V} \setminus V$, $\mathcal{U}_v$ has the countable intersection property. Then $\alpha(\psi(V)) = V$ and $\alpha(\psi(\widehat{V})) = \widehat{V}$.*

Proof. Suppose by way of contradiction that there exists $e < g \in V \setminus \alpha(\psi(V))$. Then $S(g) \in \psi(V)$ implies there exists $e < w \notin V$ such that $S(w) \subseteq S(g)$. As shown in the proof immediately above, we can assume that $w = w \vee g \notin V$ and $\Delta(w) = \Delta(g)$.

Let $e < v \in \widehat{V} \setminus V$. Since $S(v), S(w) \in \psi(V)$, $S(v \wedge w) = S(w) \cap S(v) \in \psi(V)$. So, without loss of generality, g is its projection onto $\Delta(v \wedge w)$, $w \in \widehat{V} \setminus V$, and, still, $\Delta(g) = \Delta(w)$. The importance of this is that now we can assume that $\mathcal{U}_g$ has the countable intersection property.

But, again as before, let $W_n = \{G_\delta \in \Delta(g) : g_\delta^n \leq w_\delta < g_\delta^{n+1}\}$. Again, there exists a unique $W_n \in \mathcal{U}_w$. But then $V < Vw < Vg^{n+1} = V$, and so $w \in V$. Thus no such g exists, and so $V = \alpha(\psi(V))$. Finally, as $V \subset \alpha(\psi(\widehat{V})) \subseteq \widehat{V}$ by Theorem 31 and $\alpha(\psi(\widehat{V})) \subseteq \widehat{V}$, $\alpha(\psi(\widehat{V})) = \widehat{V}$. ∎

For ease of notation (and following well established convention), if $G_\delta \in \Delta(G)$, $\widehat{G}_\delta$ will be hereafter denoted G^δ.

Proposition 33. *Let $V \in \Gamma(G)$. $\alpha(\psi(V)) \in \Gamma(G)$ if and only if either for any $e < v \in \widehat{V} \setminus V$, $\mathcal{U}_v$ has the countable intersection property, or there exists $e < v \in \widehat{V} \setminus V$ such that for all $G_\delta \in \Delta(v)$, $G^\delta/G_\delta \cong \mathbb{Z}$.*

Proof. ($\Rightarrow$) Suppose that $\alpha(\psi(V)) \in \Gamma(G)$; because $\alpha(\psi(\alpha(\psi(V)))) = \alpha(\psi(V))$, we can then assume that $V = \alpha(\psi(V))$. Also assume that for any $e < v \in \widehat{V} \setminus V$, $\mathcal{U}_v$ does not have the countable intersection property. Suppose by way of contradiction that there does not exist $e < v \in \widehat{V} \setminus V$ such that for all $G_\delta \in \Delta(v)$, G^δ/G_δ is not isomorphic to $\mathbb{Z}$.

Let $e < v \in \widehat{V} \setminus V$. Let $A = \{G_\delta \in \Delta(v) : G^\delta/G_\delta \not\cong \mathbb{Z}\}$; then $A \in \mathcal{U}_v$, as otherwise the projection x of v onto $\Delta(v) \setminus A$ is in $\widehat{V} \setminus V$, and for all $G_\delta \in \Delta(x)$, $G^\delta/G_\delta \cong \mathbb{Z}$. Without loss of generality, v is its projection onto A.

Since $\mathcal{U}_v$ does not have the countable intersection property, there exist $U_1 \supset U_2 \supset \cdots$ in $\mathcal{U}_v$ such that

$$\cap_{n=1}^{\infty} U_n = \emptyset.$$

Because for all $G_\delta \in \Delta(v)$, $G^\delta/G_\delta \not\cong \mathbb{Z}$, G^δ/G_δ is a dense subgroup of $\mathbb{R}$. Thus for any positive integer n, there exists $w_\delta \in G^\delta \setminus G_\delta$ such that $G_\delta w_\delta^n \leq G_\delta v < G_\delta w_\delta^{n+1}$.

Let $w = \bigvee_{\Delta(v)} w_\delta$, where for all $G_\delta \in U_n \setminus U_{n+1}$, $G_\delta w_\delta^n \leq G_\delta v < G_\delta w_\delta^{n+1}$. Then for any $W \in \mathcal{U}_v$, $\{G_\delta \in W : w_\delta^n \leq v_\delta\} \in \mathcal{U}_v$. So $Vw^n < Vv$ for all n, and thus $w \in V$. However, $S(w) = S(v)$ implies that $w \notin \alpha(\psi(V)) = V$. Thus our assumption was wrong and so there must exist $e < v \in \widehat{V} \setminus V$ such that for all $G_\delta \in \Delta(v)$, $G^\delta/G_\delta \cong \mathbb{Z}$.

($\Leftarrow$) We have already seen that if there exists $e < v \in \widehat{V} \setminus V$ such that $\mathcal{U}_v$ has the countable intersection property, then $\alpha(\psi(V)) = V$, and so is a value. So we must show that if there exists $e < v \in \widehat{V} \setminus V$ such that for all $G_\delta \in \Delta(v)$, $G^\delta/G_\delta \cong \mathbb{Z}$, then $\alpha(\psi(V))$ is a value.

For all $G_\delta \in \Delta(v)$, let $w_\delta \in G^\delta \setminus G_\delta$ be special such that $G_\delta w_\delta$ covers G_δ, and let $w = \bigvee_{\Delta(v)} w_\delta$. Since $S(w) = S(v)$, $w \notin \alpha(\psi(V))$. So w has a value $W \supseteq \alpha(\psi(V))$. Suppose that $\alpha(\psi(V)) \subset W$; let $e < x \in W \setminus \alpha(\psi(V))$. Then $S(x) \in \psi(V)$ and so there exists $e < y \notin V$ such that $S(y) \subseteq S(x)$. By letting v equal its projection onto $\Delta(v) \cap \Delta(y)$, we can assume that $y = v$. But then for all $G_\delta \in \Delta(v)$, $G_\delta x \geq G_\delta w$. Thus $x \geq w$ and so $x \notin W$, which contradicts our definition of x. So $\alpha(\psi(V))$ is a value of w, and thus in $\Gamma(G)$. ∎

Definition F. Let $V \in \Gamma(G)$. Define $W \in \Gamma(G)$ to be *stalk-equivalent* to V if $\alpha(\psi(W)) = \alpha(\psi(V))$, and *stalk($V$)* to be the set of regular subgroups W that are stalk-equivalent to V.

The above definition of a stalk differs from that used by Conrad in [C80] and Anderson and Kenny in [AK80]. For them, two prime subgroups P_1 and P_2 are stalk-equivalent if P_1 and P_2 contain exactly the same minimal prime subgroups.

We started the discussion of determining $\Gamma(G)$ from a plenary subset Δ by reminding the reader of the well known structure of $\Gamma(\prod_{i=1}^{\infty} \mathbb{Z})$. The following proposition shows that the arrangement of regular subgroups in $\prod_{i=1}^{\infty} \mathbb{Z}$ occurs in many other settings.

Proposition 34. *Let $V \in \Gamma(G)$. Suppose there exists $e < v \in \widehat{V} \setminus V$ such that $\mathcal{U}_v$ does not have the countable intersection property. If there exists $e < w \in \widehat{V} \setminus V$ such that $G^\delta/G_\delta \cong \mathbb{Z}$ for all $G_\delta \in \Delta(w)$, then stalk(V) contains a minimal element, namely $\alpha(\psi(V))$, and the set $\{W \in stalk(V) : W \supset \alpha(\psi(V))\}$ is an η_1-set. If no such w exists, then stalk(V) has no minimal element and is an η_1-set itself.*

Proof. Let $A = \{g \in G : \Delta(g) \subseteq S(v)\}$; then A is a closed convex ℓ-subgroup of G [BD86], and $N_v = \cap\{G_\delta : G_\delta \in \Delta(v)\} \subseteq A$. Then A/N_v is archimedean [M92] and, since G is laterally complete and special-valued, is easily shown to be isomorphic to $\prod G^\delta/G_\delta$ (where G^δ denotes the cover of G_δ). Note that $N_v \subseteq \alpha(\psi(V))$ and that $\alpha(\psi(V))/N_v$ is a minimal prime subgroup of A/N_v. The rest of the theorem now follows from what is known (see Gilman and Jerison [GJ60], Anderson and Kenny [AK80], or Darnel and Martinez [DM00]) of values of products of archimedean o-groups. ∎

Proposition 35. *Let $V \in \Gamma(G)$. Suppose there exists $e < v \in \widehat{V} \setminus V$ such that $\mathcal{U}_v$ does not have the countable intersection property. Then for any $W \in stalk(V)$ such that $\alpha(\psi(V)) \subset W$, $\widehat{W}/W \cong \mathbb{R}$.*

Proof. This was shown in [D(s)]. ∎

To review, we have seen that the class of prime subgroups of a laterally complete special-valued ℓ-group having the property that $P = \alpha(\psi(P))$ is constructible from the plenary subset of special values, and includes all closed prime subgroups, all branch primes, and all minimal prime subgroups. In addition, necessary and sufficient conditions were given for when $\alpha(\psi(P))$ is a value itself.

To conclude, the author would like to remind the reader of an alternative method devised by Anderson and Kenny [AK80] in investigating the prime subgroups of a Hahn group, $V(\Delta, \mathbb{R})$. Such Hahn groups are laterally complete and special-valued [CHH63], and their approach applies equally well to all laterally complete special-valued ℓ-groups.

To start, let $\mathcal{A}$ be the set of maximal antichains of Δ, partially ordered by $X \leq Y$ if for all $G_\alpha \in X$, there exists $G_\beta \in Y$ such that $G_\alpha \subseteq G_\beta$. Under this ordering, $\mathcal{A}$ is a lattice. For $X \leq Y$ in $\mathcal{A}$, an ultrafilter $\mathcal{U}$ on X is *compatible* with an ultrafilter $\mathcal{V}$ on Y if for any $U \in \mathcal{U}$, $\{G_\beta \in Y : G_\beta \supseteq G_\alpha \in U\} \in \mathcal{V}$, and if for any $V \in \mathcal{V}$, $\{G_\alpha \in X : G_\alpha \subseteq G_\beta \in V\} \in \mathcal{U}$. A *compatible system* of ultrafilters on $\mathcal{A}$ is a set of ultrafilters $\{\mathcal{U}(X) : X \in \mathcal{A}\}$ such that if $X \leq Y$, $\mathcal{U}(X)$ is compatible with $\mathcal{U}(Y)$. Anderson and Kenny showed that there is a bijection between minimal prime subgroups and compatible systems of ultrafilters. For a branch prime subgroup $P = \vee_{\lambda \in \Lambda}\{M_\lambda\}$, where each M_λ is a minimal prime subgroup, they showed that P could be identified with the collection of compatible systems of ultrafilters identified with each M_λ, this collection having the property that for any $X \in \mathcal{A}$, the ultrafilter on X from the compatible system of M_{λ_1} is the same as that from the system of M_{λ_2} for any $\lambda_1, \lambda_2 \in \Lambda$. Conversely, suppose that a set of compatible systems of ultrafilters has the property that for some $X \in \mathcal{A}$, every system produces the same ultrafilter on X. Anderson and Kenny showed that such a set identified a branch prime subgroup. They gave further results on classifying the stalks of primes. The prime subgroups identified by them can be shown (with some difficulty in some instances) to all be of the form $\alpha(\mathcal{F})$ for some prime ideal filter $\mathcal{F}$ on Δ.

References

[AK80]   M. Anderson and G. O. Kenny, *The root system of primes of a Hahn group.* J. Austr. Math. Soc. (Series A) **29** (1980), 17-28.

[BCD86] R. N. Ball, P. F. Conrad, and M. R. Darnel, *Above and below ℓ-subgroups of a lattice-ordered group.* Trans. Amer. Math. Soc. **297** (1986), 1-40.

[BD86]   J. P. Bixler and M. R. Darnel, *Special-valued ℓ-groups.* Alg. Univ. **22** (1986), 172-191.

[C67]   P. F. Conrad, *A characterization of lattice-ordered groups by their convex ℓ-subgroups.* J. Aust. Math. Soc. **7** (1967), 145-189.

[C80]   P. F. Conrad, *Minimal prime subgroups of lattice-ordered groups.* Czech. Math. J. **30** (1980), 280-285.

[CHH63] P. F. Conrad, J. Harvey, and W. C. Holland, *The Hahn embedding theorem for lattice-ordered groups.* Trans. Amer. Math. Soc. **108** (1963), 143-169.

[D95]   M. R. Darnel, *The Theory of Lattice-ordered Groups.* Pure and Applied Mathematics **187** (1995), Marcel Dekker.

[D(s)]   M. R. Darnel, *Measurable cardinals and a conjecture for lattice-ordered groups.* Submitted 2001.

[DM00] M. R. Darnel and J. Martinez. *Hyper-special valued lattice-ordered groups.* Forum Math. **12** (2000), 477-512.

[GJ60]   L. Gillman and M. Jerison *Rings of Continous Functions.* (1960) Van Nostrand.

[M92]   J. Martinez, *Locally conditioned radical classes of lattice-ordered groups.* Czech. Math. J. **42 (117)** (1992), 25-34.

[P61]   R. S. Pierce, *Rings of integer-valued functions.* Trans. AMS **100** (1961), 371-394.

Department of Mathematical Sciences, Indiana University South Bend,
South Bend, IN 46615, USA
mdarnel@iusb.edu

The Undecidability of the Word Problem for Distributive Residuated Lattices

Nikolaos Galatos

ABSTRACT.   Let $\mathbf{A} = \langle X \mid R \rangle$ be a finitely presented algebra in a variety $\mathcal{V}$. The algebra $\mathbf{A}$ is said to have an undecidable word problem if there is no algorithm that decides whether or not any two given words in the absolutely free term algebra $T_\mathcal{V}(X)$ represent the same element of $\mathbf{A}$. If $\mathcal{V}$ contains such an algebra $\mathbf{A}$, we say that it has an undecidable word problem. (It is well known that the word problem for the varieties of semigroups, groups and l-groups is undecidable.)

The main result of this paper is the undecidability of the word problem for a range of varieties including the variety of distributive residuated lattices and the variety of commutative distributive ones. The result for a subrange, including the latter variety, is a consequence of a theorem by Urquhart [7]. The proof here is based on the undecidability of the word problem for the variety of semigroups and makes use of the concept of an n-frame, introduced by von Neumann. The methods in the proof extend ideas used by Lipshitz and Urquhart to establish undecidability results for the varieties of modular lattices and distributive lattice-ordered semigroups, respectively.

1 Introduction

Definition 1.1. A *residuated lattice*, or *residuated lattice-ordered monoid*, is an algebra $\mathbf{l} = \langle L, \wedge, \vee, \cdot, e, \backslash, / \rangle$ such that $\langle L, \wedge, \vee \rangle$ is a lattice, $\langle L, \cdot, e \rangle$ is a monoid and multiplication is both left and right residuated, with $\backslash$ and $/$ as residuals, i.e., $a \cdot b \leq c \Leftrightarrow a \leq c/b \Leftrightarrow b \leq a \backslash c$, for all $a, b, c \in L$.

This definition of a residuated lattice is more general than the original given by Ward and Dilworth [1] in 1939. Here it is not stipulated that the monoid reduct is commutative nor that the lattice reduct has e as its top element. Residuated lattices have arisen also in logic, in connection to the Lambek calculus: see, e.g. the papers of J.M. Dunn and H. Ono in [9]. The structure theory of residuated lattices was first studied by K. Blount and C. Tsinakis in [2]. Although no further knowledge of what

231

J. Martínez (ed.), Ordered Algebraic Structures, 231–243.
© 2002 *Kluwer Academic Publishers.*

is included in this section is required for the comprehension of this paper, the reader
is advised to refer to [2] for more information about residuated lattices.

It is not hard to see that $\mathcal{RL}$, the class of all residuated lattices, is a variety and

$$x \approx x \wedge (xy \vee z)/y, \quad x(y \vee z) \approx xy \vee xz, \quad (x/y)y \vee x \approx x$$
$$y \approx y \wedge x \backslash (yx \vee z), \quad (y \vee z)x \approx yx \vee zx, \quad y(y \backslash x) \vee x \approx x$$

together with the monoid and the lattice identities form an equational basis for it.
Actually $\mathcal{RL}$ is an ideal variety, i.e., congruence relations are determined by their
e-classes. Moreover, the latter are subalgebras with specific properties described in
[2]: they have to be order convex and closed under all conjugation maps λ_a, ρ_a, where
$\lambda_a(x) = (a\backslash(xa))\wedge e$, $\rho_a(x) = ((ax)/a)\wedge e$, and a ranges over elements of the residuated
lattice.

The following lemma of [2] contains all the necessary identities for algebraic manip-
ulations in residuated lattices.

Lemma 1.2. *If x, y, z are elements of a residuated lattice, then the following properties
hold.*

1) $x(y \vee z) = xy \vee xz$ and $(y \vee z)x = yx \vee zx$.

2) $x \backslash (y \wedge z) = (x \backslash y) \wedge (x \backslash z)$ and $(y \wedge z)/x = (y/x) \wedge (z/x)$.

3) $x/(y \vee z) = (x/y) \wedge (x/z)$ and $(y \vee z) \backslash x = (y \backslash x) \wedge (z \backslash x)$.

4) $(x/y)y \leq x$ and $y(y \backslash x) \leq x$.

5) $x(y/z) \leq (xy)/z$ and $(z \backslash y)x \leq z \backslash (yx)$.

6) $(x/y)/z = x/(zy)$ and $z \backslash (y \backslash x) = (yz) \backslash x$.

7) $x \backslash (y/z) = (x \backslash y)/z$.

8) $x/e = x = e \backslash x$.

9) $e \leq x/x$ and $e \leq x \backslash x$.

10) $x(x \backslash x) = x = (x/x)x$.

11) $(x \backslash x)^2 = (x \backslash x)$ and $(x/x)^2 = (x/x)$.

12) *If the residuated lattice has a bottom element 0, then it has a top element 1, as
well, and for all x, $x0 = 0x = 0$, $x/0 = 0 \backslash x = 1$ and $1/x = x \backslash 1 = 1$.*

This paper is heavily influenced by work on similar problems. Lipshitz [5] established the undecidability of the word problem for modular lattices and Urquhart for DL-semigroups [8] and models of relevance logic [7]. Moreover, [3] contains undecidability results about relation algebras, while Freese [4] proved that the word problem for the free modular lattice on five generators is undecidable. The proofs of all the above make use of the notion of an *n-frame*, introduced by von Neumann in [10]. It is a geometric concept that was originally used in the definition of the von Staudt product of two points on a projective line. Taking advantage of the intrinsic connections between projective geometry and modular lattices, von Neumann defined this product in the latter. In other words, the notion of an n-frame can be used to define a semigroup structure in a modular lattice. Lipshitz used this fact to reduce the decidability of the word problem for modular lattices to the one for semigroups. Going one step further and using a modified version of an n-frame, Urquhart applied similar ideas to DL-semigroups. In this paper we give the definition of an n-frame and the results for modular lattices from [5], some of which will be used later on, before presenting the modified definition for residuated lattices together with the corresponding theorem.

2 Modular lattices.

We begin with a version of the original definition of von Neumann that is essentially equivalent to it.

Definition 2.1. A *modular n-frame* in a lattice **L** is an $n \times n$ matrix, $C = [c_{ij}]$, $c_{ij} \in \mathbf{L}$, (set $a_i = c_{ii}$ and $e = \bigwedge\{a_i \mid i \in \mathbb{N}_n\}$; $\mathbb{N}_n = \{1, ..., n\}$), such that:

   i) $\bigvee A_1 \wedge \bigvee A_2 = \bigvee(A_1 \cap A_2)$, for all $A_1, A_2 \subseteq \{a_1, a_2, ..., a_n\}$, where $\bigvee \emptyset = e$;

   ii) $c_{ij} = c_{ji}$, for all $i, j \in \mathbb{N}_n$;

   iii) $a_i \vee a_j = a_i \vee c_{ij}$, for all $i, j \in \mathbb{N}_n$;

   iv) $a_i \wedge c_{ij} = e$, for all distinct $i, j \in \mathbb{N}_n$;

   v) $(c_{ij} \vee c_{jk}) \wedge (a_i \vee a_k) = c_{ik}$, for all distinct triples $i, j, k \in \mathbb{N}_n$.

The following examples, taken from [3], give some idea of the motivation for the definition.

Example 2.2. Consider the real projective plane P. The lattice **L** of subspaces of P contains points, projective lines, P and $\emptyset$, ordered under inclusion. Meet is intersection of subsets of P, while the join of two projective subspaces is the least subspace containing both of them. Modularity of **L** is well known and easy to establish. A modular 3-frame, see Figure 1, will consist of essentially six points: $a_1, a_2, a_3, c_{12}, c_{13}, c_{23}$, because of (ii). The points a_1, a_2, a_3 are not collinear, by condition (i); c_{ij} has to be on

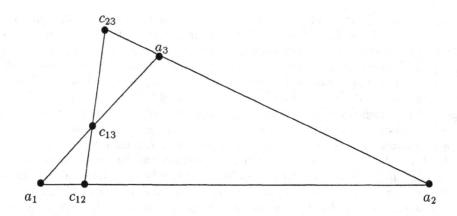

Figure 1: The geometric meaning of a modular 3-frame.

the line $a_i \vee a_j$, by (iii), while c_{12}, c_{13}, c_{23} are collinear, by condition (v); actually, c_{ik} is the point of intersection of the lines $a_i \vee a_j$ and $c_{ij} \vee c_{jk}$.

Example 2.3. Let $\mathbf{V}$ be an n-dimensional real inner product space, $\{e_i \mid i \in \mathbb{N}_n\}$ an orthonormal base of $\mathbf{V}$, $a_i = \langle e_i \rangle$, the subspace generated by e_i, and $c_{ij} = \langle e_i - e_j \rangle$. Then $[c_{ij}]$, $i, j \in \mathbb{N}_n$, is a modular n-frame in the lattice $\mathbf{L}$ of subspaces of $\mathbf{V}$.

Given a modular n-frame one can define operations of multiplication and addition on certain elements of the lattice.

Definition 2.4. Let $[c_{ij}]$ be a modular n-frame in a modular lattice $\mathbf{L}$. Define

  i) $L_{ij} = \{x \in \mathbf{L} \mid x \vee a_j = a_i \vee a_j \text{ and } x \wedge a_j = e\}$, for all distinct $i, j \in \mathbb{N}_n$;

  ii) $b \otimes_{ijk} d = (b \vee d) \wedge (a_i \vee a_k)$, for all $b \in L_{ij}, d \in L_{jk}$;

  iii) $b \odot_{ij} d = (b \otimes_{ijk} c_{jk}) \otimes_{ikj} (c_{ki} \otimes_{kij} d)$, for all $b, d \in L_{ij}$;

  iv) $b \oplus_{ij} d = [((b \vee c_{ik}) \wedge (a_j \vee a_k)) \vee ((d \vee a_k) \wedge (a_j \vee c_{ik}))] \wedge (a_i \vee a_j)$, for all $b, d \in L_{ij}$.

In [5] it is shown that the definitions of $\odot_{ij}$ and $\oplus_{ij}$ are independent of the choice of $k \in \mathbb{N}_n$, for $k \neq i, k \neq j$.

Remark 2.5. The definitions of $b \odot_{ij} d$ and $b \oplus_{ij} d$ differ from [10] and [5]. There multiplication and addition are not defined for elements of L_{ij}, but for L-numbers. An L-number α in a modular n-frame C is a set of lattice elements indexed by $\{(i,j) \mid i, j \in \mathbb{N}_n, i \neq j\}$, such that $(\alpha)_{kh} = [P(i,j,k,h)]((\alpha)_{ij})$, where $(\alpha)_{ij}$ symbolizes

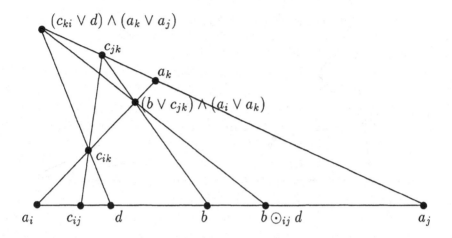

Figure 2: The geometric meaning of $\odot_{ij}$.

the (i, j)-coordinate of α and $P(i, j, k, h)$ is the composition of the two perspective isomorphisms with axes c_{jh} and c_{ik}. Lemma 6.1 of [10] guarantees that one can work with the fixed (i, j)-coordinates of L-numbers instead of them, since given $i, j \in \mathbb{N}_n$ the correspondence between α and $(\alpha)_{ij}$ is a bijection. Moreover, this bijection between L-numbers under the multiplication and addition defined in [10] and L_{ij} under $\odot_{ij}$ and $\oplus_{ij}$ is a ring isomorphism, as it can be deduced from Lemmas 6.2, 6.3, Theorem 6.1 and the appendix to Chapter 6, Part II of [10]. Freese, in [4], is the first one to use $\odot_{ij}$ and $\oplus_{ij}$, instead of multiplication and addition of L-numbers, and essentially the definition of an n-frame presented here.

In the context of the first example, L_{ij} is the set of all points x on the line $a_i \vee a_j$, $(x \vee a_i = a_i \vee a_j)$, different from a_i, $(x \wedge a_j = e)$, $b \otimes_{ijk} d$ is by definition the intersection of the lines $b \vee d$ and $a_i \vee a_j$, for $b \in L_{ij}, d \in L_{jk}$, while $b \odot_{ij} d$ and $b \oplus_{ij} d$, $b, d \in L_{ij}$ are the (von Staudt) product and sum, see Figures 2 and 3, of b and d on the line $a_i \vee a_j$, where a_i plays the role of zero, c_{ij} is the unit and a_j is infinity. Some projective geometry is required to verify this assertion.

The following theorem of [10] justifies the terminology of multiplication and addition, and validates the connection between projective geometry and modular lattices.

Theorem 2.6 [Von Neumann] *Let $C = [c_{ij}]$ be a modular n-frame in a modular lattice $\mathbf{L}$, where $n \geq 4$. Then $\mathbf{R}_{ij} = \langle L_{ij}, \oplus_{ij}, \odot_{ij}, a_i, c_{ij} \rangle$ is a ring for all distinct $i, j \in \mathbb{N}_n$. Moreover, all rings $\mathbf{R}_{ij}$ are isomorphic.*

In view of the last statement of the previous theorem the choice of indices i, j in $\mathbf{R}_{ij}$ is inessential. So, $\mathbf{R}_{12} = \langle L_{12}, \oplus_{12}, \odot_{12}, a_1, c_{12} \rangle$ is called the *ring associated with*

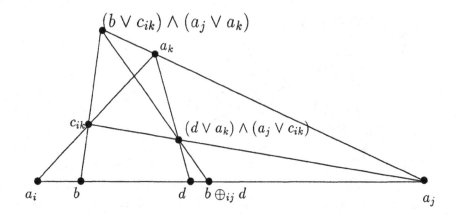

Figure 3: The geometric meaning of $\oplus_{ij}$.

the *modular n-frame C of* **L**.

For a vector space, **V**, denote by $L(\mathbf{V})$ the set of all subspaces of **V**. It is well known that $\mathbf{L}(\mathbf{V}) = \langle L(\mathbf{V}), \wedge, \vee \rangle$ is a modular lattice, where meet is intersection and the join of two subspaces is the subspace generated by their union.

The following results of [5] make use of the definition of a modular n-frame.

Lemma 2.7 [Lipshitz] *Let* **V** *be an infinite-dimensional vector space. Then,*

i) $\mathbf{L}(\mathbf{V})$ *contains a 4-frame, C, where e is the least element of* $\mathbf{L}(\mathbf{V})$ *and*

ii) *Any countable semigroup is a subsemigroup of the multiplicative semigroup of the ring associated with C.*

Theorem 2.8 [Lipshitz] *The word problem for modular lattices is undecidable.*

3   Distributive residuated lattices.

A residuated lattice is called *distributive* if its lattice reduct is distributive. Obviously the class of distributive residuated lattices is a variety and is denoted by $\mathcal{DRL}$.

We modify the definition of a modular n-frame, to suit our purposes.

Definition 3.1. A *residuated-lattice n-frame* (or just *n-frame*) in a residuated lattice **L** is an $n \times n$ matrix, $C = [c_{ij}]$, $c_{ij} \in \mathbf{L}$, (set $a_i = c_{ii}$), such that:

i) $a_i a_j = a_j a_i$, for all $i, j \in \mathbb{N}_n$;

ii) $\prod A_1 \wedge \prod A_2 = \prod (A_1 \cap A_2)$, for all $A_1, A_2 \subseteq \{a_1, a_2, ..., a_n\}$, where $\prod \emptyset = e$;

iii) $a_i^2 = a_i$, for all $i \in \mathbb{N}_n$;

iv) $c_{ij} c_{jk} \wedge a_i a_k = c_{ik}$, for all distinct triples $i, j, k \in \mathbb{N}_n$;

v) $c_{ij} = c_{ji}$, for all $i, j \in \mathbb{N}_n$;

vi) $c_{ij} a_j = a_i a_j$, for all $i, j \in \mathbb{N}_n$;

vii) $c_{ij} \wedge a_j = e$, for all distinct $i, j \in \mathbb{N}_n$.

It is clear that if multiplication is replaced by join, the conditions in the definition reduce to the ones in Definition 2.1.

Definition 3.2. i) An element a of a residuated lattice **L** is called *modular* if $c(b \wedge a) = cb \wedge a$ and $(a \wedge b)c = a \wedge bc$ for all elements b, c of L, such that $c \leq a$.

ii) An n-frame of a residuated lattice is called *modular* if $\prod A$ is modular, for all $A \subseteq \{a_1, a_2, ..., a_n\}$.

Definition 3.3. Let $[c_{ij}]$ be an n-frame in a residuated lattice **L**. Define

i) $L_{ij} = \{x \in L \mid x a_j = a_i a_j \text{ and } x \wedge a_j = e\}$, for all distinct $i, j \in \mathbb{N}_n$;

ii) $b \otimes_{ijk} d = bd \wedge a_i a_k$, for all $b \in L_{ij}, d \in L_{jk}$;

iii) $b \odot_{ij} d = (b \otimes_{ijk} c_{jk}) \otimes_{ikj} (c_{ki} \otimes_{kij} d)$, for all $b, d \in L_{ij}$ and for all distinct triples i, j, k.

The definition of $\odot_{ij}$ doesn't depend on the choice of k, as is shown in the lemma below.

Lemma 3.4. *Let* $c = [c_{ij}]$ *be a modular 4-frame in a residuated lattice* **L**.

i) *If* $b \in L_{ij}$, *then* $b \leq a_i a_j$;

ii) *If* $b \in L_{ij}$ *and* $d \in L_{jk}$, *then* $b \otimes_{ijk} d \in L_{ik}$, *for all distinct triples* $i, j, k \in \mathbb{N}_n$;

iii) *If* $b \in L_{ij}$, $d \in L_{jk}$ *and* $f \in L_{kl}$, *then* $(b \otimes_{ijk} d) \otimes_{ikl} f = b \otimes_{ijl} (d \otimes_{jkl} f)$, *for all distinct quadruples* $i, j, k, l \in \mathbb{N}_n$;

iv) *If* $b, d \in L_{ij}$, *then* $(b \otimes_{ijk} c_{jk}) \otimes_{ikj} (c_{ki} \otimes_{kij} d) = (b \otimes_{ijl} c_{jl}) \otimes_{ilj} (c_{li} \otimes_{lij} d)$, *for all distinct quadruples* $i, j, k, l \in \mathbb{N}_4$.

Proof. i) $b = be \leq ba_j = a_ia_j$.

ii) We first show that $(b \otimes_{ijk} d)a_k = a_ia_k$.

$$
\begin{aligned}
(b \otimes_{ijk} d)a_k &= (a_ia_k \wedge bd)a_k, \\
&= bda_k \wedge a_ia_k, && a_ia_k \text{ is modular and } a_k \leq a_ia_k, \text{ since } e \leq a_i \\
&= ba_ja_k \wedge a_ia_k, && da_k = a_ja_k, \text{ since } d \in L_{jk} \\
&= a_ia_ja_k \wedge a_ia_k, && ba_j = a_ia_j, \text{ since } b \in L_{ij} \\
&= a_ia_k, && \text{(ii) of Def. 3.1}
\end{aligned}
$$

To prove $b \otimes_{ijk} d \in L_{ij}$ we also need to show that $(b \otimes_{ijk} d) \wedge a_k = e$.

$$
\begin{aligned}
(b \otimes_{ijk} d) \wedge a_k &= bd \wedge a_ia_k \wedge a_k \\
&= bd \wedge a_k, && a_k \leq a_ia_k, \text{ since } e \leq a_i \\
&\leq bd \wedge a_ja_k, && a_k \leq a_ja_k, \text{ since } e \leq a_i \\
&= (b \wedge a_ja_k)d, && a_ja_k \text{ is modular and } d \leq a_ja_k, \text{ by (i)} \\
&= (b \wedge a_ja_k \wedge a_ia_j)d, && b \wedge a_ia_j = b, \text{ by (i)} \\
&= (b \wedge a_j)d, && a_ja_k \wedge a_ia_j = a_j, \text{ by (ii) of Def. 3.1} \\
&= d, && b \wedge a_j = e, \text{ since } b \in L_{ij}
\end{aligned}
$$

So, $(b \otimes_{ijk} d) \wedge a_k = b \otimes_{ijk} d \wedge a_k \wedge a_k \leq d \wedge a_k = e$, since $d \in L_{jk}$. Moreover,

$$
e = ee \wedge ee \wedge e \leq bd \wedge a_ia_k \wedge a_k = (b \otimes_{ijk} d) \wedge a_k.
$$

Thus, $(b \otimes_{ijk} d) \wedge a_k = e$.

iii) Since $b \in L_{ij}$, $d \in L_{jk}$ and $f \in L_{kl}$, by (ii) we get, $b \otimes_{ijk} d \in L_{ik}$ and $d \otimes_{jkl} f \in L_{jl}$; thus, $(b \otimes_{ijk} d) \otimes_{ikl} f$, $b \otimes_{ijl} (d \otimes_{jkl} f) \in L_{il}$.

$$
\begin{aligned}
(b \otimes_{ijk} d) \otimes_{ikl} f &= (bd \wedge a_ia_k)f \wedge a_ia_l \\
&= (bd \wedge a_ia_ja_k \wedge a_ia_ka_l)f \wedge a_ia_l, && \text{(ii) of Def. 3.1} \\
&= (bd \wedge a_ia_ka_l)f \wedge a_ia_l, && bd \leq a_ia_ja_ja_k = a_ia_ja_k, \\
&&& \text{by (i), since } b \in L_{ij}, d \in L_{jk} \\
&&& \text{and } a_j^2 = a_j \\
&= bdf \wedge a_ia_ka_l \wedge a_ia_l, && a_ia_ka_l \text{ is modular and} \\
&&& f \leq a_ka_l \leq a_ia_ka_l, \\
&&& \text{since } f \in L_{kl} \\
&= bdf \wedge a_ia_l, && \text{(ii) of Def. 3.1}
\end{aligned}
$$

Similarly, $b \otimes_{ijl} (d \otimes_{jkl} f) = bdf \wedge a_ia_l$, so

$$
(b \otimes_{ijk} d) \otimes_{ikl} f = b \otimes_{ijl} (d \otimes_{jkl} f).
$$

iv) First note that condition (iv) of the definition of an n-frame can be written as

$c_{rs} = c_{rt} \otimes_{rts} c_{ts}$, for all distinct triples $r, t, s \in \mathbb{N}_n$.

$$
\begin{aligned}
(a \otimes_{ijk} c_{jk}) \otimes_{ikj} (c_{ki} \otimes_{kij} b) &= (a \otimes_{ijk} (c_{jl} \otimes_{jlk} c_{lk})) \otimes_{ikj} (c_{ki} \otimes_{kij} b) \\
&= ((a \otimes_{ijl} c_{jl}) \otimes_{ilk} c_{lk}) \otimes_{ikj} (c_{ki} \otimes_{kij} b) \\
&= (a \otimes_{ijl} c_{jl}) \otimes_{ilj} (c_{lk} \otimes_{lkj} (c_{ki} \otimes_{kij} b)) \\
&= (a \otimes_{ijl} c_{jl}) \otimes_{ilj} ((c_{lk} \otimes_{lki} c_{ki}) \otimes_{lij} b) \\
&= (a \otimes_{ijl} c_{jl}) \otimes_{ilj} (c_{li} \otimes_{lij} b)
\end{aligned}
$$

Thus the definition of $\odot_{ij}$ is independent of k. ∎

Lemma 3.5. *Let $c = [c_{ij}]$ be a modular 4-frame in a residuated lattice $\mathbf{L}$.*

i) *If $b, d \in L_{ij}$, then $b \odot_{ij} d \in L_{ij}$, for all distinct $i, j \in \mathbb{N}_n$.*

ii) *If $b, d, f \in L_{12}$, then $(b \odot_{12} d) \odot_{12} f = b \odot_{12} (d \odot_{12} f)$.*

Proof. i) Since $b \in L_{ij}, c_{jk} \in L_{jk}$ and $c_{ki} \in L_{ki}, d \in L_{ij}$, we have $b \otimes_{ijk} c_{jk} \in L_{ik}$ and $c_{ki} \otimes_{kij} d \in L_{kj}$. So,

$$
b \odot_{ij} d = (b \otimes_{ijk} c_{jk}) \otimes_{ikj} (c_{ki} \otimes_{kij} d) \in L_{ij}.
$$

ii)
$$
\begin{aligned}
(b \odot_{12} d) \odot_{12} f &= \{[(b \otimes_{123} c_{23}) \otimes_{132} (c_{31} \otimes_{312} d)] \otimes_{123} c_{23}\} \otimes_{132} (c_{31} \otimes_{312} f) \\
&= \{[(b \otimes_{124} c_{24}) \otimes_{142} (c_{41} \otimes_{412} d)] \otimes_{123} c_{23}\} \otimes_{132} (c_{31} \otimes_{312} f) \\
&= \{(b \otimes_{124} c_{24}) \otimes_{143} [(c_{41} \otimes_{412} d) \otimes_{423} c_{23}]\} \otimes_{132} (c_{31} \otimes_{312} f) \\
&= (b \otimes_{124} c_{24}) \otimes_{142} \{[(c_{41} \otimes_{412} d) \otimes_{423} c_{23}] \otimes_{432} (c_{31} \otimes_{312} f)\} \\
&= (b \otimes_{124} c_{24}) \otimes_{142} \{[c_{41} \otimes_{413} (d \otimes_{123} c_{23})] \otimes_{432} (c_{31} \otimes_{312} f)\} \\
&= (b \otimes_{124} c_{24}) \otimes_{142} \{c_{41} \otimes_{412} [(d \otimes_{123} c_{23}) \otimes_{132} (c_{31} \otimes_{312} f)]\} \\
&= (b \otimes_{123} c_{23}) \otimes_{132} \{c_{31} \otimes_{312} [(d \otimes_{123} c_{23}) \otimes_{132} (c_{31} \otimes_{312} f)]\} \\
&= b \odot_{12} (d \odot_{12} f)
\end{aligned}
$$

A fact that establishes the associativity of $\odot_{12}$. ∎

Corollary 3.6. *Let $C = [c_{ij}]$ be a modular residuated-lattice 4-frame in a residuated lattice $\mathbf{L}$. Then, $\mathbf{S}_{12} = \langle L_{12}, \odot_{12} \rangle$ is a semigroup, called the semigroup associated with the 4-frame C.*

Lemma 3.7. *Let $\mathbf{L}$ be a distributive residuated lattice, with a top element, T, and a bottom element, B. If $a, \tilde{a}, \in L$, $a^2 \leq a$, $a\tilde{a} \leq \tilde{a}$, $\tilde{a}a \leq \tilde{a}$, $a \wedge \tilde{a} = B$ and $a \vee \tilde{a} = T$, then, a is modular.*

Proof. Let $b, c \in L, c \leq a$. Then, $(a \wedge b)c \leq ac \leq a^2 \leq a$ and $(a \wedge b)c \leq bc$; thus, $(a \wedge b)c \leq a \wedge bc$. On the other hand,

$$
\begin{aligned}
a \wedge bc &= a \wedge (b \wedge T)c = a \wedge (b \wedge (a \vee \tilde{a}))c \\
&= a \wedge ((b \wedge a) \vee (b \wedge \tilde{a}))c = a \wedge ((b \wedge a)c \vee (b \wedge \tilde{a})c) \\
&\leq a \wedge ((b \wedge a)c \vee \tilde{a}a) \leq (a \wedge (b \wedge a)c) \vee (a \wedge \tilde{a})
\end{aligned}
$$

$$\leq \quad (b \wedge a)c \vee B = (b \wedge a)c$$

Thus, $a \wedge bc = (b \wedge a)c$. Similarly, we get the other condition $a \wedge cb = c(a \wedge b)$.   ∎

Let $\mathbf{V}$ be a vector space. For $A, B \in L_{\mathbf{V}} = \mathcal{P}(V)$, the power set of $\mathbf{V}$, let

$$A \wedge B = A \cap B, \quad A \vee B = A \cup B,$$

$$AB = \{a + b \mid a \in A, b \in B\},$$

$$A \backslash B = B/A = \{c \mid \{c\}A \subseteq B\} \quad \text{and} \quad e = \{0_{\mathbf{V}}\}.$$

It is easy to see that $\mathbf{L_V} = \langle L_{\mathbf{V}}, \wedge, \vee, \cdot, e, \backslash, / \rangle$ is a distributive residuated lattice. Moreover, $L(\mathbf{V})$ is a subset of $L_{\mathbf{V}}$, but $\mathbf{L(V)}$ is not a sublattice of the lattice reduct of $\mathbf{L_V}$. Nevertheless, a subset A of V is in $L(\mathbf{V})$ if and only if $e \leq A$ and $AA = A$. Additionally, $\wedge_{L(\mathbf{V})} = \wedge_{\mathbf{L_V}}$ and $\vee_{L(\mathbf{V})} = \cdot_{\mathbf{L_V}}$.

Definition 3.8. If $\mathbf{S} = \langle S, \bullet \rangle$, $S = \langle x_1, x_2, ..., x_n \mid r_1^{\bullet}(\overline{x}) = s_1^{\bullet}(\overline{x}), ..., r_k^{\bullet}(\overline{x}) = s_k^{\bullet}(\overline{x}) \rangle$, is a finitely presented semigroup and $\mathcal{V}$ is a variety of residuated lattices, let $\mathbf{L(S, \mathcal{V})}$ be the residuated lattice in $\mathcal{V}$ with the presentation described below:

Generators:

$$x_1', x_2', ..., x_n', \ c_{ij} \ (i, j \in \mathbb{N}_4), \ \top, \bot \ \text{and} \ \overline{\prod A} \ (A \in \mathcal{A}(C) = \mathcal{P}(\{a_1, a_2, a_3, a_4\})).$$

Relations:

   i) Equations (i)-(vii) of Definition 3.1 (for $n = 4$);

   ii) $x_i' a_2 = a_1 a_2$ and $x_1' \wedge a_2 = e$, for all $i \in \mathbb{N}_n$;

   iii) $r_i^{\odot_{12}}(\overline{x}') = s_i^{\odot_{12}}(\overline{x}')$, for all $i \in \mathbb{N}_k$, where $t^{\odot_{12}}$ denotes the evaluation of t in the semigroup associated with the 4-frame $[c_{ij}]$;

   iv) $\bot^2 = \bot$, $\top^2 = \top = \top/\bot = \bot/\bot = \bot\backslash\top = \bot\backslash\bot$, $\bot \leq e \leq \top$ and $\bot \leq x \leq \top$, $\bot x = x\bot = \bot$, for every generator x;

   v) $x^2 \leq x$, $x\tilde{x} \leq \tilde{x}$, $\tilde{x}x \leq \tilde{x}$, $x \wedge \tilde{x} = \bot$ and $x \vee \tilde{x} = \top$, for all x of the form $\prod A, A \in \mathcal{A}(C)$.

Let $R(\overline{x})$ denote the conjunction $\bigwedge_{i \in \mathbb{N}_k} r_i(\overline{x}) = s_i(\overline{x})$ of the relations of $\mathbf{S}$ and $R'(\overline{x}', C, \overline{\mathcal{A}}(C), \bot, \top)$ of the relations of $\mathbf{L(S, \mathcal{V})}$. ($\overline{x}$ denotes $(x_1, x_2, ..., x_n)$)

Lemma 3.9. For every semigroup $\mathbf{S}$, $\mathbf{L(S, \mathcal{V})}$ has a bounded lattice reduct and $\bot, \top$ are the bottom and top elements.

Proof. We will prove that $\bot \leq w \leq \top$, for every word w in the generators. We first prove that "$\bot \leq w$ and $\bot w = w\bot = \bot$" for every word, w, using induction on the complexity of w. Properties stated in Lemma 1.2 will be used without reference. The statement is true for the generators and for e, by (iv) in the relations of $\mathbf{L}(\mathbf{S}, \mathcal{V})$. If the statement is true for words u, v, i.e. $\bot \leq u, v$ and $\bot u = u\bot = \bot v = v\bot = \bot$ then:

- $\bot(u \vee v) = \bot u \vee \bot v = \bot$ and $(u \vee v)\bot = \bot$. Also, $\bot \leq u \vee v$.

- $\bot \leq u \wedge v$ and $\bot = \bot\bot \leq \bot(u \wedge v) \leq \bot u \wedge \bot v = \bot$, while $(u \wedge v)\bot = \bot$ is proven in a similar way.

- $\bot \leq \bot\bot \leq uv$ and $\bot uv = \bot v = \bot$, while the other products equal $\bot$, also.

- Since $\bot v = \bot \leq u$, we have $\bot \leq u/v$; thus $\bot = \bot\bot \leq \bot(u/v)$ and $\bot \leq (u/v)\bot$.

Moreover, $\bot(u/v) \leq (\bot u)/v = \bot/v \leq \bot/\bot = \top$, so $u/v \leq \bot \backslash \top = \bot \backslash \bot = \bot/\bot$; hence $\bot(u/v) \leq \bot$ and $(u/v)\bot \leq \bot$. Thus, $\bot(u/v) = \bot$ and $(u/v)\bot = \bot$. For left division we work analogously. Consequently, $\bot$ is the bottom element of $\mathbf{L}$ and, by (12) of Lemma 1.2, $\top = \bot/\bot$ is the top element of $\mathbf{L}$. ∎

The following lemma is a well known fact from the theory of Universal Algebra.

Lemma 3.10. *Let $A = \langle \overline{x} | R(\overline{x}) \rangle$, be a finite presentation of an algebra $\mathbf{A}$ in a variety $\mathcal{V}$ where $\overline{x} = (x_1, ..., x_n)$, $n \in \mathbf{N}$ is the sequence of generators and $R(\overline{x})$ the conjunction of the relations. Also, let r, s be n-ary semigroup terms; then the following are equivalent:*

i) *$\mathbf{A}$ satisfies $r^{\mathbf{A}}(\overline{x}) = s^{\mathbf{A}}(\overline{x})$.*

ii) *For every algebra $\mathbf{B}$ in $\mathcal{V}$, if there exist elements $y_1, ..., y_n \in B$, such that $R(\overline{y})$ holds in $\mathbf{B}$, then $\mathbf{B}$ satisfies $r^{\mathbf{B}}(\overline{y}) = s^{\mathbf{B}}(\overline{y})$.*

Proof. For the non-trivial direction, note that the natural epimorphism from the free algebra of $\mathcal{V}$ on $\overline{x}$ to the subalgebra of $\mathbf{B}$ generated by $\overline{y}$, $F_{\mathcal{V}}(\overline{x}) \longrightarrow Sg_{\mathbf{B}}(\overline{y})$, $x_i \mapsto y_i$, factors through $\mathbf{F}_{\mathcal{V}}(\overline{x})/R(\overline{x}) \cong \mathbf{A}$. So, $f : A \longrightarrow Sg_{\mathbf{B}}(\overline{y}) \subseteq B$, $x_i \mapsto y_i$ is a homomorphism. Since $r^{\mathbf{A}}(\overline{x}) = s^{\mathbf{A}}(\overline{x})$, we get

$$r^{\mathbf{B}}(\overline{y}) = r^{\mathbf{B}}(f(\overline{x})) = f(r^{\mathbf{A}}(\overline{x})) = f(s^{\mathbf{A}}(\overline{x})) = s^{\mathbf{B}}(f(\overline{x})) = s^{\mathbf{B}}(\overline{y})$$

in $\mathbf{B}$. ∎

Lemma 3.11. *Let $\mathbf{S}$ be a semigroup, r, s semigroup terms and $\mathcal{V}$ a variety of distributive residuated lattices. If $\mathbf{S}$ satisfies $r^{\bullet}(\overline{x}) = s^{\bullet}(\overline{x})$ then $\mathbf{L}(\mathbf{S}, \mathcal{V})$ satisfies $r^{\odot_{12}}(\overline{x}') = s^{\odot_{12}}(\overline{x}')$.*

Proof. $C = [c_{ij}]$ is a 4-frame in $\mathbf{L}(\mathbf{S}, \mathcal{V})$, by (i) of $R'(\overline{x}', C, \tilde{\mathcal{A}}(C), \perp, \top)$ and $x'_i \in L_{12}$, by (ii). Moreover, by (v) and Lemma 3.7, $\prod A$ is modular, for all $A \in \mathcal{A}(C)$, hence C is modular. By Corollary 3.6, $\langle L_{12}, \odot_{12} \rangle$ is a semigroup and, by (iii), it satisfies $R(\overline{x}')$; thus, by Lemma 3.10, it also satisfies $r^{\odot_{12}}(\overline{x}') = s^{\odot_{12}}(\overline{x}')$.   ∎

We can now prove the main theorem.

Theorem 3.12. *Let $\mathcal{V}$ be a variety of distributive residuated lattices, containing $\mathbf{L}_V$, for some infinite-dimensional vector space $\mathbf{V}$. Then, there is a finitely presented residuated lattice in $\mathcal{V}$, with undecidable word problem.*

Proof. Let $\mathbf{S} = \langle S, \bullet \rangle$, $S = \langle x_1, x_2, ..., x_n | r_1^{\bullet}(\overline{x}) = s_1^{\bullet}(\overline{x}), ..., r_k^{\bullet}(\overline{x}) = s_k^{\bullet}(\overline{x}) \rangle$, be a finitely presented semigroup with undecidable word problem (see [6]) and consider $\mathbf{L}(\mathbf{S}, \mathcal{V})$.

We will show that, for every pair r, s of semigroup words, $\mathbf{S}$ satisfies $r^{\bullet}(\overline{x}) = s^{\bullet}(\overline{x})$ if and only if $\mathbf{L}(\mathbf{S}, \mathcal{V})$ satisfies $r^{\odot_{12}}(\overline{x}') = s^{\odot_{12}}(\overline{x}')$. Since one direction follows from Lemma 3.11, suppose that $\mathbf{S}$ does not satisfy $r(\overline{x}) = s(\overline{x})$. By Lemma 2.7, $\mathbf{S}$ is embeddable, via f, say, into the multiplicative semigroup of the ring, $\mathbf{R}$, associated with a modular 4-frame $\hat{C}$ in the modular lattice $\mathbf{L}(\mathbf{V})$. So, $r^{\mathbf{R}}(f(\overline{x})) = s^{\mathbf{R}}(f(\overline{x}))$ is false in $\mathbf{R}$, where $f(\overline{x}) = (f(x_1), ..., f(x_n))$, thus also false in $\mathbf{L}(\mathbf{V})$, if viewed as a lattice equation. Since, as noted before, $\wedge_{\mathbf{L}(\mathbf{V})} = \wedge_{\mathbf{L}_V}$ and $\vee_{\mathbf{L}(\mathbf{V})} = \cdot_{\mathbf{L}_V}$, $\mathbf{L}_V$ fails $r^{\mathbf{R}}(f(\overline{x})) = s^{\mathbf{R}}(f(\overline{x}))$, where the latter is considered a residuated lattice equation $(r^{\mathbf{L}_V}(f(\overline{x})) = s^{\mathbf{L}_V}(f(\overline{x})))$. On the other hand, if we view the above mentioned modular 4-frame as a residuated lattice 4-frame and take $\emptyset$ as $\perp$, V as $\top$ and $V - x$ as $\overline{x}$, for all $\overline{x} \in \tilde{\mathcal{A}}(\hat{C})$, it follows that $\mathbf{L}_V$ satisfies $R'(f(\overline{x}), \hat{C}, \tilde{\mathcal{A}}(\hat{C}), \perp, \top)$. Indeed, (i) and (iv) of $R'(f(\overline{x}), \hat{C}, \tilde{\mathcal{A}}(\hat{C}), \perp, \top)$ are obvious, while (ii) is true, since $f(x_i)$ is a member of the multiplicative semigroup of $\mathbf{R}$ and this semigroup plays the role of L_{12}. Condition (iii) holds because it holds in $\mathbf{S}$ for $\overline{x}$ and (v) is very easy to check. So, for $\overline{y} = f(\overline{x})$, $\mathbf{L}_V$ satisfies $R'(\overline{y}, C, \tilde{\mathcal{A}}(C), \perp, \top)$, but not $r^{\mathbf{L}_V}(\overline{y}) = s^{\mathbf{L}_V}(\overline{y})$; hence, by Lemma 3.10, $\mathbf{L}(\mathbf{S}, \mathcal{V})$ fails $r^{\odot_{12}}(\overline{x}') = s^{\odot_{12}}(\overline{x}')$.

If the word problem for $\mathbf{L}(\mathbf{S}, \mathcal{V})$ were decidable then the one for $\mathbf{S}$ would be decidable, too. Thus, $\mathcal{V}$ has an undecidable word problem.   ∎

Recall that a quasi-equation is a universally quantified implication in which the assumption is a conjunction of equations and the conclusion is a single equation.

Corollary 3.13. *If $\mathcal{V}$ is a variety such that $\mathsf{HSP}(\mathbf{L_V}) \subseteq \mathcal{V} \subseteq \mathcal{DRL}$, for some infinite-dimensional vector space $\mathbf{V}$, then $\mathcal{V}$ has an undecidable quasi-equational theory.*

Corollary 3.14. *The word problem and quasi-equational theory for (commutative) distributive residuated lattices is undecidable.*

As pointed out by one of the referees, results in [7] inply the consequence of Theorem 3.12 in the commutative case. Moreover, the result for $\mathcal{DRL}$ alone can be proved in a much more simple and direct way; [11] contains a discussion on decidability and residuated lattices. Thus the novelty of the result in this paper lies in the non-commutative varieties different from $\mathcal{DRL}$.

The equational theory of $\mathcal{RL}$ is known to be decidable (see [11]). It is an open problem, though, whether the same is true for $\mathcal{DRL}$ or for other subvarieties of $\mathcal{RL}$.

References

[1] M. Ward and R. P. Dilworth, *Residuated Lattices.* Trans. of the AMS **45** (1939), 335–354.

[2] K. Blount and C. Tsinakis, *The structure of residuated lattices.* Preprint.

[3] H. Andréka, S. Givant and I. Németi, *Decision Problems for Equational Theories of Relation Algebras.* Memoirs of the AMS **126**(604) (1997).

[4] Ralph Freese, *Free modular lattices.* Trans. of the AMS **261** (1980), 81-91.

[5] L. Lipshitz, *The undecidability of the word problems for projectice geometries and modular lattices.* Trans. of the AMS **193** (1974), 171-180.

[6] G. Rozenberg and A. Salomaa, *Cornerstones of Undecidability.* Intl. Ser. in Computer Science; (1994) Prentice Hall.

[7] A. Urquhart, *The undecidability of entailment and relevant implication.* Journal of Symbolic Logic **49**(4) (1984), 1059-1073.

[8] A. Urquhart, *Decision problems for distributive lattice-ordered semigroups.* Alg. Univ. **33** (1995), 399-418.

[9] P. Schroeder and K. Dosen (Eds.), *Substructural Logics.* Clarendon Press, Oxford, (1993).

[10] J. von Neumann, *Continuous Geometry.* (1960) Princeton University Press.

[11] P. Jipsen and C. Tsinakis, *A survey of residuated lattices.* In these Proceedings, 19-56.

Department of Mathematics, Vanderbilt University, Nashville, TN 37240, USA
ngalatos@math.vanderbilt.edu

Least Integer Closed Groups[1]

Anthony W. Hager, Chawne M. Kimber [2] and Warren Wm. McGovern

This paper is dedicated to Paul Conrad
on the occasion of his 80th birthday.

ABSTRACT.   An a-closure of a lattice-ordered group is an extension which is maximal with respect to preserving the lattice of convex ℓ-subgroups under contraction. We describe the a-closures of some local singular archimedean lattice-ordered groups with designated weak unit. In particular, we provide explicit descriptions of all of the a-closures of groups that are singularly convex, such as the group $C(X, \mathbb{Z})$ of continuous integer-valued functions on a zero-dimensional space.

1   Preliminaries

An (abelian) *lattice-ordered group (ℓ-group)* is a group $(G, +, \leq, \vee, \wedge)$ with a partial ordering such that $g \leq h \Rightarrow g + k \leq h + k$ for all $g, h, k \in G$, and with respect to which G is a lattice (that is, for all $g, h \in G$, the supremum $g \vee h$ and infimum $g \wedge h$ exist). An *ℓ-subgroup* is a subgroup that is also a sublattice. G^+ is the positive cone of the group and $|g| = g \vee 0 + (-g) \vee 0$. The ℓ-group G is a *(real) vector lattice* if for every $g \in G^+$, we have that $rg \in G$ for every $0 \leq r \in \mathbb{R}$. G is *archimedean* if for every $g, h \in G^+$ there exists a natural number n such that ng is not less than or equal to h. See Chapter 10 of [9] for properties of archimedean ℓ-groups. For instance, Theorem 53.3 is the result that all archimedean ℓ-groups are abelian.

An ℓ-subgroup $H \leq G$ is *convex* if $0 \leq g \leq h \in H$ implies that $g \in H$. Let $\mathfrak{C}(G)$ denote the lattice of all convex ℓ-subgroups of G. The convex ℓ-subgroup generated by an element $g \in G$ is denoted $G(g)$. If $P \in \mathfrak{C}(G)$ and for every $g, h \in G^+$, we have that

---

[1] AMS Subject Classifications. 06F25, 20F60

Keywords. lattice-ordered groups, archimedean groups, singular groups, hyperarchimedean, rings of continuous functions, groups of continuous functions

[2] This author was partially supported by a doctoral fellowship from the Florida Education Fund and was Van Vleck Visiting Assistant Professor of Mathematics at Wesleyan University during the development of this work. She thanks both institutions for their generosity.

J. Martínez (ed.), Ordered Algebraic Structures, 245–260.
© 2002 *Kluwer Academic Publishers.*

$g \wedge h \in P \Rightarrow g \in P$ or $h \in P$, then we call P a *prime* subgroup. By Zorn's Lemma, minimal prime subgroups exist.

A positive element $u \in G^+$ is a *weak unit* if $\{g \in G \mid |g| \wedge u\} = \{0\}$. We work in the category **W** consisting of archimedean ℓ-groups with designated weak unit (G, u) and lattice-preserving group homomorphisms that also preserve the unit. The ℓ-group $(C(X), 1)$ of real-valued continuous functions on a completely regular space is a **W**-object under pointwise operations. In the main, we are interested in particular subgroups: $C(X, \mathbb{Z})$ is the group of all the integer-valued continuous functions; $S(X, \mathbb{R})$ consists of continuous functions with finite range; and $S(X, \mathbb{Z})$ contains the integer-valued continuous functions with finite range. Note that for any $r \in \mathbb{R}$, we use $\mathbf{r}$ to represent the function on X that is constantly equal to r.

Let (G, u) be in **W**. By Zorn's Lemma, there exist convex ℓ-subgroups of G that are maximal with respect to not containing u. We call such subgroups *values* of u. Let YG be the set of values of the designated unit u. Then YG is a compact Hausdorff space in the hull-kernel topology.

Define

$$D(X) = \{ f : X \to \mathbb{R} \cup \{\pm\infty\} \mid f \text{ continuous and } f^{-1}\mathbb{R} \text{ dense} \}.$$

$D(X)$ is a lattice under pointwise operations. However the pointwise sum of two elements need not exist: for $f, g \in D(X)$, the sum $f + g$ is defined on the dense set $f^{-1}\mathbb{R} \cap g^{-1}\mathbb{R}$, though it may not extend to a continuous function on X. Recall that a completely regular space X is *quasi-F* if for every dense cozero set $A \subseteq X$, every bounded $\mathbb{R}$-valued continuous function on A has a continuous extension to X. It is shown in [17] that $D(X)$ is a group under pointwise addition (thus an ℓ-group) precisely when X is a quasi-F space. However, $D(X)$ may contain a sublattice H that is a group under the operation: $(f+g)(x) = f(x)+g(x)$ for all x in some dense subset of X. When this is the case, we call H an ℓ-*group in* $D(X)$. The following representation theorem has many incarnations, among them [20] and Theorem 2.7 of [15].

Theorem 1.1. [Yosida Embedding Theorem] *Let (G, u) be in* **W**. *Then there is an ℓ-isomorphism of G onto $\widehat{G}$, an ℓ-group in $D(YG)$ such that $\widehat{G}$ separates the points of YG and $u \mapsto \mathbf{1}$.*

Henceforth, we identify any object (G, u) of **W** with its image in $D(YG)$.

An element $s \in G^+$ is called *singular* if $0 \le g \le s \Rightarrow g \wedge (s - g) = 0$. We call an ℓ-group G *singular* if for each $g \in G^+$, there exists a singular s such that $s \le g$. In particular, (G, u) is in **W** and u is singular, then G is called singular. Define

$$\mathbf{W_s}G = \{g \in G \mid p \in g^{-1}\mathbb{R} \Rightarrow g(p) \in \mathbb{Z}\}.$$

In [13], the authors demonstrate that $\mathbf{W_s}$ is a monocoreflection of **W** into the (full) subcategory of singular groups in **W**; $\mathbf{W_s}G \le G$ is the maximum subgroup of G that is singular in **W**. We will call $\mathbf{W_s}G$ the *singular part* of G.

G is a *large* ℓ-subgroup of H if for every $h \in H^+$ there is $g \in G^+$ and a natural number n such that $g \le nh$. Let $g, h \in G^+$. We say that g and h are *a-equivalent* and write $g \sim_a h$ if there exist natural numbers n, m such that $g \le nh$ and $h \le mg$. If G is an ℓ-subgroup of H, then H is an *a-extension* of G if every positive element of H is a-equivalent to a positive element of G. In this case, we write $G \le_a H$. By Theorem 2.1 of [6], $G \le_a H$ if and only if the map $K \mapsto K \cap G$ gives a lattice isomorphism $\mathfrak{C}(H) \to \mathfrak{C}(G)$. If G has no proper a-extensions, then we say G is *a-closed*. In Example 6.4 of [6], Conrad produces two nonisomorphic abelian a-closures of an ℓ-group. Therefore, a-closures are not unique. For many other examples of this, see [11].

The first of the following properties is a generalization of Theorem 1.1 of [7]; the second property follows from Theorem 2.1 of [6].

Proposition 1.2. *Let G be an ℓ-group.*

(i) *Let $G \le_a H$. If G is archimedean then H is archimedean.*

(ii) *If (G, u) is an object of $\mathbf{W}$ and $G \le_a H$, then (G, u) is a $\mathbf{W}$-subobject of (H, u), $YH = YG$ and $G^* \le_a H^*$.*

Proof. 1. Let G be an ℓ-group. Assume that H is a non-archimedean a-extension of G. Let $h_1, h_2 \in H^+$ such that $0 \le h_2 - nh_1$ for every natural number n. There exist $g_1, g_2 \in G$ and $m_1, m_2 \in \mathbb{N}$ such that $g_1 \le m_1 h_1$ and $h_2 \le m_2 g_2$. Then $0 \le m_1 m_2 g_2 - ng_1$ for every natural number n and hence G is not archimedean.

2. First, we show that u is a weak unit in H. Let $h \in H^+$. There exists $g \in G^+$ and natural numbers m, n such that $h \le mg$ and $g \le nh$. If $u \wedge h = 0$, then $0 \le u \wedge g \le u \wedge (nh) = 0$. Thus, $g = 0 = h$, as desired.

Second, we must show that $YH = YG$. The contraction mapping $\varphi : YH \to YG$ is surjective by Proposition 12.11 of [9]. By Theorem 2.1 of [7], φ is one-to-one and therefore a homeomorphism.

Let $h \in (H^*)^+$ and let $g \in G$ be such that $g \sim_a h$. Then $0 \le g \le mh \le nu$ for some $m, n \in \mathbb{N}$. Therefore, $g \in G^*$ and hence $G^* \le_a H^*$. ∎

The task we undertake is to describe some of the a-closures of ℓ-groups that are singular in $\mathbf{W}$. First, we focus on the subclass of $\mathbf{W}$ in which these a-closures must lie. Second, we describe those ℓ-groups that are least integer closed (defined below). Third, we discuss those ℓ-groups that are strongly least integer closed (defined below) and show that all singularly convex groups (defined below) are in this class. As an example, we explicitly compute all of the a-closures of the singularly c^3 groups and, in particular, of the group $C(X, \mathbb{Z})$.

Unless otherwise stated, all groups are in $\mathbf{W}$ and are identified with their Yosida representations.

2 Bounded away ℓ-groups

In [7], Conrad introduced the class of *hyperarchimedean* ℓ-groups (there called *epiarchimedean*): the ℓ-groups for which every ℓ-homomorphic image is archimedean. In **W**, these are precisely the groups in which $G = G(u)$ and for every $g \in G^+$, there exists $r \in \mathbb{R}$ such that $0 < g(p) \Rightarrow 0 < r \leq g(p)$; see Theorem 2.3 below and [11]. Note that with Proposition 1.2 it is evident that

Proposition 2.1. [[7], Theorem 1.1] *An a-extension of a hyperarchimedean group is also hyperarchimedean.*

Removing the $G = G(u)$ condition in the definition of hyperarchimedean ℓ-groups allows for some interesting results.

Definition 2.2. From [18]: An element $g \in D(X)$ is *bounded away from* 0 if there exists $r \in \mathbb{R}$ such that $0 < g(p) < \infty \Rightarrow 0 < r \leq g(p)$. Call a **W**-object (G, u) *bounded away* if in $D(YG)$ each of the elements of G is bounded away from 0. Let S be a subgroup of $\mathbb{R}$ containing 1. We say that G is *S-bounded away* if every $g \in G$ and for each $s \in S$ there exists $r \in \mathbb{R}$ (depending on g and s) such that if $0 < s < g(p) < \infty$ then $s < r \leq g(p)$.

Since elements of singular groups are integer-valued, all singular groups are bounded away. To each element of other kinds of groups, we associate an integer-valued function by composing elements $g \in G$ with the "least integer function" on $\mathbb{R}$. That is, for $g \in G$ and $n \in \mathbb{Z}$, define $[g](p) = [g(p)] = n$ if $p \in g^{-1}(n-1, n]$ and $[g](p) = g(p)$, if $g(p) = \pm\infty$. The theorem below specifies when one can expect $[g]$ to be continuous.

Theorem 2.3. *Let (G, u) be in **W** and let $G^* = G(u)$. The following are equivalent:*

(i) *G is bounded away.*

(ii) *G is $\mathbb{Z}$-bounded away.*

(iii) *G is $\mathbb{Q}$-bounded away.*

(iv) *For every $g \in G$, the set $g(YG) \cap \mathbb{Q}$ is a closed discrete subset of $\mathbb{Q}$.*

(v) *For every $g \in G$ and every $q \in \mathbb{Q}$, then set $g^{-1}\{q\}$ is open.*

(vi) *For every $g \in G$, the zeroset $Z(g) = \{p \in YG \mid g(p) = 0\}$ is clopen in YG.*

(vii) *For every $g \in G$, the function $[g]$ is continuous on YG.*

(viii) *Every **W**-homomorphic image of G is bounded away.*

(ix) *G^* is hyperarchimedean.*

(x) G^* is the maximum hyperarchimedean convex ℓ-subgroup of G.

(xi) Every proper prime subgroup of G not containing G^* is a value of u.

(xii) Every value of u is a minimal prime subgroup of G.

(xiii) $G^* = \{g \in G \mid \forall h \in G, \exists m \in \mathbb{N}, |g| \wedge m|h| = |g| \wedge (m+1)|h| \}$

(xiv) $G^* = \cap_{g \in G^+} G(g) \oplus g^\perp$, where $g^\perp = \{h \in G \mid |g| \wedge |h| = 0\}$.

(xv) For every $g \in G^+$, there is a natural number n such that

$$1 \wedge ng \wedge (1 - ng) \leq 0 \quad \text{or} \quad 1 \wedge ng \wedge (1 - ng)^+ = 0.$$

(xvi) For every $g \in G^+$ there is a natural number n such that $(ng - 1)^{+\perp} = g^\perp$.

Proof. The equivalence of $(5), (11), (12), (15)$ and (16) is by Theorem 1.1 of [7]; the eqivalence of $(9), (13)$ and (14) is by Theorem 2.1 from [19]. It is evident that $(5) \Rightarrow (6), (8) \Rightarrow (1)$ and $(9) \Leftrightarrow (10)$.

$(1) \Leftrightarrow (2)$: Proceed via translation. That is, G is bounded away from 0 if and only if for every $g \in G$ and for every $n \in \mathbb{N}$, $g - \mathbf{n}$ is bounded away from 0. This is equivalent to the statement that for every $g \in G$ and for every $n \in \mathbb{N}$ there exists $r \in \mathbb{R}$ such that $n < g(p) \Rightarrow 0 < n < n + r \leq g(p)$, which is exactly the definition of $\mathbb{Z}$-bounded away.

$(1) \Leftrightarrow (3)$: It is easy to verify that G is bounded away if and only if the divisible hull dG is bounded away; see Section 4 of [7] and the following. Every $g \in G$ is bounded away from 0 if and only if for every $g \in G$ and for every $0 < q \in \mathbb{Q}$ there exists $r \in \mathbb{R}$ such that $0 < g(p) \Rightarrow 0 < qr \leq qg(p)$, which is the statement that dG is bounded away.

Using this, we verify that (1) is equivalent to (3), as above. G is bounded away from 0 if and only if for every $g \in G$ and for every $0 \leq q \in \mathbb{Q}$, $g - \mathbf{q}$ is bounded away from 0 in dG. This is equivalent to the statement that for every $g \in G$ and for every $0 \leq q \in \mathbb{Q}$ there exists $r \in \mathbb{R}$ such that $q < g(p) \Rightarrow 0 < q < q + r \leq g(p)$, which is exactly the definition of $\mathbb{Q}$-bounded away.

$(1) \Leftrightarrow (4)$: Let $\{q_j\}_{j=1}^\infty \subseteq g(YG) \cap \mathbb{Q}$ be a convergent sequence. If $q_j \to q = \frac{m}{n} \in \mathbb{Q}$, then $ng - mu$ is not bounded away from zero. Thus, $g \in G$ is bounded away if and only if $g(YG) \cap \mathbb{Q}$ contains no convergent sequence and its limit. This is equivalent to saying that the set is closed and discrete.

$(1) \Leftrightarrow (9)$: By the definitions and Theorem 1.1 of [6].

$(6) \Rightarrow (5)$: Let $q = \frac{m}{n} \in \mathbb{Q}$, then $Z(ng - \mathbf{m}) = g^{-1}\{q\}$ is clopen by condition (6).

$(6) \Leftrightarrow (7)$: $Z(g - \mathbf{n})$ is open for every integer n and every $g \in G$ if and only if $g^{-1}\{n\}$ is open for every integer n and every $g \in G$. This is the same as the statement that $g^{-1}(n - 1, n]$ is clopen for every n and every g, which precisely says that every $[g]$ is continuous.

$(9) \Leftrightarrow (8)$: Let $\varphi : G \to H$ be a surjective $\mathbf{W}$-morphism. Since G^* is hyperarchimedean, we know that $\varphi(G^*) = H^*$ is hyperarchimedean. Therefore, H is bounded away by $(1) \Leftrightarrow (9)$.

(6) $\Leftrightarrow$ (2): Condition (6) is equivalent to the property that for every $g \in G$ and for every integer n the set $g^{-1}\{n\}$ is open. This, in turn, is equivalent to: for each $g \in G$ and for each $n \in \mathbb{Z}$ there is an open neighborhood $U_n \subseteq \mathbb{R}$ such that $g(YG) \cap U_n = \{n\}$. The latter property is equivalent to (2).   ∎

Corollary 2.4. *If G is bounded away, then every a-extension of G is bounded away.*

Proof. Let G be bounded away. Assume that $G \leq_a H$, let $0 < h \in H^*$ and obtain $g \in G^*$ such that $g \sim_a h$. Then $Z(g) = Z(h)$. Since $Z(g)$ is clopen by Theorem 2.3, so is $Z(h)$. Therefore, H is bounded away.   ∎

Proposition 2.5. *The following are equivalent.*

(i) *G is $\mathbb{R}$-bounded away.*

(ii) *For each $g \in G$ and $r \in \mathbb{R}$, there is an open neighborhood $r \in V_r \subset \mathbb{R}$ such that $|g(YG) \cap V_r| < \omega$.*

(iii) *For each $g \in G$ and $r \in \mathbb{R}$, the set $g^{-1}\{r\}$ is open.*

(iv) *For each $g \in G$ and $p \in YG$ such that $g(p) \in \mathbb{R}$, there is an open neighborhood $p \in U_p \subseteq YG$ such that $|g(U_p)| = 1$.*

(v) *For each $g \in G$ and $p \in YG$ such that $g(p) \in \mathbb{R}$, there is an open neighborhood $p \in U_p \subseteq YG$ such that $|g(U_p)| < \omega$.*

(vi) *For each $g \in G$, the set $g(YG) \cap \mathbb{R}$ is closed, countable and discrete in $\mathbb{R}$.*

If G is a vector lattice, then G is bounded away if and only if G satisfies the above conditions.

Proof. The proof of the equivalence of (1) and (6) is the same as the proof of (1) $\Leftrightarrow$ (4) in Theorem 2.3.

(1) $\Leftrightarrow$ (2): An element $g \in G$ is bounded away from every $r \in \mathbb{R}$ if and only if no $r \in \mathbb{R}$ is a cluster point of the range $g(YG)$, which is precisely condition (2).

(4) $\Leftrightarrow$ (5): Clearly (4) $\Rightarrow$ (5). Conversely, let $g \in G$ and $p \in YG$. Then there is a neighborhood U'_p such that $g(U'_p) = \{r_1, \ldots r_n\}$, where $g(p) = r_1$. Then there exists $\varepsilon > 0$ such that $g(U'_p) \cap (r_1 - \varepsilon, r_1 + \varepsilon) = \{r_1\}$. Let $U_p = g^{-1}(r_1 - \varepsilon, r_1 + \varepsilon) \cap U'_p$. Then $g(U_p) = \{r_1\}$.

(3) $\Leftrightarrow$ (4): The set $g^{-1}\{r\}$ is open if and only if for every $p \in g^{-1}\{r\}$ there exists a neighborhood $U_p \subseteq g^{-1}\{r\}$.

(1) $\Leftrightarrow$ (3): An element $g \in G$ is bounded away from $r \in \mathbb{R}$ if and only if $g - r \in D(YG)$ is bounded away from 0. We have already seen that this is the same as saying that $g^{-1}\{r\} = Z(g - r)$ is open.

It is always the case that $\mathbb{R}$-bounded away implies bounded away. If G is a vector lattice, then for all $r \in \mathbb{R}$, the constant function $\mathbf{r}$ is in G. Hence, if G is a bounded away vector lattice, $g - \mathbf{r} \in G$ is bounded away from 0 and, therefore, G must be $\mathbb{R}$-bounded away. ∎

Example 2.6. A hyperarchimedean ℓ-group that is not $\mathbb{R}$-bounded away: This example is 6.4 of [6]. Let $\alpha\mathbb{N}$ denote the one-point compactification of the discrete space $\mathbb{N}$, in which α is the point at infinity. Then $S(\alpha\mathbb{N}, \mathbb{Z})$ is the ℓ-group of eventually constant sequences of integers, under pointwise addition and the pointwise ordering. Define a new function by $b(n) = \pi + \frac{1}{n}$ and $b(\alpha) = \pi$. Then the ℓ-group $S(\alpha\mathbb{N}, \mathbb{Z}) + b\mathbb{Q}$ is hyperarchimedean. The element b fails condition (2) of Proposition 2.5, so the group is not $\mathbb{R}$-bounded away. See [11] for some generalizations of this example.

Example 2.7. A vector lattice that is not bounded away: Let $C(\mathbb{N})$ be the vector lattice of continuous real-valued functions on the discrete space $\mathbb{N}$, under pointwise addition and the pointwise ordering. In the Yosida representation $C(\mathbb{N}) \hookrightarrow D(\beta\mathbb{N})$, the Stone-extension of the function $f(n) = \frac{1}{n}$ fails to be bounded away from zero. Therefore, $C(\mathbb{N})$ is not bounded away. In fact, this example indicates how one shows that $C(X)$ is bounded away if and only if X is finite.

We note a fairly close connection with a class of ℓ-groups studied by Marlow Anderson. Let G be an ℓ-group (not necessarily in **W**), and let $\mathcal{M}(G)$ be the set of prime subgroups of G that are both maximal and minimal. Then G is called *locally flat* if $\cap\mathcal{M}(G) = \{0\}$; see [1].

Theorem 2.8. [[1], Theorem 2.2] *Let G be an ℓ-group. Then G is locally flat if and only if G can be embedded as a large ℓ-subgroup of $C(X)$, for some space X such that the zerosets of G form a clopen base for the closed sets of X.*

A value V of u is called *real* if the totally ordered group G/V is archimedean, thus embeddable in $\mathbb{R}$ by Hölder's Theorem.

Proposition 2.9. *A **W**-object (G, u) is locally flat if G is bounded away and the set of real values of u is dense in YG.*

Proof. Let X be the set of real values of u. Since X is dense in YG, the canonical embedding, i.e., the Yosida representation restricted to X, given by

$$G \hookrightarrow \prod_{V \in X} G/V \subseteq C(X)$$

is large. Since G is bounded away, the zerosets of G form a clopen base for the closed sets of YG and hence restrict to a clopen base for X. Thus, G is locally flat by Theorem 2.8. ∎

The converse of Proposition 2.9 is false: If Y is an infinite zero-dimensional space with a dense set X of P-points, then $C(Y)$ is not bounded away, but is locally flat since $C(Y)|_X$ is a representation as in Theorem 2.8. (See [1], 4.4.)

3   Least integer closed ℓ-groups

Definition 3.1. We call G *least integer closed* (resp., *weakly least integer closed*) if for all $g \in G$ we have $[g] \in G$ (resp., if there exists $g' \in G$ and a dense set $U \subseteq YG$ such that $[g]|_U = g'|_U$).

Proposition 3.2. *Hyperarchimedean groups are least integer closed and least integer closed groups are bounded away.*

Proof. If G is hyperarchimedean, then for every $g \in G$, we know that $[g] \in S(YG, \mathbb{Z}) \leq G$; thus, every hyperarchimedean group is least integer closed. In view of Theorem 2.3, a least integer closed group is bounded away. ∎

The examples following Theorem 3.4 illustrate that the converses of the above are not true.

Lemma 3.3. *Let G be bounded away. Then for each $g \in G^+$, we have that $[g]$ is continuous, $g \leq [g]$ and there is a positive integer m such that $[g] \leq mg$. Thus, every least integer closed group is an a-extension of its singular part.*

Proof. By Theorem 2.3, all of the functions $[g]$ are continuous since G is bounded away. Since $g \in G^+$ is bounded away from 0, there exists $r = \min(g(YG) \setminus \{0\})$. Let $m \in \mathbb{N}$ such that $mr > 1$. Then $g \leq [g]$ and $[g] \leq mg$. Finally, if G is least integer closed, then for each $g \in G^+$, we have $g \sim_a [g] \in \mathbf{W_s}G$, as desired. ∎

Theorem 3.4. *The following are equivalent.*

(i) *G is least integer closed.*

(ii) *G is weakly least integer closed and bounded away.*

(iii) *G is weakly least integer closed and $\mathbf{W_s}G \leq_a G$.*

(iv) *$G = \mathbf{W_s}G + G^*$ and G is bounded away.*

(v) *$G = \mathbf{W_s}G + G^*$ and $\mathbf{W_s}G \leq_a G$.*

Proof. That (1) $\Rightarrow$ (2) is clear; (2) $\Rightarrow$ (1) is a consequence of the equivalence of conditions (1) and (7) of Theorem 2.3. Lemma 3.3 is the interesting part of the proof of (1) $\Rightarrow$ (3). The following finishes the proof.
  (3) $\Rightarrow$ (5) : By the preceding argument, $G = \mathbf{W_s}G + G^*$.
  (5) $\Rightarrow$ (4) : By Corollary 2.4, G is bounded away.
  (4) $\Rightarrow$ (1) : Since G is bounded away, G^* is hyperarchimedean and, hence, G^* is also least integer closed. Then for $g = f + h \in \mathbf{W_s}G + G^*$, we have $[g] = f + [h] \in G$. ∎

Each of the properties (weakly least integer closed, $\mathbf{W_s}G \leq_a G$, and $G = \mathbf{W_s}G + G^*$) appearing in Theorem 3.4 merits and will receive some concentrated attention from the authors in future papers. In particular, we have observed (as did a referee) that there is a nice connection between weakly least integer closed groups and projectable groups and this is addressed in [12]. Examples 3.7–3.10 demonstrate that the three classes introduced in the preceeding theorem are distinct. The next observation is motivating, both for the introduction of the class of least integer closed groups, and for the material of sections 4 and 5 below.

Proposition 3.5. *Let X be a zero-dimensional Tychonoff space. If $C(X,\mathbb{Z}) \leq_a H$, then H is least integer closed and thus $H = C(X,\mathbb{Z}) + H^*$ where $S(X,\mathbb{Z}) \leq_a H^*$.*

Proof. First note that $Y = YC(X,\mathbb{Z})$ is the maximal zero-dimensional compactification of X (see [13]), so $X \subseteq Y$. Now suppose that $C(X,\mathbb{Z}) \leq_a H$. By part (2) of Proposition 1.2, $YH = Y$ and $S(X,\mathbb{Z}) \leq_a H^*$ since $C^*(X,\mathbb{Z}) = S(X,\mathbb{Z})$. Let $h \in H$. By Corollary 2.4, H is bounded away, so $[h]$ is continuous on Y. Thus $[h]|_X$ is continuous. Since $h \sim_a f \in C(X,\mathbb{Z})$, the element h is real-valued on X and so is $[h]$. This shows that $[h] \in C(X,\mathbb{Z})$, so H is least integer closed and $\mathbf{W_s}H = C(X,\mathbb{Z})$. Now apply Theorem 3.4. ∎

In 5.4 below, we prove a converse to Proposition 3.5: If $S(X,\mathbb{Z}) \leq_a K$, then $C(X,\mathbb{Z}) + K$ is an ℓ-group a-extending $C(X,\mathbb{Z})$. We turn now to examples distinguishing the properties in Theorem 3.4.

Example 3.6. The following is an example of a vector lattice that is bounded away but is not an a-extension of its singular part. By Theorem 3.4, the group also is not weakly least integer closed. Let $X = \alpha\mathbb{N}$ and define $f, g \in D(X)$ by $f(n) = n^3$, $g(n) = n^{2-\frac{1}{n}}$ and $f(\alpha) = g(\alpha) = \infty$. Let G be the vector lattice spanned by $S(X,\mathbb{R}), f$ and g. Then $\mathbf{W_s}G = S(X,\mathbb{Z}) + f\mathbb{Z}$. If $\mathbf{W_s}G \leq_a G$ then since g is unbounded, we must have $g \sim_a k + mf \in S(X,\mathbb{Z}) + f\mathbb{Z}$, for some nonzero integer m. Note that $k + mf \sim_a f$. Yet, if $f \sim_a g$ then there exists an integer s such that $f \leq sg$, whence $s \geq n^{1+\frac{1}{n}}$ for all n. This is a contradiction. Thus, G is not an a-extension of $\mathbf{W_s}G$.

Example 3.7. Recall that every least integer closed group is bounded away and an a-extension of its singular part. This example shows that the converse does not hold. Let $X = \alpha\mathbb{N}$ and define $f \in D(X)$ by $f(n) = 2n + \chi_E(n)$ and $f(\alpha) = \infty$, where χ_E is the characteristic function of the set of even numbers in $\mathbb{N}$. If H is the divisible hull of the ℓ-group generated by $S(X,\mathbb{R})$ and f, then H is bounded away and

$$\mathbf{W_s}H = S(X,\mathbb{Z}) + f\mathbb{Z} \leq_a H.$$

Since $[\frac{1}{2}f] \notin \mathbf{W_s}H$, the group H is not least integer closed. Moreover, there is no $f' \in H$ such that $f'|_U = [\frac{1}{2}f]|_U$, for any dense set U; so H is not weakly least integer closed either.

Example 3.8. The group $C(\mathbb{N})$ is weakly least integer closed by Theorem 3.10, but, as noted in Example 2.7, it is not bounded away. Therefore $C(\mathbb{N})$ is not least integer closed by Proposition 3.2. To see this explicitly, note that if $h(n) = \frac{1}{n}$ then $[h]$ is the characteristic function of $\mathbb{N}$, which is not continuous.

Example 3.9. Note that every element of the group $G = C(\alpha\mathbb{N})$ is bounded, so $G = \mathbf{W_s}G + G^*$, but G is not bounded away.

The following result from a paper in preparation by the authors gives a hint of the interesting properties of weakly least integer closed groups. Recall that a space X is *basically disconnected* if $cl_X \operatorname{coz}(f)$ is open for every $f \in C(X)$.

Theorem 3.10. *Let X be a Tychonoff space. The following are equivalent.*

(i) $C(X)$ *is weakly least integer closed.*

(ii) $D(X)$ *is an ℓ-group which, as a $\mathbf{W}$-object, is weakly least integer closed.*

(iii) X *is basically disconnected.*

4   Strongly least integer closed ℓ-groups

Let X be zero-dimensional. In Proposition 3.5, we showed that any a-extension of $C(X, \mathbb{Z})$ has the form $C(X, \mathbb{Z}) + K$, where K is an a-extension of $S(X, \mathbb{Z})$. The goal of this section and the next is to generalize this result.

Theorem 4.1. *The following are equivalent:*

(i) G *is bounded away and*

$$G \leq_a H \Rightarrow \mathbf{W_s}G \leq_a \mathbf{W_s}H, \quad G^* \leq_a H^* \text{ and } H = \mathbf{W_s}H + H^*.$$

(ii) G *is bounded away and* $(G \leq_a H \Rightarrow H = \mathbf{W_s}H + H^*)$.

(iii) *Every a-extension of G is least integer closed.*

(iv) *Every a-extension of G is bounded away and*

$$G \leq_a H \Rightarrow \mathbf{W_s}G \leq_a \mathbf{W_s}H, \quad G^* \leq_a H^* \text{ and } H = \mathbf{W_s}H + H^*.$$

(v) G *is least integer closed and* $(G \leq_a H \Rightarrow H = \mathbf{W_s}H + H^*)$.

Proof. (1) $\Rightarrow$ (2): clear.

(2) $\Rightarrow$ (3): Follows from Corollary 2.4 and Theorem 3.4.

(3) $\Rightarrow$ (4): Let $G \leq_a H$. By Proposition 3.2, H is bounded away; it follows from Proposition 1.2 that $G^* \leq_a H^*$. We see that $H = \mathbf{W_s}H + H^*$ by Theorem 3.4. Lemma 3.3 shows that $\mathbf{W_s}G \leq_a G$. Therefore, $\mathbf{W_s}G \leq_a H$ and we conclude that $\mathbf{W_s}G \leq_a \mathbf{W_s}H$.

(4) $\Rightarrow$ (5): This follows from Theorem 3.4.

(5) $\Rightarrow$ (1): By Proposition 3.2, G is bounded away; it follows that H is bounded away by Corollary 2.4 and, therefore, least integer closed by Theorem 3.4. Lemma 3.3 shows that $\mathbf{W_s}G \leq_a \mathbf{W_s}H$. ∎

Definition 4.2. An ℓ-group G is *strongly least integer closed* (or *SLIC*) if it satisfies one, hence all, of the conditions of Theorem 4.1.

In section 5, we identify a fairly wide class of ℓ-groups which are stongly least integer closed. For now, we note the following.

Proposition 4.3. *Hyperarchimedean ℓ-groups are strongly least integer closed.*

Proof. Since an a-extension of a hyperarchimedean group must be hyperarchimedean, this follows from Proposition 3.2 and Theorem 3.4. ∎

Corollary 4.4. *If $\mathbf{W_s}G$ is strongly least integer closed and $\mathbf{W_s}G \leq_a G$, then G is strongly least integer closed.*

Proof. By condition (4) of Theorem 4.1. ∎

Corollary 4.5. *If G is strongly least integer closed, then $\mathbf{W_s}G \leq_a G$.*

Proof. G is least integer closed, by Theorem 4.1. Therefore, $\mathbf{W_s}G \leq_a G$ by Theorem 3.4. ∎

5   Singularly convex ℓ-groups

We now present the desired generalization of Proposition 3.5.

Definition 5.1. A singular ℓ-group G is *singularly a-closed* if it has no proper a-extension which is singular.

Example 5.2. As mentioned before, the group $\mathbf{W_s}H$ in Example 3.7 is singular, and singularly a-closed, yet it is not strongly least integer closed since $\mathbf{W_s}H \leq_a H \neq \mathbf{W_s}H + H^*$.

Example 5.3. Let G be the ℓ-group generated by $S(\alpha\mathbb{N}, \mathbb{Z})$ and the function f defined by $f(x) = 2n + \pi\chi_E(n)$ if $x = n$ and $f(\alpha) = \infty$, where χ_E is the characteristic function of the set of even numbers. Then G is not strongly least integer closed, since no a-extension can contain $[f]$ and G is not an a-extension of $\mathbf{W_s}G$. Yet, $\mathbf{W_s}G = S(\alpha\mathbb{N}, \mathbb{Z})$ is singularly a-closed and strongly least integer closed by Corollary 5.8 and Corollary 5.9.

Theorem 4.1 indicates that finding the a-closures of a strongly least integer closed group G should split into two steps. We must first find the maximal hyperarchimedean extensions of G^* and, secondly, find the singular a-closures of $\mathbf{W_s}G$. These are relatively difficult tasks to carry out. The first step is demonstrated in great detail in [11] for $G = S(\alpha\mathbb{N}, \mathbb{Z})$. Below, we show that for certain other, (possibly unbounded) singular groups, this first step suffices also: we hold the singular part of a singularly a-closed group fixed so that we need only a-extend the bounded part of the group to compute the a-closures.

Recall that $f \in D(YG)$ is *locally in* G if for every $p \in YG$ there is a neighborhood U_p and an element $g \in G$ such that $f|_{U_p} = g|_{U_p}$. If G contains all the elements of $D(YG)$ that are locally in G, then we call G *local*. For further discussion of local groups, see [14] and [13].

Theorem 5.4. *Let G be singular and local and let $G^* \leq_a K$. Then $G + K$ is a least integer closed ℓ-group and $G \leq_a G + K$.*

Proof. Since both G and K are groups, $G + K$ is a group. We must show that it is a lattice. Let $f = g + k \in G + K$. The set $\text{pos}(f) \equiv \{p \mid f(p) \geq 0\} = f^{-1}[0, \infty]$ is always closed. In this situation, it is also open: Since K is bounded, there exists an integer M such that $|k| < M$. Then

$$\text{pos}(f) = \cup_{|n| \leq M}(\{p \mid k(p) \geq -n\} \cap g^{-1}\{n\}) \cup g^{-1}[M, \infty],$$

where each set appearing in this expression is open since G is singular and K is hyperarchimedean. Thus, $\text{pos}(f)$ is clopen.

Let $A = \text{pos}(f)$ and let $\chi_A \in S(YG, \mathbb{Z}) \subseteq G^*$ be the corresponding characteristic function. Since the unit of K is strong,

$$k\chi_A = (k \vee 0) \wedge (M\chi_A) - ((-k) \vee 0) \wedge (M\chi_A) \in K;$$

note that, in general, a $\mathbf{W}$-object will be local if its unit is strong. Since G is local, $g\chi_A \in G$. Thus $f \vee 0 = f\chi_A = g\chi_A + k\chi_A \in G + K$ and, therefore, $G + K$ is a lattice.

$G + K$ is least integer closed since for $f = g + k \in G + K$, we have that $[f] = g + [k] \in G + S(YG, \mathbb{Z}) \subseteq G$. By Lemma 3.3, we have that $G \leq_a G + K$ since $G + K$ is least integer closed and $\mathbf{W_s}(G + K) = G$. ∎

Theorem 5.5. *Let G be local and singularly a-closed. If $G^* \leq_a K$ is an a-closure, then $G + K$ is an a-extension maximal with respect to being least integer closed. In particular, if G is also strongly least integer closed, then $G + K$ is an a-closure of G.*

Proof. $G \leq_a G + K$ by Theorem 5.4. Assume that $G + K \leq_a H$ where H is least integer closed. Let $h \in \mathbf{W_s}H^+$ and obtain $g + k \in G + K$ such that $h \sim_a g + k$. Since G is local, $g + [k] \in G$ and hence, $h \sim_a g + k \sim_a g + [k] \in G$. Thus, $G \leq_a \mathbf{W_s}H$. We then conclude that $\mathbf{W_s}H = G$ since G is singularly a-closed. In addition, $H^* = K$ since K is a-closed and $K = (G + K)^* \leq_a H^*$. Thus $H = \mathbf{W_s}H + H^* = G + K$, as desired. The final statement of the theorem follows from the above since every a-extension of a strongly least integer closed group is least integer closed by Theorem 4.1. ∎

We now consider modifications to our purposes of the classes of c^3, and convex, **W**-objects discussed in [16], [2], [3], [4] and [5], among other places. (See these papers for remarks on the efficacy of the classes.)

[13] defines G to be *singularly c^3* if G is singular and for each $g_1, g_2, \ldots \in G$, each $f \in C(\cap_{n \in \omega} g_n^{-1}\mathbb{R}, \mathbb{Z})$ extends over YG to an element of G. This means that G is the direct limit of the system

$$\{ C(U, \mathbb{Z}) \mid U = \cap_{n \in \omega} g_n^{-1}\mathbb{R}, \text{ and } g_n \in G \}$$

with bonding maps given by restriction of functions. This direct limit is the union of the $C(U, \mathbb{Z})$ modulo the equivalence $f_1 \sim f_2$ if $f_1 = f_2$ on the intersection of the domains. So we abbreviate the condition to $G = \cup_U C(U, \mathbb{Z})$.

We define G to be *singularly convex* if G is singular and a convex subset of $D(YG, \mathbb{Z})$. One can easily verify the following.

Proposition 5.6.

1. If X is zero-dimensional (not necessarily compact), then $C(X, \mathbb{Z})$ is singularly c^3.

2. Singularly c^3 implies singularly convex.

Theorem 5.7. *If G is singularly convex, then G is local, singularly a-closed and strongly least integer closed.*

Proof. Let G be singularly convex. If $f \in D(YG)$ is locally in G, then by compactness, $YG = \cup_{i=1}^n U_i$ for open U_i, with $g_i \in G$ for which $f|_{U_i} = g|_{U_i}$. Then $f \in D(YG, \mathbb{Z})$ and $|f| \leq \vee |g_i| \in G$, whence $f \in G$. Now suppose $G \leq_a H$ with H singular. Using (2) of Proposition 1.2, $H \subseteq D(YG, \mathbb{Z})$, and since $h \in H$ implies that $|h| \leq mg$ for some m and g, $h \in G$ by convexity. Once again let $G \leq_a H$. If $h \in H^+$ and $h \sim_a g \in G^+$ then by Theorem 2.3 we know that $[h] \in D(YG)$, $[h] \sim_a g$ and for some integer m we have $0 \leq [h] \leq mg$. Since G is singularly convex, $[h] \in G$. Thus, H is least integer closed. ∎

Corollary 5.8. *Every singularly c^3 group is strongly least integer closed. In fact, if $G = \cup_U C(U, \mathbb{Z})$, then $G \leq_a H$, if and only if $H = G + \cup_U K_U$ for groups $K_U = H^* \cap S(U, \mathbb{Z})$, where $S(U, \mathbb{Z}) \leq_a K_U$ for each U.*

Corollary 5.9.

1. Let G be singularly convex and $G \leq H$. Then $G \leq_a H$ if and only if $H = G + K$ where $G^* \leq_a K$. Such an H is a-closed if and only if K is a-closed.

2. If $\mathbf{W_s}H$ is singularly convex and if H is an a-extension of $\mathbf{W_s}H$, then H is a-closed if and only if H^* is a-closed.

Proof. These are merely restatements of Theorem 5.5 in the light of Theorem 5.7. ∎

Open questions:

(i) Must every singularly a-closed, local group that is also strongly least integer closed be singularly convex?

(ii) For what groups is G a-closed if and only if $\mathbf{W_s}G$ is singularly a-closed and G^* is a-closed?

(iii) For what class of groups is the vector lattice hull an a-closure?

(iv) Is an a-extension of a local group necessarily local?

(v) What conditions on an archimedean f-ring will guarantee that every a-closure is also an f-ring?

References

[1] M. Anderson, *Locally flat vector lattices.* Canad. J. Math. **32** no. 4, (1980), 924-936.

[2] E. Aron, *Embedding lattice-ordered algebras in uniformly closed algebras.* (1971) University of Rochester Dissertation.

[3] E. Aron and A. Hager, *Convex vector lattices and ℓ-algebras.* Topology and Appl. **12** (1981), 1-10.

[4] R. Ball and A. Hager, *One the localic Yosida representation of an archimedean lattice-ordered group with weak order unit.* J. Pure and Applied Algebra, **70** (1991), 17-43.

[5] R. Ball and A. Hager, *Algebraic extensions of an archimedean lattice-ordered group, I.* J. Pure and Applied Algebra, **85** (1993), 1-20.

[6] P. Conrad, *Archimedean extensions of lattice-ordered groups.* J. Indian Math. Soc. **30** (1966), 131-160.

[7] P. Conrad, *Epi-archimedean groups*, Czech. Math. J. **24** (99) (1974), 192-218.

[8] P. Conrad, M. Darnel and D. G. Nelson, *Valuations of lattice-ordered groups.* J. Algebra **192** no. 1, (1997), 380-411.

[9] M. Darnel, *Theory of Lattice-Ordered Groups.* Pure and Appl. Math. **187** (1995) Marcel Dekker, New York.

[10] L. Gillman and M. Jerison, *Rings of Continuous Functions.* (1960) D. Van Nostrand Publ. Co.

[11] A. Hager and C. Kimber, *Some examples of hyperarchimedean lattice-ordered groups.* Submitted.

[12] A. Hager, C. Kimber and W. McGovern *Pseudo-least integer closed groups.* In progress.

[13] A. Hager and J. Martinez, *Archimedean singular lattice-ordered groups.* Alg. Univer. **40** (1998), 119-147.

[14] A. Hager and L. Robertson, *Extremal units in an archimedean Riesz space.* Rend. Sem. Mat. Univ. Padova **59** (1978), 97-115.

[15] A. Hager and L. Robertson, *Representing and ringifying a Riesz space.* Symp. Math. **21** (1977), 411-431.

[16] M. Henriksen, J. Isbell, and D. Johnson, *Residue class fields of lattice-ordered algebras.* Fund. Math. **50** (1961), 107-117.

[17] M. Henriksen and D. Johnson, *On the structure of a class of Archimedean lattice-ordered algebras.* Fund. Math. **50** (1961), 73-94.

[18] C. Kimber and W. McGovern, *Bounded away lattice-ordered groups.* (1998) Manuscript.

[19] J. Martinez, *The hyper-archimedean kernel sequence of a lattice-ordered group.* Bull. Austral. Math. Soc. **10** (1974), 337-340.

[20] K. Yosida, *On the representation of the vector lattice.* Proc. Imp. Acad. Tokyo **18** (1942), 339-343.

Department of Mathematics, Wesleyan University, Middletown, CT 06459-0128, USA
ahager@wesleyan.edu

Department of Mathematics, Lafayette College, Easton, PA 18042, USA
kimberc@lafayette.edu

Department of Mathematics and Statistics, Bowling Green State University,
Bowling Green, OH 43403, USA
warrenb@bgnet.bgsu.edu

Structure Spaces of Maximal ℓ-Ideals of Lattice-Ordered Rings

J. Ma and P. Wojciechowski[1]

Dedicated to Paul Conrad
on the occasion of his 80th birthday.

ABSTRACT.   Let R and S be two unital ℓ-reduced ℓ-semisimple lattice-ordered rings with $1 > 0$. It is shown that if there exists an ℓ-group isomorphism between the underlying lattice-ordered groups of R and S which preserves the identity element 1, then the structure spaces of maximal ℓ-ideals of R and S endowed with the hull-kernel topology are homeomorphic.

Let X be a compact Hausdorff space and $C(X)$ the ring of real continuous functions on X. Then $C(X)$ is an f-ring with the coordinate-wise order. In 1947, Kaplansky proved the classical result that if $C(X)$ and $C(Y)$ are isomorphic as lattices for two compact Hausdorff spaces X and Y, then X and Y are homeomorphic [7]. In 1968, by extending the arguments used by Kaplansky, Subramanian proved the generalization for f-rings: if two unital commutative ℓ-semisimple f-rings are isomorphic as lattices, then their structure spaces of maximal ℓ-ideals are homeomorphic [11]. Generally, this is not true for lattice-ordered rings (Example 1). In this paper, by extending the arguments used by Subramanian, it is shown that if there exists a unit-preserving ℓ-group isomorphism between two unital ℓ-reduced ℓ-semisimple ℓ-rings with $1 > 0$, then their structure spaces of maximal ℓ-ideals are homeomorphic (Theorem 11). We also include the above result for f-rings as a consequence without assuming commutativity (Corollary 15). Moreover, it is shown that the structure space of a unital ℓ-reduced ℓ-ring with $1 > 0$ is homeomorphic to the structure space of its ℓ-subring consisting of all elements whose absolute values are f-elements (Theorem 16).

First we review some definitions and results on lattice-ordered rings (ℓ-rings) and lattices we will use later. For the general theory on ℓ-rings and lattices, we refer the reader to Fuchs [5] and Birkhoff [2].

---

[1] The authors are very grateful to Professor Jorge Martinez and the referee for their valuable comments and suggestions.

J. Martínez (ed.), Ordered Algebraic Structures, 261–274.
© 2002 Kluwer Academic Publishers.

Let R be an ℓ-ring. An (left, right) ℓ-*ideal* I of R is a (left, right) ring ideal and a sublattice of R which is *convex* in the sense that $0 \leq y \leq x \in I$ implies $y \in I$, for all $x, y \in R$. A proper ℓ-ideal I of R is called ℓ-*prime* if $JK \subseteq I$ implies $J \subseteq I$ or $K \subseteq I$, for all ℓ-ideals J, K of R. An ℓ-ring R is called ℓ-*semisimple* if the intersection of all its maximal ℓ-ideals is zero, and R is called ℓ-*reduced* if R contains no nonzero positive nilpotent element. Let I be an ℓ-ideal of an ℓ-ring R, and let $x \in R$. Then $x(I)$ denotes the homomorphic image of x in R/I. In this paper, the *structure space* of an ℓ-ring R, denoted by $\mathrm{Max}_\ell(R)$, is the set of all its maximal ℓ-ideals endowed with the hull-kernel topology. Thus the *closure* of a set $\{M_\alpha : \alpha \in \Gamma\}$ in $\mathrm{Max}_\ell(R)$ is

$$\{M \in \mathrm{Max}_\ell(R) : \cap_{\alpha \in \Gamma} M_\alpha \subseteq M\}.$$

An ℓ-ring is called *unital* if it has an identity element, denoted by 1. It is clear that if R is unital, then $\mathrm{Max}_\ell(R)$ is compact.

Given an ℓ-ring R, the set $R^+ = \{a \in R : a \geq 0\}$ is called the *positive cone* of R. An element $a \in R^+$ is called an f-*element* if $b \wedge c = 0$ implies $ab \wedge c = ba \wedge c = 0$, for all $b, c \in R$. Let

$$T = T(R) = \{a \in R : |a| \text{ is an } f\text{-element of } R\},$$

where $|a| = a \vee -a$ for each $a \in R$. Then $T(R)$ is a convex ℓ-subring of R, and R is called an f-*ring* precisely when $R = T(R)$. Let

$$T^\perp = T(R)^\perp = \{r \in R : |r| \wedge |a| = 0, \forall a \in T(R)\}.$$

We note that if R is unital with $1 > 0$, then

$$T^\perp = \{r \in R : |r| \wedge 1 = 0\}.$$

Also let

$$U(T) = \{r \in R : |r| > T(R)\},$$

where $|r| > T(R)$ means $|r| \geq a$, for each $a \in T(R)$. By [3, Lemma 6.2], if $T(R) \neq 0$ is totally ordered, then we have the decomposition $R = U(T) \cup (T \oplus T^\perp)$, where the direct sum is regarded as the direct sum of ℓ-subgroups T and $T^\perp$, and $U(T) \cap (T \oplus T^\perp) = \emptyset$.

Let L be a lattice. A *lattice-prime ideal* P of L is a nonempty proper subset of L having the following properties.
(1) for all $a, b \in P$, $a \vee b \in P$,
(2) $b \leq a, a \in P$ and $b \in L$ implies $b \in P$,
(3) for all $a, b \in L$, $a \wedge b \in P$ implies $a \in P$ or $b \in P$.

Now we give an example to show that even if two unital commutative ℓ-semisimple ℓ-rings with $1 > 0$ are isomorphic as ℓ-groups, their structure spaces of maximal ℓ-ideals may not be homeomorphic. Recall that a *lattice-ordered field* (ℓ-*field*) is an ℓ-ring and a field.

Example 1. Let $\mathbb{Q}$ be the field of rationals, and let R be the direct sum of two copies of $\mathbb{Q}$ (so R is an f-ring). Let $S = \mathbb{Q}(\sqrt{2}) = \mathbb{Q} + \mathbb{Q}\sqrt{2}$ with coordinate-wise order. Then S is an ℓ-field. If we define $\phi\colon R \to S$ by $\phi((a,b)) = a + b\sqrt{2}$, then it is easy to check that ϕ is an ℓ-isomorphism between the underlying ℓ-groups of R and S (namely, ϕ is a group isomorphism and also a lattice isomorphism from R to S). It is also clear that R contains two maximal ℓ-ideals.

We notice that the ϕ in Example 1 does not preserve identity element 1. In the following it is shown that if there exists a unit-preserving ℓ-group isomorphism between two unital ℓ-reduced ℓ-semisimple ℓ-rings with $1 > 0$, then their structure spaces of maximal ℓ-ideals are homeomorphic. Essentially we follow Subramanian's proof [11], which is motivated by Kaplansky's proof [7].

The following lemma shows that in a unital ℓ-ring with $1 > 0$ certain ℓ-ideals contain f-elements.

Lemma 2. *Let R be a unital ℓ-ring with $1 > 0$, and let I, J be two ℓ-ideals of R with $R = I + J$. Then there exist two f-elements x and y of R such that*
(a) $y \in I$ and $x \in J$;
(b) $1(I) = x(I)$ and $1(J) = y(J)$;
(c) $0 \le x \le 1$, $0 \le y \le 1$, and $x \wedge y = 0$ (hence $xy = yx = 0$). If I and J are proper, then $x \notin I$ and $y \notin J$.

Proof. Since $R = I + J$, we have $1 = i + j$ for some $i \in I$ and $j \in J$. Since $1 > 0$, $1 = |1| = |i + j| \le |i| + |j|$, and hence $1 = i_1 + j_1$, where $0 \le i_1 \le |i|$ and $0 \le j_1 \le |j|$, so $i_1 \in I$ and $j_1 \in J$. Let $z = i_1 \wedge j_1$, $x = j_1 - z$, and $y = i_1 - z$. Then $x \wedge y = 0$, $x \in J$, $y \in I$, and $z \in I \cap J$. Since $1 = x + y + 2z$, we have $1(I) = x(I) + y(I) + 2z(I) = x(I)$ and $1(J) = x(J) + y(J) + 2z(J) = y(J)$. Since $0 \le x, y \le 1$, they are f-elements. ∎

Lemma 2 can be stated more generally. Let R be a unital ℓ-ring with $1 > 0$. If $R = K + J$ for two left or right ℓ-ideals of R, then $1 = i + j$ with $0 \le i \in K$ and $0 \le j \in J$, and i and j are f-elements since $0 \le i, j \le 1$. This fact will be often used later.

We now record some consequences of Lemma 2.

Corollary 3. *Let R be a unital ℓ-ring with $1 > 0$, and let Φ_1 and Φ_2 be two nonempty disjoint closed sets in $\mathrm{Max}_\ell(R)$. Then there exist two f-elements x and y in R such that*
(a) for each $M \in \Phi_1$, $y \in M$ and for each $N \in \Phi_2$, $x \in N$;
(b) for each $M \in \Phi_1$, $1(M) = x(M)$ and for each $N \in \Phi_2$, $1(N) = y(N)$;
(c) $0 < x < 1, 0 < y < 1$, and $x \wedge y = 0$ (hence $xy = yx = 0$).

Proof. Let I_k denote the intersection of all maximal ℓ-ideals in Φ_k ($k = 1, 2$). Then I_1 and I_2 are proper. If $R \ne I_1 + I_2$, then there exists a maximal ℓ-ideal M such that

$I_1 + I_2 \subseteq M$, so $I_1 \subseteq M$ and $I_2 \subseteq M$. Since Φ_1 and Φ_2 are closed, $M \in \Phi_1$ and $M \in \Phi_2$, which contradicts that Φ_1 and Φ_2 are disjoint. Thus $R = I_1 + I_2$ and the Corollary follows from Lemma 2. ∎

Corollary 4. *Let R be a unital ℓ-ring with $1 > 0$, and let Φ_1 and Φ_2 be two nonempty disjoint closed sets in $\text{Max}_\ell(R)$. Then for any two elements $a, b \in R$, there exists $c \in R$ such that $c(M) = (a - 1)(M)$, for each $M \in \Phi_1$ and $c(N) = (b + 1)(N)$, for each $N \in \Phi_2$.*

Proof. Let x and y be the same as in Corollary 3. Given $a, b \in R$, let $c = x(a-1) + y(b+1)$. Since for each $M \in \Phi_1, y(M) = 0$ and $x(M) = 1(M)$,

$$c(M) = x(M)(a-1)(M) + y(M)(b+1)(M) = (a-1)(M).$$

Similarly, since for each $N \in \Phi_2, x(N) = 0$ and $y(N) = 1(N)$,

$$c(N) = x(N)(a-1)(N) + y(N)(b+1)(N) = (b+1)(N).$$

∎

Let F be an f-ring and let J be an ℓ-prime ℓ-ideal of F. Then $ab \in J$ implies $a \in J$ or $b \in J$ for all $a, b \in F$ [6, Corollary 4.6, p. 175]. Generally, this is not true for an ℓ-ring [4, Example 2.3]. The following result, however, shows that if I is an ℓ-prime ℓ-ideal of a unital ℓ-reduced ℓ-ring R with $1 > 0$, then $xy \in I$ implies $x \in I$ or $y \in I$ for all $x, y \in T(R)$. This result will be often used later.

Lemma 5. *Let R be a unital ℓ-reduced ℓ-ring with $1 > 0$, and let I be an ℓ-prime ℓ-ideal of R. If $xy \in I$ or $x \wedge y \in I$, where x and y are in $T(R)$, then $x \in I$ or $y \in I$.*

Proof. By [8, Lemma 4], $T(R/I)$ is a totally ordered domain. Let $xy \in I$, and $x, y \in T(R)$. Then $x(I)y(I) = 0$ in R/I and $x(I), y(I) \in T(R/I)$. Thus $x(I) = 0$ or $y(I) = 0$ since $T(R/I)$ is a domain, so $x \in I$ or $y \in I$. If $x \wedge y \in I$, then $x(I) \wedge y(I) = 0$ in R/I, and hence $x \in I$ or $y \in I$, since $T(R/I)$ is totally ordered. ∎

The Lemma 5 is not true if R is not ℓ-reduced as shown in the following example. This matrix example serves as a counterexample for many cases in which ℓ-rings are not ℓ-reduced.

Example 6. Let L be a totally ordered field, and let L_n ($n \geq 2$) be the $n \times n$ matrix ring over L. Then L_n may be lattice-ordered by letting a matrix be positive exactly when each of its entries is positive, that is, the positive cone is $(L^+)_n$. Since 0 and L_n are the only ideals of L_n, 0 is an ℓ-prime ℓ-ideal of $(L_n, (L^+)_n)$. It is straightforward to verify that $T(L_n) = \{D \in L_n : D \text{ is a diagonal matrix}\}$. Let e_{ij} be the matrix units. Then for any $i \neq j$, e_{ii} and e_{jj} are nonzero f-elements with $e_{ii}e_{jj} = 0$.

For the rest of this paper we always assume that R is a unital ℓ-reduced ℓ-semisimple ℓ-ring with $1 > 0$. A lattice-prime ideal P of R is called *dominated*, if $x \in P$ and $(y - x)^+ \wedge 1 = 0$ imply $y \in P$, for all $x, y \in R$, or equivalently, if $x \in P$ and $y \leq x + z$, where $z \wedge 1 = 0$, then $y \in P$. A lattice-prime ideal P of R is called *associated* with a maximal ℓ-ideal M if $x \in P$ and $(y - x)^- \wedge 1 \notin M$ imply $y \in P$.

The next three lemmas establish the relation between dominated lattice-prime ideals and maximal ℓ-ideals. The Lemma 7 shows how to construct dominated lattice-prime ideals associated with a given maximal ℓ-ideal.

Lemma 7. *Let M be a maximal ℓ-ideal of R, and let a be a fixed element of R. Define $P = \{r \in R : (r - a)^+ \wedge 1 \in M\}$. Then P is a dominated lattice-prime ideal of R associated with M.*

Proof. P is nonempty and proper since $a \in P$ and $a + 1 \notin P$. We first show that P is a lattice-prime ideal. Let $r, s \in P$. Since

$$[(r \vee s) - a]^+ \wedge 1 = [(r - a)^+ \vee (s - a)^+] \wedge 1 = [(r - a)^+ \wedge 1] \vee [(s - a)^+ \wedge 1] \in M,$$

we have $r \vee s \in P$. Let $t \leq s \in P$ for some $t \in R$. Then $(t - a)^+ \wedge 1 \leq (s - a)^+ \wedge 1 \in M$ implies $(t - a)^+ \wedge 1 \in M$, so $t \in P$. Now let $u \wedge v \in P$, for some $u, v \in R$. Since

$$[(u - a)^+ \wedge 1] \wedge [(v - a)^+ \wedge 1] = [(u - a)^+ \wedge (v - a)^+] \wedge 1 = (u \wedge v - a)^+ \wedge 1 \in M,$$

by Lemma 5, we have $(u - a)^+ \wedge 1 \in M$ or $(v - a)^+ \wedge 1 \in M$ since each maximal ℓ-ideal is ℓ-prime, so $u \in P$ or $v \in P$. Thus P is a lattice-prime ideal of R. To see that P is dominated, let $x \in P$ and $(y - x)^+ \wedge 1 = 0$ for some $y \in R$. Since

$$\begin{aligned} (y - a)^+ \wedge 1 &= (y - x + x - a)^+ \wedge 1 \leq [(y - x)^+ + (x - a)^+] \wedge 1 \\ &\leq [(y - x)^+ \wedge 1] + [(x - a)^+ \wedge 1] \in M, \end{aligned}$$

$(y - a)^+ \wedge 1 \in M$, and hence $y \in P$. Thus P is a dominated lattice-prime ideal. Finally let $x \in P$ and $(y - x)^- \wedge 1 \notin M$ for $x, y \in R$. Then $(y - x)^+ \wedge 1 \in M$ by Lemma 5, since $[(y - x)^- \wedge 1] \wedge [(y - x)^+ \wedge 1] = 0$, and hence, as above,

$$(y - a)^+ \wedge 1 \leq [(y - x)^+ \wedge 1] + [(x - a)^+ \wedge 1] \in M,$$

so $(y - a)^+ \wedge 1 \in M$ and $y \in P$. Thus P is a dominated lattice-prime ideal of R associated with M. ∎

Given a maximal ℓ-ideal M, there may exist more than one dominated lattice-prime ideal associated with M by Lemma 7. However, the following lemma shows that every dominated lattice-prime ideal is associated with exactly one maximal ℓ-ideal.

Lemma 8. *Every dominated lattice-prime ideal P of R is associated with exactly one maximal ℓ-ideal of R.*

Proof. Suppose that P is not associated with any maximal ℓ-ideal. Then for each maximal ℓ-ideal M_α there exist $x_\alpha, y_\alpha \in R$ such that $x_\alpha \in P, y_\alpha \notin P$ and yet $(y_\alpha - x_\alpha)^- \wedge 1 \notin M_\alpha$. For $x, y \in R$, let $\delta(x, y)$ denote the open set in $\text{Max}_\ell(R)$ that consists of all maximal ℓ-ideals not containing $(y - x)^- \wedge 1$. Then we have a family of open sets $\delta(x_\alpha, y_\alpha)$ which covers $\text{Max}_\ell(R)$. Since $\text{Max}_\ell(R)$ is compact, there exists a finite subcover $\delta(x_i, y_i), i = 1, \ldots, n$. Consider now $\delta(\vee_{i=1}^n x_i, \wedge_{i=1}^n y_i)$. Since $(\wedge_{i=1}^n y_i - \vee_{i=1}^n x_i)^- \geq (y_j - x_j)^- \geq 0$ for each $j = 1, \ldots, n$, for each maximal ℓ-ideal M_α, $(\wedge_{i=1}^n y_i - \vee_{i=1}^n x_i)^- \wedge 1 \notin M_\alpha$, and hence $(\wedge_{i=1}^n y_i - \vee_{i=1}^n x_i)^+ \wedge 1 \in M_\alpha$ for each $M_\alpha \in \text{Max}_\ell(R)$, by Lemma 5. Since R is ℓ-semisimple, $(\wedge_{i=1}^n y_i - \vee_{i=1}^n x_i)^+ \wedge 1 = 0$, and hence, since P is dominated and $\vee_{i=1}^n x_i \in P, \wedge_{i=1}^n y_i \in P$, so $y_j \in P$, for some $1 \leq j \leq n$, which is a contradiction. Thus P is at least associated with one maximal ℓ-ideal of R.

Now suppose P is associated with two different maximal ℓ-ideals of R, say, M and N. Choose $a \in P, b \notin P$. Since the sets $\{M\}$ and $\{N\}$ in $\text{Max}_\ell(R)$ are disjoint and closed, by Corollary 4, there exists an element $c \in R$ such that $c(M) = (a - 1)(M)$ and $c(N) = (b + 1)(N)$. Thus $(c - a)^-(M) = 1(M)$ and $(b - c)^-(N) = 1(N)$, so $(c - a)^- \wedge 1 \notin M$ and $(b - c)^- \wedge 1 \notin N$. Thus $c \in P$, and hence $b \in P$, which is a contradiction. Thus P is associated with exactly one maximal ℓ-ideal of R. ∎

It is natural to ask when two dominated lattice-prime ideals are associated with the same maximal ℓ-ideal. This question is answered in the following result.

Lemma 9. *Two dominated lattice-prime ideals P_1 and P_2 of R are associated with the same maximal ℓ-ideal of R if and only if there is a dominated lattice-prime ideal P such that $P \subseteq P_1 \cap P_2$.*

Proof. "⇒" Let P_1 and P_2 be associated with one maximal ℓ-ideal M. Choose $x \in P_1$ and $y \in P_2$. Write $a = (x \wedge y) - 1$ and $P = \{r \in R : (r - a)^+ \wedge 1 \in M\}$. From Lemma 7, P is a dominated lattice-prime ideal associated with M. We show that $P \subseteq P_1 \cap P_2$. Let $r \in P$. Since

$$(r - a)^+ \wedge 1 = [r - (x \wedge y) + 1]^+ \wedge 1 = [(r - x + 1)^+ \vee (r - y + 1)^+] \wedge 1$$
$$= [(r - x + 1)^+ \wedge 1] \vee [(r - y + 1)^+ \wedge 1],$$

we have $(r - x + 1)^+ \wedge 1 \in M$ and $(r - y + 1)^+ \wedge 1 \in M$. Since $(r - x)^+ \leq (r - x + 1)^+$ and $(r - y)^+ \leq (r - y + 1)^+$, we have $(r - x)^+ \wedge 1 \in M$ and $(r - y)^+ \wedge 1 \in M$. If $(r - x)^- \wedge 1 \in M$, then, since $(r - x)^+(M) \wedge 1(M) = 0$ and $(r - x)^-(M) \wedge 1(M) = 0$, $(r - x)^+(M)$ and $(r - x)^-(M) \in T(R/M)^\perp$, so $(r - x)(M) \in T(R/M)^\perp$. From $R/M = U(R/M) \cup (T(R/M) \oplus T(R/M)^\perp)$, we have

$$(r - x + 1)^+(M) = [(r - x + 1)(M)]^+ = [(r - x)(M) + 1(M)]^+$$
$$= [(r - x)(M)]^+ + [1(M)]^+ = (r - x)^+(M) + 1(M).$$

Thus, $1(M) \in T(R/M)^{\perp}$ since $(r - x + 1)^+(M)$, $(r - x)^+(M) \in T(R/M)^{\perp}$, which is a contradiction. Thus $(r - x)^- \wedge 1 \notin M$, so $r \in P_1$. Similarly, $(r - y)^- \wedge 1 \notin M$, so $r \in P_2$. Thus $P \subseteq P_1 \cap P_2$.

"$\Leftarrow$" Let P be a dominated lattice-prime ideal contained in $P_1 \cap P_2$, and P, P_1, and P_2 be associated with maximal ℓ-ideals M, M_1, and M_2, respectively. If $M \neq M_1$, then $\{M\}$ and $\{M_1\}$ are disjoint closed sets in $\mathrm{Max}_\ell(R)$. Choose $a \in P$ and $b \notin P_1$. There exists $c \in R$ such that $c(M) = (a - 1)(M)$ and $c(M_1) = (b + 1)(M_1)$ by Corollary 4. Thus $c \in P$ since $(c - a)^- \wedge 1 \notin M$ and P is associated with M, and then $b \in P_1$ since $(b - c)^- \wedge 1 \notin M_1$, and P_1 is associated with M_1 and $c \in P \subseteq P_1$. This contradicts $b \notin P_1$. Thus $M = M_1$. Similarly, $M = M_2$. Thus $M_1 = M_2$. ∎

Next, we characterize the closure of a nonempty set in $\mathrm{Max}_\ell(R)$ by using dominated lattice-prime ideals.

Lemma 10. Let a be a fixed element in R. Let $\Phi = \{M_\alpha : \alpha \in \Gamma\}$ be any nonempty set in $\mathrm{Max}_\ell(R)$. Then a maximal ℓ-ideal M belongs to the closure of Φ if and only if there exists a dominated lattice-prime ideal P associated with M such that it contains the intersection $A(\Phi)$ of all the dominated lattice-prime ideals which contain a and are also associated with some member in Φ.

Proof. Let M be in the closure of $\Phi = \{M_\alpha : \alpha \in \Gamma\}$. By Lemma 7,

$$P = \{r \in R : (r - a)^+ \wedge 1 \in M\} \quad \text{and} \quad P_\alpha = \{r \in R : (r - a)^+ \wedge 1 \in M_\alpha\}$$

are dominated lattice-prime ideals associated with M and M_α, respectively. Obviously, $a \in P_\alpha$ for each $\alpha \in \Gamma$. Now if $r \in A(\Phi)$, then $r \in P_\alpha$ for each $\alpha \in \Gamma$, so $(r-a)^+ \wedge 1 \in M_\alpha$ for each $M_\alpha \in \Phi$, and hence $(r - a)^+ \wedge 1 \in \cap_{\alpha \in \Gamma} M_\alpha \subseteq M$. Thus $r \in P$, so $A(\Phi) \subseteq P$.

Conversely, suppose M does not belong to the closure of Φ. Let P be any dominated lattice-prime ideal associated with M. Consider some $b \notin P$. Since $\{M\}$ and the closure of Φ are two disjoint closed sets in $\mathrm{Max}_\ell(R)$, by Corollary 4, there is an element $c \in R$ such that $c(M_\alpha) = (a - 1)(M_\alpha)$ for all $M_\alpha \in \Phi$ and $c(M) = (b + 1)(M)$. Thus $(c - a)^- \wedge 1 \notin M_\alpha$ for all M_α in Φ, and hence $c \in A(\Phi)$ by the definition of $A(\Phi)$. From $(b - c)^- \wedge 1 \notin M$, we have $c \notin P$ since $b \notin P$. Thus no dominated lattice-prime ideal associated with M can contain $A(\Phi)$. ∎

Now we are ready to prove the main result in this paper.

Theorem 11. Let R and S be two unital ℓ-reduced ℓ-semisimple ℓ-rings with $1 > 0$. If there exists an ℓ-group isomorphism between R and S which preserves identity element 1, then $\mathrm{Max}_\ell(R)$ and $\mathrm{Max}_\ell(S)$ are homeomorphic.

Proof. Let φ be an ℓ-group isomorphism from R to S with $\varphi(1) = 1$. We show first that for a subset P of R, P is a dominated lattice-prime ideal of R if and only if $\varphi(P)$ is a dominated lattice-prime ideal of S. Suppose that P is a dominated lattice-prime ideal of R. Since φ is an ℓ-group isomorphism, $\varphi(P)$ is a lattice-prime ideal of S. To see that $\varphi(P)$ is dominated, let $x, y \in S$ such that $x \in \varphi(P)$ and $(y - x)^+ \wedge 1 = 0$. Let $x = \varphi(a)$ and $y = \varphi(b)$, where $a \in P, b \in R$. Since

$$\varphi[(b - a)^+ \wedge 1] = \varphi[(b - a)^+] \wedge \varphi(1) = [\varphi(b - a)]^+ \wedge 1 = (y - x)^+ \wedge 1 = 0,$$

$(b - a)^+ \wedge 1 = 0$, so $b \in P$, and hence $y = \varphi(b) \in \varphi(P)$. Thus $\varphi(P)$ is dominated. Similarly, if $\varphi(P)$ is a dominated lattice-prime ideal of S, then P is a dominated lattice-prime ideal of R. Therefore, φ induces a one-to-one correspondence between the set of all dominated lattice-prime ideals of R and the set of all dominated lattice-prime ideals of S.

Two dominated lattice-prime ideals are called *equivalent* if they are associated with the same maximal ℓ-ideal. By Lemma 8, we have an equivalence relation on the set of all dominated lattice-prime ideals. Let $[P]$ denote the equivalence class containing the dominated lattice-prime ideal P. We define $\psi : \mathrm{Max}_\ell(R) \to \mathrm{Max}_\ell(S)$,

$$M \mapsto [P] \mapsto [\varphi(P)] \mapsto N,$$

where $M \in \mathrm{Max}_\ell(R), P$ is a dominated lattice-prime ideal of R associated with M, and $\varphi(P)$ is associated with $N \in \mathrm{Max}_\ell(S)$. Let P_1 and P_2 be two dominated lattice-prime ideals of R associated with the same maximal ℓ-ideal of R. Then there exists a dominated lattice-prime ideal P of R such that $P \subseteq P_1 \cap P_2$ by Lemma 9, so $\varphi(P) \subseteq \varphi(P_1) \cap \varphi(P_2)$, and hence, by Lemma 9 again, $\varphi(P_1)$ and $\varphi(P_2)$ are dominated lattice-prime ideals of S associated with the same maximal ℓ-ideal of S. Thus ψ is well-defined. By Lemmas 7, 8, and 9, ψ is one-to-one and onto.

Let b be a fixed element of S, and let $b = \varphi(a)$, where $a \in R$. For a nonempty set Φ in $\mathrm{Max}_\ell(R)$, $\psi(\Phi)$ is a nonempty set in $\mathrm{Max}_\ell(S)$. Let $A(\psi(\Phi))$ be the intersection of all the dominated lattice-prime ideals of S which contain b and are also associated with some member in $\psi(\Phi)$ and let $A(\Phi)$ be the intersection of all the dominated lattice-prime ideals of R which contain a and are also associated with some member in Φ. We claim that $\varphi(A(\Phi)) = A(\psi(\Phi))$. Let $y = \varphi(x) \in \varphi(A(\Phi))$, where $x \in A(\Phi)$, and let I ba a dominated lattice-prime ideal of S which contains b and is associated with some $N \in \psi(\Phi)$. Let $I = \varphi(P)$ and $N = \psi(M)$. Then P is a dominated lattice-prime ideal of R which contains a and is associated with $M \in \Phi$. Thus $A(\Phi) \subseteq P$, so $x \in P$ and $y = \varphi(x) \in I = \varphi(P)$. Therefore, $y \in A(\psi(\Phi))$, and hence $\varphi(A(\Phi)) \subseteq A(\psi(\Phi))$. Now let $z \in A(\psi(\Phi))$, and let $z = \varphi(w)$, where $w \in R$. We show that $w \in A(\Phi)$. Let P_1 be a dominated lattice-prime ideal of R which contains a and is associated with some $M_1 \in \Phi$ and let $I_1 = \varphi(P_1)$, $N_1 = \varphi(M_1)$. Then I_1 is a dominated lattice-prime ideal of S which contains b and is associated with $N_1 \in \psi(\Phi)$, so $A(\psi(\Phi)) \subseteq I_1$, and hence $z = \varphi(w) \in I_1 = \varphi(P_1)$, therefore, $w \in P_1$. Thus $w \in A(\Phi)$.

Now let Φ be a closed set in $\mathrm{Max}_\ell(R)$. We show that $\psi(\Phi)$ is closed in $\mathrm{Max}_\ell(S)$. If $N \in \mathrm{Max}_\ell(S)$ is in the closure of $\psi(\Phi)$, then by Lemma 10, there exists a dominated lattice-prime ideal I of S associated with N such that I contains $A(\psi(\Phi))$. Let $I = \varphi(P)$

and $N = \psi(M)$. Then P is a dominated lattice-prime ideal of R associated with M. Since $\varphi(A(\Phi)) = A(\psi(\Phi)) \subseteq I = \varphi(P)$, $A(\Phi) \subseteq P$. Thus, by Lemma 10, $M \in \Phi$, and hence $N = \psi(M) \in \psi(\Phi)$, so $\psi(\Phi)$ is closed, therefore, ψ^{-1} is continuous. By a similar argument, it can be shown that ψ is also continuous. Thus $\mathrm{Max}_\ell(R)$ and $\mathrm{Max}_\ell(S)$ are homeomorphic.   ∎

The conditions that R, S are ℓ-reduced and $1 > 0$ cannot be dropped in Theorem 11, as shown in the following examples.

Example 12. Let A be the direct sum of two copies of the ℓ-field $\mathbb{Q}(\sqrt{2})$ defined as in Example 1. Then A is a unital ℓ-reduced ℓ-semisimple ℓ-ring with $1 > 0$. Let $(\mathbb{Q}_2, (\mathbb{Q}^+)_2)$ be the lattice-ordered 2×2 matrix ring defined as in Example 6. Then $(\mathbb{Q}_2, (\mathbb{Q}^+)_2)$ is a unital ℓ-semisimple ℓ-ring with $1 > 0$, but $(\mathbb{Q}_2, (\mathbb{Q}^+)_2)$ is not ℓ-reduced. Define $f : \mathbb{Q}_2 \to A$ by

$$f \begin{pmatrix} a & b \\ c & d \end{pmatrix} = (a + b\sqrt{2}, d + c\sqrt{2}).$$

Then f is an ℓ-group isomorphism that preserves 1. Since A contains two maximal ℓ-ideals and $\mathbb{Q}_2$ contains one maximal ℓ-ideal, their structure spaces cannot be homeomorphic. Therefore, the condition that R and S are ℓ-reduced cannot be droped from Theorem 11.

Example 13. Consider the 2×2 matrix ring $\mathbb{Q}_2$ and the field $\mathbb{Q}(\sqrt{2})$ again. This time we give them different lattice orders from those in Example 12. The lattice orders that we are going to use were described by Weinberg in [12] and [13].
    We first consider $\mathbb{Q}_2$. Let

$$f_1 = \begin{pmatrix} 1 & 2 \\ 0 & 0 \end{pmatrix}, f_2 = \begin{pmatrix} 0 & 0 \\ 1 & 1 \end{pmatrix}, f_3 = \begin{pmatrix} 2 & 2 \\ 0 & 0 \end{pmatrix}, f_4 = \begin{pmatrix} 0 & 0 \\ 1 & 2 \end{pmatrix}.$$

Then $\mathbb{Q}_2 = \mathbb{Q}f_1 + \mathbb{Q}f_2 + \mathbb{Q}f_3 + \mathbb{Q}f_4$. Define the positive cone as

$$(\mathbb{Q}_2)^+ = \mathbb{Q}^+ f_1 + \mathbb{Q}^+ f_2 + \mathbb{Q}^+ f_3 + \mathbb{Q}^+ f_4.$$

Then $\mathbb{Q}_2$ becomes an ℓ-ring since the product of any two matrices in $\{f_1, f_2, f_3, f_4\}$ is positive. Since none of f_1, f_2, f_3, f_4 is nilpotent, $\mathbb{Q}_2$ is ℓ-reduced and since $1 = -(f_1 + f_2) + (f_3 + f_4)$, 1 is not positive.
    Next, consider $\mathbb{Q}(\sqrt{2})$. Let $x = 2 + \sqrt{2}$ and $y = 1 + \sqrt{2}$. Then $\mathbb{Q}(\sqrt{2}) = \mathbb{Q}x + \mathbb{Q}y$. Since $x^2 = 2x + 2y$, $y^2 = x + y$, and $xy = yx = x + 2y$, $\mathbb{Q}(\sqrt{2})$ becomes an ℓ-field if we define the positive cone as

$$(\mathbb{Q}(\sqrt{2}))^+ = \mathbb{Q}^+ x + \mathbb{Q}^+ y.$$

Note that 1 is not positive since $1 = x - y$.

Now let A be the direct sum of two copies of the ℓ-field $\mathbb{Q}(\sqrt{2})$ with the above lattice order. Then A is a unital ℓ-reduced ℓ-semisimple ℓ-ring in which 1 is not positive. Define $f : \mathbb{Q}_2 \to A$ by

$$f(\alpha_1 f_1 + \alpha_2 f_2 + \alpha_3 f_3 + \alpha_4 f_4) = (\alpha_3 x + \alpha_1 y, \alpha_4 x + \alpha_2 y),$$

where $\alpha_i \in \mathbb{Q}, 1 \leq i \leq 4$. Then f is an ℓ-group isomorphism that preserves 1. Therefore, the condition that 1 is positive cannot be omitted in Theorem 11.

There are many unital ℓ-reduced ℓ-semisimple ℓ-rings with $1 > 0$ that are not f-rings. Below we list some of them. An ℓ-ring is called ℓ-*simple* if the only ℓ-ideals are 0 and itself, and an ℓ-ring is called an ℓ-*domain* if $ab = 0$ implies $a = 0$ or $b = 0$ for all $a \geq 0$ and $b \geq 0$. An ℓ-ring is ℓ-semisimple if and only if it is a subdirect product of ℓ-simple ℓ-rings, and an ℓ-ring is ℓ-reduced if and only if it is a subdirect product of ℓ-domains [10, Corollary 1]. Thus, ℓ-reduced ℓ-semisimple ℓ-rings could be built using ℓ-simple ℓ-domains via subdirect product, or simply via direct sum.

Some examples of unital ℓ-simple ℓ-domains with $1 > 0$ are provided below. Recall that an ℓ-algebra over a totally ordered field L is called *Archimedean over L* if it has no nonzero bounded subspaces.

Example 14. (I) [1, Example 10, p. 49]. Let L be a totally ordered field, and let $L(G)$ be the group algebra of the finite group G over L. Lattice order $L(G)$ by defining $\sum \alpha_i g_i \geq 0$, where $g_i \in G$, $\alpha_i \in L$, if and only if all $\alpha_i \geq 0$. Then $L(G)$ is a unital finite-dimensional Archimedean ℓ-algebra over L with $1 > 0$. By [1, Remark 6, p. 49], $L(G)$ is ℓ-simple. It is straightforward to check that $L(G)$ is an ℓ-domain.

The above example has some general meaning. Let A be a unital commutative finite-dimensional Archimedean ℓ-algebra over L with $1 > 0$. Given an element $a \in A$, let $\ell(1-a) = \{x \in A : (1-a)x = 0\}$ be the annihilator of $(1-a)$. Then $\ell(1-a)$ is an ideal of A. If there exists an element $e \in A^+$, $e \neq 1$ and $e^n = 1$ for some $n \geq 2$ such that the dimension of $\ell(1-e)$ over L is 1, then A is isomorphic to the ℓ-algebra $L(G)$ defined above, where G is a finite cyclic group [9, Theorem 3].

(II) [10, Example 7]. Let S be a unital totally ordered domain and let $A = S[x]$ be the polynomial ring in the indeterminate x. For each $n \geq 1$, define the positive cone of A as

$$P_n = \left\{ \sum_{i=0}^{k} \alpha_i x^i : k \geq 2n \text{ and } \alpha_k > 0, \text{ or } k < 2n \text{ and each } \alpha_i \geq 0 \right\}.$$

Then (A, P_n) is an ℓ-simple ℓ-domain with $1 > 0$.

We now consider unital ℓ-semisimple f-rings. Recall that a positive element e in an ℓ-ring is called a *weak unit* by Birkhoff and Pierce if $e \wedge a = 0$ implies $a = 0$ for any

$a \geq 0$. In a unital ℓ-ring A with $1 > 0$, 1 is a weak unit if and only if A is an *almost f-ring*, that is, the identity $x^+ x^- = 0$ holds in A, or equivalently, $x \wedge y = 0$ implies $xy = 0$ [1, Theorem 15, p. 60]. In an ℓ-reduced unital ℓ-ring with $1 > 0$, 1 is a weak unit if and only if it is an f-ring [1, Corollary 1, p. 61].

Now if F is a unital ℓ-semisimple f-ring, then F contains no nonzero nilpotent element since all nilpotent elements in F form a nilpotent ℓ-ideal of F [1, Theorem 16, p. 63], and 1 is a weak unit in F. Since 1 is a weak unit, every lattice-prime ideal of F is dominated, and hence, by the same argument in Theorem 11 we obtain the following corollary, which was considered by Subramanian in [11] for commutative f-rings.

Corollary 15. *Let F_1 and F_2 be two unital ℓ-semisimple f-rings. If they are isomorphic as lattices, then $\mathrm{Max}_\ell(F_1)$ and $\mathrm{Max}_\ell(F_2)$ are homeomorphic.*

Let A be a unital ℓ-ring with $1 > 0$. It is interesting and important to investigate the relation between ℓ-ideals of A and ℓ-ideals of $T(A)$. We observe that if I is an ℓ-prime ℓ-ideal of A, then $I \cap T(A)$ is not necessarily an ℓ-prime ℓ-ideal of $T(A)$. For instance, considering the matrix ring $(L_n, (L^+)_n)$ in Example 6, 0 is an ℓ-prime ℓ-ideal of L_n, but 0 is not ℓ-prime in $T(L_n)$ since $T(L_n)$ is isomorphic to the direct sum of n copies of L. However if A is also ℓ-reduced, then $I \cap T(A)$ is an ℓ-prime ℓ-ideal of $T(A)$ since if $xy \in I \cap T(A)$ for some $x, y \in T(A)$, then $x \in I$ or $y \in I$ by Lemma 5, so $x \in I \cap T(A)$ or $y \in I \cap T(A)$.

The important relation between maximal ℓ-ideals of A and maximal ℓ-ideals of $T(A)$ is established in the next result.

Theorem 16. *Let A be a unital ℓ-reduced ℓ-ring with $1 > 0$. Then $\mathrm{Max}_\ell(A)$ and $\mathrm{Max}_\ell(T(A))$ are homeomorphic.*

Proof. Let M be a maximal ℓ-ideal of A. Then $M \cap T(A)$ is an ℓ-prime (actually prime) ℓ-ideal of $T(A)$ by Lemma 5. Since in an f-ring the ℓ-ideals that contain an ℓ-prime ℓ-ideal form a chain, there exists a unique maximal ℓ-ideal of $T(A)$ that contains $M \cap T(A)$. We denote this unique maximal ℓ-ideal of $T(A)$ by $M(T)$. Now we define $\varphi : \mathrm{Max}_\ell(A) \to \mathrm{Max}_\ell(T(A))$ by $\varphi(M) = M(T)$.

Let $M, N \in \mathrm{Max}_\ell(A)$ and $\varphi(M) = \varphi(N)$. If $M \neq N$, then $A = M + N$, and hence there exist $0 \leq i \in M \cap T(A)$ and $0 \leq j \in N \cap T(A)$ such that $1 = i + j$, so $1 = i + j \in \varphi(M) = \varphi(N)$, which is a contradiction. Thus $\varphi(M) = \varphi(N)$ implies $M = N$, so φ is one-to-one.

Now let I be a maximal ℓ-ideal of $T(A)$. Then I is ℓ-prime in $T(A)$, so $T(A)^+ \setminus I$ is closed under multiplication since $T(A)$ is an f-ring. Let K be an ℓ-ideal of A which is maximal with respect to the property that $K \cap (T(A)^+ \setminus I) = \emptyset$. Then K is an ℓ-prime ℓ-ideal of A [4, 2.7], and it is clear that $K \cap T(A) \subseteq I$. Let M be a maximal ℓ-ideal of A containing K. Then $M \cap T(A)$ and I must be comparable since $K \cap T(A)$ is ℓ-prime in $T(A)$, and hence $M \cap T(A) \subseteq I$. Thus $I = M(T)$, so φ is onto.

Let $\Phi = \{M_\alpha : \alpha \in \Gamma\}$ be a closed set in $\mathrm{Max}_\ell(A)$. We show that $\varphi(\Phi) = \{M_\alpha(T) : \alpha \in \Gamma\}$ is closed in $\mathrm{Max}_\ell(T(A))$. Let I be a maximal ℓ-ideal of $T(A)$ that contains $\cap_{\alpha \in \Gamma} M_\alpha(T) \supseteq \cap_{\alpha \in \Gamma}(M_\alpha \cap T(A))$. If $I \notin \varphi(\Phi)$, then $M_\alpha \cap T(A) \not\subseteq I$ for each $\alpha \in \Gamma$, and hence $(M_\alpha \cap T(A)) + I = T(A)$ for each $\alpha \in \Gamma$, since I is maximal. Thus for each $\alpha \in \Gamma$, there exist $0 \le x_\alpha \in (M_\alpha \cap T(A)) \setminus I$ and $0 \le y_\alpha \in I \setminus (M_\alpha \cap T(A))$ such that $1 = x_\alpha + y_\alpha$. Let $x_\alpha \wedge y_\alpha = z_\alpha$. Then $0 \le z_\alpha \in (M_\alpha \cap T(A)) \cap I$ for each $\alpha \in \Gamma$. Since $y_\alpha \notin M_\alpha$ and $z_\alpha \in M_\alpha$, $y_\alpha - z_\alpha \notin M_\alpha$, for each $\alpha \in \Gamma$.

Let

$$\delta(y_\alpha - z_\alpha) = \{M \in \mathrm{Max}_\ell(A) : (y_\alpha - z_\alpha) \notin M\}, \text{ for each } \alpha \in \Gamma.$$

Then we obtain an open cover $\{\delta(y_\alpha - z_\alpha) : \alpha \in \Gamma\}$ for Φ, and hence a finite subcover $\delta(y_i - z_i), i = 1, \dots, n$, can be extracted out of this cover because the compactness of $\mathrm{Max}_\ell(A)$ implies that Φ is compact. For every $M_\alpha \in \Phi$, there exists $y_j - z_j \notin M_\alpha$, for some $1 \le j \le n$. Since $(x_j - z_j) \wedge (y_j - z_j) = 0, (x_j - z_j) \in M_\alpha$ by Lemma 5, so $\wedge_{i=1}^n (x_i - z_i) \in M_\alpha$ for each $\alpha \in \Gamma$. Thus

$$\wedge_{i=1}^n (x_i - z_i) \in \cap_{\alpha \in \Gamma}(M_\alpha \cap T(A)) \subseteq \cap_{\alpha \in \Gamma} M_\alpha(T) \subseteq I,$$

so $x_k - z_k \in I$ for some $1 \le k \le n$ since I is maximal in $T(A)$, and hence $x_k \in I$, which is a contradiction. Therefore, $I \in \varphi(\Phi)$.

Now let $\Psi = \{I_\alpha : \alpha \in \Gamma\} \subseteq \mathrm{Max}_\ell(T(A))$ be closed. We verify that $\varphi^{-1}(\Psi)$ is closed in $\mathrm{Max}_\ell(A)$. Let $M \in \mathrm{Max}_\ell(A)$ be in the closure of $\varphi^{-1}(\Psi)$. If $\cap_{\alpha \in \Gamma} I_\alpha \not\subseteq M(T)$. Then $(\cap_{\alpha \in \Gamma} I_\alpha) + M(T) = T(A)$, so $1 = x + y$, where $0 \le x \in (\cap_{\alpha \in \Gamma} I_\alpha) \setminus M(T)$ and $0 \le y \in M(T) \setminus (\cap_{\alpha \in \Gamma} I_\alpha)$. Since $x \in I_\alpha$ for each $\alpha \in \Gamma$, we have $y \notin I_\alpha$ for each $\alpha \in \Gamma$. Let $x \wedge y = z$. Then $(x - z) \wedge (y - z) = 0$. Let $I_\alpha = M_\alpha(T) \supseteq M_\alpha \cap T(A)$, where M_α is a maximal ℓ-ideal of A for each $\alpha \in \Gamma$. Then $(y - z) \notin I_\alpha$ for each $\alpha \in \Gamma$ implies $(y - z) \notin M_\alpha$ for each $\alpha \in \Gamma$ since $y - z$ is an f-element, so $(x - z) \in M_\alpha$, for each $\alpha \in \Gamma$, by Lemma 5. But $\cap_{\alpha \in \Gamma} M_\alpha \subseteq M$ since $\varphi^{-1}(\Psi) = \{M_\alpha : \alpha \in \Gamma\}$, so $x - z \in M$, and hence $x - z \in M(T)$ and $x \in M(T)$, which is a contradiction. Thus $\cap_{\alpha \in \Gamma} I_\alpha \subseteq M(T)$, and hence $M(T) \in \Psi$ since Ψ is closed, so $M \in \varphi^{-1}(\Psi)$. Therefore $\varphi^{-1}(\Psi)$ is closed in $\mathrm{Max}_\ell(A)$.   ∎

We record two consequences of Theorem 16. Let F be a unital f-ring. Then each ℓ-prime ℓ-ideal I of F is contained in a unique maximal ℓ-ideal of F since F/I is totally ordered. The same result holds for a unital ℓ-reduced ℓ-ring with $1 > 0$ as shown below.

Corollary 17. *In a unital ℓ-reduced ℓ-ring with $1 > 0$ each ℓ-prime ℓ-ideal is contained in a unique maximal ℓ-ideal.*

Proof. Let A be a unital ℓ-reduced ℓ-ring with $1 > 0$, and let I be an ℓ-prime ℓ-ideal of A. Then I is contained in a maximal ℓ-ideal of A. Suppose that M and N are maximal ℓ-ideals containing I. Then $I \cap T(A)$ is contained in $M \cap T(A)$ and $N \cap T(A)$. Let $M(T)$ and $N(T)$ be defined as in Theorem 16. Then $M(T)$ and $N(T)$ are comparable since they both contain $I \cap T(A)$ which is ℓ-prime in $T(A)$, and hence $M(T) = N(T)$. Therefore, $M = N$. Thus I is contained in a unique maximal ℓ-ideal. ∎

Let F be a unital f-ring. In [6], Johnson showed that the maximal ℓ-ideals of F and maximal left (right) ℓ-ideals of F coincide [6, Corollary 4.7, p. 187 and Corollary 2.2, p. 213]. Certainly this is not true for every unital ℓ-ring with $1 > 0$. For instance, in Example 6, the only maximal ℓ-ideal of L_n is 0, but there are more than one maximal left (right) ℓ-ideal in L_n. The following result, however, shows that in a unital ℓ-reduced ℓ-ring with $1 > 0$, the maximal ℓ-ideals and maximal left (right) ℓ-ideals are in one-to-one correspondence.

Corollary 18. *Let A be a unital ℓ-reduced ℓ-ring with $1 > 0$. Then every maximal ℓ-ideal of A is contained in a unique maximal left (right) ℓ-ideal of A, and each maximal left (right) ℓ-ideal of A contains a maximal ℓ-ideal of A.*

Proof. Let M be a maximal ℓ-ideal of A. Then M is contained in a maximal left ℓ-ideal by Zorn's lemma. Now suppose that M is contained in two different maximal left ℓ-ideals L_1 and L_2. Then we have $M \cap T(A) \subseteq L_1 \cap T(A)$ and $L_2 \cap T(A)$. Since $L_1 \cap T(A)$ and $L_2 \cap T(A)$ are left ℓ-ideals of $T(A)$, they are contained in some maximal left ℓ-ideals of $T(A)$, and since every maximal left (right) ℓ-ideal in $T(A)$ is a maximal ℓ-ideal, there exist maximal ℓ-ideals I_1 and I_2 of $T(A)$ such that $L_1 \cap T(A) \subseteq I_1$ and $L_2 \cap T(A) \subseteq I_2$. Since $M \cap T(A)$ is contained in I_1 and I_2, $I_1 = I_2$. Now $A = L_1 + L_2$ implies $1 = i + j$, where $0 \leq i \in L_1$ and $0 \leq j \in L_2$. Since i and j are f-elements, $i \in L_1 \cap T(A) \subseteq I_1$ and $j \in L_2 \cap T(A) \subseteq I_2$, and hence $1 = i + j \in I_1 = I_2$, which is a contradiction. Thus each maximal ℓ-ideal M of A is contained in a unique maximal left ℓ-ideal.

Now let L be a maximal left ℓ-ideal of A. Then, by the same argument as above, $L \cap T(A) \subseteq M(T)$ for some maximal ℓ-ideal M of A, where $M(T)$ is defined as in Theorem 16. If $M \not\subseteq L$, then $A = M + L$, and hence $1 = i + j$, where $0 \leq i \in M$ and $0 \leq j \in L$, so $1 \in M(T)$, which is a contradiction. Thus $M \subseteq L$. ∎

Note that in a unital ℓ-reduced ℓ-ring with $1 > 0$, maximal ℓ-ideals and left (right) maximal ℓ-ideals are generally different [10, Example 2], although they are in one-to-one correspondence by Corollary 18.

References

[1] G. Birkhoff and R. S. Pierce, *Lattice-ordered rings*. An. Acad. Brasil. Ci. **28** (1956), 41-69.

[2] G. Birkhoff, *Lattice Theory*. Rev. ed., AMS Colloquium Publications, **25** (1948).

[3] P. Conrad, *Some structure theorems for lattice-ordered groups.* Trans. AMS **99** (1961), 212-240.

[4] J. E. Diem, *A radical for lattice-ordered rings.* Pacific J. Math. **25** (1968), 71-82.

[5] L. Fuchs, *Partially Ordered Algebraic Systems.* (1963) Pergamon Press.

[6] D. G. Johnson, *A structure theory for a class of lattice-ordered rings.* Acta Math. **104** (1960), 533-565.

[7] I. Kaplansky, *Lattices of continuous functions.* Bull. AMS **53** (1947), 617-623.

[8] J. Ma, *On lattice-ordered rings with chain conditions.* Comm. in Algebra **25** (1997), 3483-3495.

[9] J. Ma, *Lattice-ordered matrix algebras with the usual lattice order.* J. of Algebra **228** (2000), 406-416.

[10] S. A. Steinberg, *Radical theory in lattice-ordered rings.* Symposia Math. **21** (1977), 379-400.

[11] H. Subramanian, *Kaplansky's Theorem for f-rings.* Math. Ann. **179** (1968), 70-73.

[12] E. C. Weinberg, *On the scarcity of lattice-ordered matrix rings.* Pacific J. Math. **19** (1966), 561-571.

[13] E. C. Weinberg, *Lectures on ordered groups and rings.* Lecture notes (University of Illinois, Urbana, 1968).

Department of Mathematical Sciences, University of Houston Clear Lake,
Houston, TX 77058, USA
ma@cl.uh.edu

Department of Mathematical Sciences, The University of Texas at El Paso,
El Paso, TX 79968, USA
piotr@math.utep.edu

Polar Functions, I:
The Summand-Inducing Hull of an Archimedean ℓ-Group with Unit.

Jorge Martínez

ABSTRACT.   This is an introduction to the concept of a polar function on **W**, the category of archimedean ℓ-groups with designated unit, and to that of its dual, a covering function on compact spaces.

Let $\mathcal{X}$ be a subalgebra of the boolean algebra of polars $\mathcal{P}(G)$ of the **W**-object G. An essential extension H of G is said to be an $\mathcal{X}$-splitting extension of G if the extension of each $K \in \mathcal{P}(G)$ to H is a cardinal summand. The least $\mathcal{X}$-splitting extension $G[\mathcal{X}]$ of G is studied here. Dually, one considers a compact Hausdorff space, and a subalgebra $\mathfrak{K}$ of $\mathfrak{R}(X)$, the boolean algebra of all regular closed sets. A $\mathfrak{K}$-cover Y of X is represented by an irreducible map $g : Y \longrightarrow X$ subject to the condition that $\mathrm{cl}_Y g^{-1}(\mathrm{int}_X A)$ is clopen, for each $A \in \mathfrak{K}$. There is a minimum $\mathfrak{K}$-cover. To each subalgebra $\mathcal{X}$ of polars of G there corresponds canonically a subalgebra $\mathfrak{K}$ of regular closed sets of the Yosida space YG of G, in such a way that the Yosida space of $G[\mathcal{X}]$ is the minimum $\mathfrak{K}$-cover of YG. This general setup is applied to some well known situations, to recapture constructions such as the projectable hull. On the topological side one may recover the cloz cover of a compact space.

A function which assigns to each G a subalgebra $\mathcal{X}(G)$ of polars of G is called a polar function. The dual notion for compact spaces is the covering function: assigning to each space X a subalgebra $\mathfrak{K}(X)$ of regular closed sets. By transfinitely iterating the basic constructions of least $\mathcal{X}$-splitting extensions and minimum $\mathfrak{K}$-covers, one obtains their idempotent closures, $\mathcal{X}^\flat$ and $\mathfrak{K}^\flat$, respectively. These closures give rise to, respectively, hull classes of archimedean ℓ-groups and covering classes of compact spaces.

The objectives of this article are to establish the existence and describe the construction of the least extension of an archimedean ℓ-group with designated weak unit, with respect to making cardinal summands out of a distinguished boolean subalgebra of polars. There is a topological counterpart to this subject matter; it is carried out on the Yosida spaces of these objects. For a compact space and a distinguished boolean subalgebra of regular closed sets, one considers the least irreducible pre-image in which the regular inverse image of each of these regular closed sets is clopen. This study was motivated and developed, in part and in parallel, by the work in [HM∞a]; that study

275

J. Martínez (ed.), Ordered Algebraic Structures, 275–299.
© 2002 *Kluwer Academic Publishers.*

deals with covers of compact spaces in complete generality, while, on the other hand, the algebraic focus there is on hulls which produce uniformly complete real f-algebras.

It should be noted that much of the algebraic work can be generalized to non-archimedean lattice-ordered groups and to f-rings as well. Indeed, in ongoing work - [HM∞b] – summand-inducing hulls are considered in the setting of semiprime rings. If one looks for the origins of these preoccupations with hulls of lattice-ordered groups, then the place to begin is in Conrad's [C73]. There one will find references to other papers on lateral completion and the like, as well as to Chambless' [Ch71], which we shall have occasion to mention later on. Our technique resembles most that of Bleier, in [Bl74], which deals with the projectable and strongly projectable hulls. In fact, Bleier's achievement is more general, in that he sets his discussion up for all lattice-ordered groups which are representable as subdirect products of totally ordered groups. Some general machinery is described in [Bl74], although the full potential of it is unrealized. As to the topological matters, we will point out, albeit only in general terms, how this work relates to that of [HVW89], but so far as we are aware the linking up of the algebraic and topological frameworks is new.

1   Introduction

The ambient category here is $\mathbf{W}$, of all archimedean lattice-ordered groups with designated weak unit $u > 0$. The morphisms are the ℓ-homomorphisms which preserve the designated units. When we have an ℓ-subgroup $G \leq H$ we say that H *is an extension of* G if the inclusion is a monomorphism of $\mathbf{W}$, and, thus, G inherits the weak unit of H. We use $\mathcal{C}(G)$ for the lattice of all convex ℓ-subgroups of the $\mathbf{W}$-object G. $\mathcal{P}(G)$ will denote the boolean algebra of all polars of G; we review the subject of polars below. $\mathcal{X}$ will, generically, stand for a subalgebra of $\mathcal{P}(G)$. It will be assumed that $\mathcal{X}$ always contains all cardinal summands of G. This is just a convenience for now, but in §5 we will actually need this provision.

The main source for basic material from the theory of lattice-ordered groups – abbreviated ℓ-*groups* – is [D95]; occasionally we shall also use material from [BKW77]. The reader should refer to these for any unexplained terminology in ℓ-group theory. One should single out the essential hull here; we recall the principal facts about it, as our discourse involves it in a prominent way.

Definition & Remarks 1.1. (a) Recall then that if $G \leq H$ is an extension in $\mathbf{W}$, we say that G is *essential in* H – or that H is an *essential* extension of G – if for each $0 < h \in H$ there is a $g > 0$ in G such that $g \leq nh$, for a suitable positive integer n.

(b) Suppose that G is an ℓ-group. The *polar of* $S \subseteq G$ is

$$S^{\perp} = \{\, g \in G \,:\, |g| \wedge |s| = 0,\, \forall\, s \in S \,\}.$$

A *polar* is a subgroup of this form; equivalently, one of the form $K = K^{\perp\perp}$. Note that every polar is a convex ℓ-subgroup. Since we are going to be computing polars in different groups, we shall subscript $S^{\perp\sigma}$ to indicate the group in which the polar is calculated, unless there is no possibility of confusion. Let $\mathcal{P}(G)$ stand for the set of all polars of G; it is well known that $\mathcal{P}(G)$ is a complete boolean algebra for which infimum is set-theoretic intersection.

Note that for an extension in $\mathbf{W}$, $G \leq H$, H is an essential extension of G if and only if the trace map $P \mapsto P \cap G$ is an isomorphism of boolean algebras from $\mathcal{P}(H)$ onto $\mathcal{P}(G)$; this is proved in [BKW77], Theorem 11.1.15. We shall frequently be concerned with the inverse of this map, namely the assignment $K \mapsto K^{\perp_H \perp_H}$.

A *cardinal summand* K of G is a polar for which $G = K \oplus K^{\perp}$. Note that the subset $\mathcal{S}(G)$ of all cardinal summands of G is a subalgebra of $\mathcal{P}(G)$. We shall frequently drop the adjective *cardinal* and simply speak of *summands*. Let us underscore the assumption about the subalgebras to be considered here: all subalgebras of $\mathcal{P}(G)$ under discussion are assumed to contain $\mathcal{S}(G)$.

(c) In [C71] Conrad developed the essential hull of an archimedean ℓ-group. We need not recall the particulars here, except to note that for each $\mathbf{W}$-object G there is a $\mathbf{W}$-object G^e which is an essential extension of G and such that if $G \leq H$ is essential then the inclusion of G in H extends to a one-to-one ℓ-homomorphism of H into G^e. In particular, G^e is *essentially closed*; that is, it has no proper essential extensions of its own in $\mathbf{W}$.

Also worth a specific mention are the following two concepts which will be revisited often. What exactly is meant by a "hull" in this context we shall explain in 5.12.

Definition & Remarks 1.2. A $\mathbf{W}$-object G is *strongly projectable* if every polar of G is a summand. G is *projectable* if $G = a^{\perp\perp} \oplus a^{\perp}$, for each $a \in G$. There is, for each $\mathbf{W}$-object G a projectable and a strongly projectable hull. In the literature, the projectable (resp. strongly projectable) hull of G is characterized as the least ℓ-group which is projectable (resp. strongly projectable), and contains G as an essential ℓ-subgroup. The existence of these is well established. For a systematic development of them, the reader is referred to [C73] and also to [Ch71], for a more revealing account. Closest to our development is the construction of [Bl74].

Before exploring new material we should briefly review the Yosida Representation Theorem. For a reference we recommend [HR77]. As is the custom, $C(X)$ denotes the ring of all continuous real valued functions defined on the topological space X. It is a ring with respect to pointwise operations, and, likewise, an object in $\mathbf{W}$, the designated unit being the constant function 1. The standard reference for material on rings of continuous functions is still [GJ76].

Definition & Remarks 1.3. YG stands for the *Yosida space* of the **W**-object G; that is to say, the space of values of the designated unit, with the hull-kernel topology. It is well known that YG is a compact Hausdorff space.

Now for any compact Hausdorff space X, $D(X)$ shall denote the set of all continuous functions $f : X \longrightarrow \mathbb{R} \cup \{\pm\infty\}$, where the range is the extended reals with the usual topology, such that $f^{-1}\mathbb{R}$ is a dense subset of X. $D(X)$ is a lattice under pointwise operations, but not a group or ring under the obvious pointwise operations, unless some assumptions are made about X. We need not raise those issues here. What is needed is an understanding of the term "ℓ-group in $D(X)$": a subset $H \subseteq D(X)$ which is an ℓ-group such that under the lattice operations it is a sublattice of $D(X)$, and, for each $h, k \in H$, $(h + k)(x) = h(x) + k(x)$ on a dense subset of X. Now we are able to state the Yosida Representation Theorem:

> For each **W**-object G with designated unit $u > 0$, there is an ℓ-isomorphism ϕ of G onto an ℓ-group G' in $D(YG)$ such that $\phi(u) = 1$.

The Yosida space YG has a canonical role in the above theorem when one takes into account separation of points. In fact, the isomorphism ϕ mentioned in the statement of the theorem has the feature that if $x \neq y$ in X, then there is a $g \in G$ such that $\phi(g)(x) \neq \phi(g)(y)$; this is *separation of points*. The Yosida space is unique (up to homeomorphism) in the following sense:

> Suppose that Z is a compact Hausdorff space and that there is an ℓ-isomorphism θ of G onto an ℓ-group H in $D(Z)$ which separates the points of Z and such that $\theta(u) = 1$, then there is a homeomorphism $t : Z \longrightarrow YG$ such that, for each $g \in G$ and $z \in Z$,
>
> $$\phi(g)(tz) = \theta(g)(z).$$

Finally, if the unit u of G is strong, the image of the Yosida Representation lies in $C(YG)$.

For the record, βX denotes the Stone-Čech compactification of the Tychonoff space X. It is also convenient to introduce $\mathfrak{B}(X)$ for the algebra of clopen sets of X. $S(X)$ stands for the subalgebra of $C(X)$ consisting of all continuous functions of finite range. Having concluded a review of the barest necessities we proceed to the subject at hand.

2   Making Summands out of Polars

Definition & Remarks 2.1. Consider an intermediate **W**-object $G \leq H \leq G^e$; as has been noted, this is tantamount to assuming that H is an essential extension of G.

Since G^e is strongly projectable (see [C71]), $K^{\perp G^e}$ is a cardinal summand of G^e, for each $K \in \mathcal{P}(G)$. Suppose now that $\mathcal{X} \leq \mathcal{P}(G)$ is a subalgebra such that $K^{\perp_H \perp_H}$ is a cardinal summand of H, for each $K \in \mathcal{X}$. Then H is called an $\mathcal{X}$-splitting extension of G. We see that G^e is a $\mathcal{X}$-splitting extension, for each subalgebra $\mathcal{X}$. We show in Theorem 2.4 that each **W**-object G has a least $\mathcal{X}$-splitting extension.

In §4 of [Bl74] Bleier defines what is essentially the notion of an $\mathcal{X}$-splitting extension. However, as he is only interested in the projectable hull in that section, the full potential of the construct remains unexplored. In particular, the subtlety of transfinitely creating an idempotent polar function, as defined and worked out in §5 of this paper, goes unnoticed in earlier treatments, chiefly because the projectable and strongly projectable hulls do not have to be so iterated.

As a first step toward Theorem 2.4, we record the following lemma. If $g \in G$ and $K \in \mathcal{P}(G)$ then $g[K]$ denotes the projection of g on the summand $K^{\perp G^e \perp G^e}$ in G^e.

Lemma 2.2. *Suppose $K \in \mathcal{P}(G)$ and H is an extension of G in G^e. Then $K^{\perp_H \perp_H}$ is a summand of H if and only if, for each $h \in H$, $h[K] \in H$.*

Proof. The sufficiency is clear, so we move on to prove the necessity. Suppose that $K^{\perp_H \perp_H}$ is a summand of H. Pick $h \in H$, and write $h = x + y$, with $x \in K^{\perp_H \perp_H}$ and $y \in K^{\perp_H}$. Note that $K^{\perp_H \perp_H} \leq K^{\perp G^e \perp G^e}$ and $K^{\perp_H} \leq K^{\perp G^e}$, so that $y \wedge h[K] = 0$, while $x \wedge (h - h[K]) = 0$. Then it is easy to see that, in fact, $x = h[K]$, proving the lemma. ∎

One small corollary of the preceding lemma, for later use:

Corollary 2.3. *Suppose that $G \leq H_1 \leq H_2 \leq G^e$, and that the common designated unit of these is a strong order unit of H_2. If H_1 is an $\mathcal{X}$-splitting extension of G, then so is H_2.*

Proof. Suppose that $u > 0$ is the designated unit. Let $K \in \mathcal{X}$; by the preceding lemma, $u[K] \in H_1 \leq H_2$. On the other hand, for each $h \geq 0$ in H_2, $h \leq nu$, for a suitable positive integer n. Finally,

$$h = h \wedge nu = (h \wedge nu[K]) + (h \wedge (nu - nu[K])),$$

and the first term in the displayed sum is $h[K]$, and it is in H_2. ∎

As the reader will appreciate, the first part of the next theorem follows easily from the preceding lemma.

Theorem 2.4. *For each **W**-object G and each subalgebra $\mathcal{X}$ of $\mathcal{P}(G)$, there is a least $\mathcal{X}$-splitting extension of G, denoted $G[\mathcal{X}]$. The elements of $G[\mathcal{X}]$ are the finite sums of the form*

$$\sum_{i=1}^{n} g_i[K_i],$$

where each $g_i \in G$ and each $K_i \in \mathcal{X}$ and $K_i \cap K_j = \{0\}$, whenever $i \neq j$.

Proof. Suppose that $\{ H_\lambda : \lambda \in \Lambda \}$ is the set of all $\mathcal{X}$-splitting extensions of G in G^e, and set $H \equiv \cap_{\lambda \in \Lambda} H_\lambda$. We prove that H is $\mathcal{X}$-splitting; this suffices. To this end we employ Lemma 2.2. Pick $h \in H$ and consider $h[K]$, with $K \in \mathcal{X}$. According to the lemma, $h[K] \in H_\lambda$, for each $\lambda \in \Lambda$, and, consequently, $h \in H$.

For the rest, let's temporarily denote the set of all sums

$$\sum_{i=1}^{n} g_i[K_i],$$

with each $g_i \in G$ and each $K_i \in \mathcal{X}$ by $\hat{G}$. We will show that every member of $\hat{G}$ can be disjointified, as specified in the theorem. In any case, it should be clear that $\hat{G}$, as defined, is a subgroup of each $\mathcal{X}$-splitting extension of G, and that $G \leq \hat{G}$. By showing that each such sum can be disjointified, it will be easy to demonstrate that $\hat{G}$ is an ℓ-subgroup, and that it is $\mathcal{X}$-splitting.

The proof of disjointification is an induction argument. We will initiate that, and leave the inducting to the reader. A straighforward calculation is involved. Suppose that $a, b \in G$ and $K, L \in \mathcal{X}$. Then

$$a[K] + b[L] = a[K \cap L^{\perp}] + (a + b)[K \cap L] + b[L \cap K^{\perp}].$$

Note that since $\mathcal{X}$ is a subalgebra, the righthand side of the identity belongs to $\hat{G}$. This shows that a sum of two projections can be disjointified.

Now, if

$$a = \sum_{i=1}^{n} g_i[K_i] \in \hat{G},$$

disjointified, as prescribed in the statement of the theorem, then

$$a \vee 0 = \sum_{i=1}^{n} (g_i \vee 0)[K_i],$$

proving that $a \vee 0 \in \hat{G}$. Finally, if $L \in \mathcal{X}$, then

$$a[L] = \sum_{i=1}^{n} g_i[K_i \cap L] \in \hat{G},$$

showing that $\hat{G}$ is $\mathcal{X}$-splitting. Thus, $\hat{G} = G[\mathcal{X}]$, and the proof is complete. ∎

For archimedean f-rings with identity the above construct behaves well.

Corollary 2.5. *Suppose that A is an archimedean f-ring with identity. For each subalgebra $\mathcal{X}$ of $\mathcal{P}(A)$, $A[\mathcal{X}]$ is an ℓ-subring of A^e.*

Proof. What should be noted here is that A^e is an f-ring which contains A as an ℓ-subring, and that polars are ring ideals; (see [C71]). Then the claim is immediate from the description of $A[\mathcal{X}]$ in Theorem 2.4.   ∎

In advance of §3 and §4, there is one more issue, concerning Yosida spaces, which ought to be aired now. The proof of Corollary 2.3 should already alert us to it.

Definition & Remarks 2.6. (a) If $g > 0$ is in the ℓ-group G, and $0 \leq a \leq g$ such that $a \wedge (g - a) = 0$, we call a a *component* of g. It is easy to see that the set $\operatorname{comp}(g)$ of all components of g is a boolean algebra, and a sublattice of the underlying lattice of G.

Let $Y(g)$ denote the space of all values of g under the hull-kernel topology. Note that if $a \in \operatorname{comp}(g)$ then $\operatorname{coz}(a) \subseteq Y(g)$ is a clopen set. Conversely, using the fact that the sets $\operatorname{coz}(g)$ $(g \in G)$ form a base for the topology on $Y(g)$, one easily shows that each clopen set of $Y(g)$ is of the form $\operatorname{coz}(a)$, for a suitable component a of g.

(b) Recall that $\mathcal{S}(G)$ stands for the subalgebra of all summands of G. Observe that if $S \in \mathcal{S}(G)$ then by projecting the designated unit $u > 0$ on S, one gets a component of u, namely $u[S]$. The map $S \mapsto u[S]$ is a boolean embedding of $\mathcal{S}(G)$ into $\operatorname{comp}(u)$. In general, the map fails to be surjective; see (c) below. On the other hand, if u is a strong order unit, then $S \mapsto u[S]$ is surjective, giving us a natural isomorphism between $\mathcal{S}(G)$, $\operatorname{comp}(u)$ and $\mathfrak{B}(YG)$.

This map is also surjective if G is projectable.

(c) The following simple example should be instructive. Consider $G = S(\beta\mathbb{N})$, the group of all sequences of finite range. Let $H_1 = C(\beta\mathbb{N})$, the group of all bounded sequences. Let H_2 be the ℓ-subgroup of $C(\mathbb{N})$ generated by H_1 and the identity function. Notice that H_2 is not projectable. With the constant function 1 as the designated unit, the range of the map $K \mapsto K^{\perp H_2 \perp H_2}$ (with $K \in \mathcal{P}(G) = \mathcal{S}(G)$) is $\mathcal{P}(H_2)$, whereas $\mathcal{S}(H_2)$ is the subalgebra of the principal polars of eventually constant sequences. In other words, H_2 is not a $\mathcal{S}(G)$-splitting extension of G. However, H_1 is a $\mathcal{S}(G)$-splitting extension of G, thus pointing out that Corollary 2.3 is false if the "strong unit" hypothesis is removed.

Observe that $YG = YH_1 = YH_2 = \beta\mathbb{N}$. Also, in H_2, the map $S \mapsto 1[S]$ of (b) is not surjective.

Let us now consider some applications.

Examples 2.7. We fix a **W**-object G.

(a) Let $\mathcal{X} = \mathcal{P}(G)$; suppose that $H \leq G^e$ is an extension of G. Observe that the assignment $K \mapsto K^{\perp_H \perp_H}$ is a boolean isomorphism from $\mathcal{P}(G)$ onto $\mathcal{P}(H)$. Therefore, to say that H is a $\mathcal{P}(G)$-splitting extension means that $H = P \oplus P^{\perp_H}$, for each $P \in \mathcal{P}(H)$; that is to say, H is strongly projectable. Thus, $G[\mathcal{P}(G)]$ is simply the strongly projectable hull of G, as discussed in [C73] and [Bl74].

(b) Let $\mathcal{P}_\omega(G)$ be the subalgebra of $\mathcal{P}(G)$ generated by the lattice of principal polars, $\mathcal{P}_0(G)$. Note that G is projectable if and only if $\mathcal{P}_\omega(G) \leq \mathcal{S}(G)$.

To put the matter formally, $H \leq G^e$ is an $\mathcal{P}_\omega(G)$-splitting extension of G if and only if $H = g^{\perp_H \perp_H} \oplus g^{\perp_H}$, for each $g \in G$. We verify that $G[\mathcal{P}_\omega(G)]$ is projectable, and thereby settle that it is the projectable hull of G, as described in [C73]. We strongly recommend that the reader compare this account of the projectable hull with those of [Ch71] and, especially, [Bl74].

Pick $x \in G[\mathcal{P}_\omega(G)]$ and express it according to the provisions of Theorem 2.4,

$$x = \sum_{i=1}^{n} g_i[K_i],$$

where each $g_i \in G$ and each $K_i \in \mathcal{P}_\omega(G)$ and $K_i \cap K_j = \{0\}$, whenever $i \neq j$. Evidently, $x^{\perp\perp}$ is the disjoint supremum of the $g_i[K_i]^{\perp\perp}$ (computed in $G[\mathcal{P}_\omega(G)]$), and for each i,

$$g_i[K_i]^{\perp\perp} = (g_i^{\perp_G \perp_G} \cap K_i)^{\perp\perp}.$$

By design this polar is a summand. It follows that $G[\mathcal{P}_\omega(G)]$ is projectable.

(c) Suppose that α denotes an uncountable, regular cardinal. Let $\mathcal{P}_\alpha^\alpha(G)$ stand for the α-generated polars having an α-generated complement. (Note: a polar P of G is *α-generated* if it can be generated by fewer than α elements. That is to say, there is a subset $S \subseteq G$, such that $|S| < \alpha$ and $P = S^{\perp\perp}$.) It is easy to see that $\mathcal{P}_\alpha^\alpha(G)$ is a subalgebra of $\mathcal{P}(G)$. We shall abbreviate "$\mathcal{P}_\alpha^\alpha(G)$-splitting" to *$\alpha$-splitting*, as well as use $G[\alpha]$ in place of $G[\mathcal{P}_\alpha^\alpha(G)]$. $G[\alpha]$ already occurs in [HM01], in the context of the ring of α-quotients of an f-ring.

As with most issues involving **W**, there is a topological side to this discussion, owing to the Yosida Representation Theorem. We shall now investigate this topological counterpart.

3   Covers of Spaces with Prescribed Clopen Sets

All topological spaces are assumed to be compact and Hausdorff. We begin with a brief review of covers of compact spaces. The reader is referred to [H89], where the context is more general than this one. Chapter 6 of [PW89] contains a good account of this subject matter. Our $\mathfrak{K}$-covers are the covers described in [HVW89] partly through an ultrafilter construction.

Definition & Remarks 3.1. (a) Fix a space X. Recall that a continuous surjection $g : Y \longrightarrow X$ is said to be *irreducible* if X is not the image under g of any proper closed subset of Y. The irreducible surjections $g : Y \longrightarrow X$ and $h : Z \longrightarrow X$ are *equivalent* if there is a homeomorphism $\overline{g} : Y \longrightarrow Z$ such that $h \cdot \overline{g} = g$. Clearly, equivalence of irreducible maps onto X is an equivalence relation, and it is on the set of equivalence classes, $\mathrm{Cov}(X)$, that we define a partial ordering. Actually, we present a quasi-ordering on the class of all irreducible coverings, which, modulo the equivalence, is the partial ordering we want. Thenceforth, we blur distinctions between a map and the equivalence class which contains it.

Again, suppose that $g : Y \longrightarrow X$ and $h : Z \longrightarrow X$ are irreducible surjections. We put $h \leq g$ if there is a continuous surjection $u : Y \longrightarrow Z$, such that $h \cdot u = g$. Such a map u is, necessarily, irreducible.

When we speak of a *cover* of X we mean, depending on the context, an irreducible surjection onto X or else the equivalence class containing it.

(b) It is well known that an irreducible surjection $g : Y \longrightarrow X$ induces a isomorphism between the boolean algebras $\mathfrak{R}(Y)$ and $\mathfrak{R}(X)$ of regular closed sets; this appears as 6.5(a) in [PW89]. From $\mathfrak{R}(Y) \longrightarrow \mathfrak{R}(X)$ the assignment in question is simply $A \mapsto g(A)$, and the inverse isomorphism is $B \mapsto \mathrm{cl}_Y g^{-1}(\mathrm{int}_X B)$. We designate the latter $g^{\leftarrow} : \mathfrak{R}(X) \longrightarrow \mathfrak{R}(Y)$. We are about to consider a particular subalgebra $\mathfrak{K}$ of $\mathfrak{R}(X)$ and the covers of X, $g : Y \longrightarrow X$, such that each set in $g^{\leftarrow}(\mathfrak{K})$ is clopen. Alternatively, we consider the irreducible surjections $g : Y \longrightarrow X$ such that each $B \in \mathfrak{K}$ is the image of a clopen set.

If $g : Y \longrightarrow X$ is in $\mathrm{Cov}(X)$ and each $A \in g^{\leftarrow}(\mathfrak{K})$ is clopen, we say that Y is a $\mathfrak{K}$-*cover* of X.

In this section we shall prove (Theorem 3.5) that each compact X has a minimum $\mathfrak{K}$-cover. By way of contrast in the approach, in [HVW89], the authors construct the cover whose existence we shall demonstrate, and, in Theorem 2.11 of that article, they show that it is least. Their cover is referred to as a *Wallman cover*, and it is constructed by considering the space of ultrafilters of member of $\mathfrak{K}$, with a complication in the construction when the sets in $\mathfrak{K}$ do not form a base for the closed sets of X. We refer the reader to [HVW89] for details.

To establish that there is a least $\mathfrak{K}$-cover, we actually need to reconnect with the algebra, in **W**. Early in §1 we adopted the convention that all designated subalgebras of polars should contain all summands. Likewise, here, we shall assume that each subalgebra $\mathfrak{K}$ of $\mathfrak{R}(X)$ contains all the clopen sets of X.

Definition & Remarks 3.2. Once again, X is a fixed space. We remind the reader that $C(X)$ is considered a member of **W**, with the constant function 1 as designated unit. There is a well known correspondence between the polars of $C(X)$ and the regular closed subsets of X, which we now describe. Suppose that $A \in \mathfrak{R}(X)$. Put

$$\tilde{A} \equiv \{ f \in C(X) : \mathrm{coz}(f) \subseteq A \}.$$

It is well known that $\tilde{A} \in \mathcal{P}(C(X))$. For the record,

$$\tilde{A}^{\perp} = \{\, f \in C(X) \,:\, \mathrm{coz}(f) \cap A = \emptyset \,\}.$$

Note that, in the correspondence $A \mapsto \tilde{A}$, clopen sets correspond to cardinal summands. To see that, refer to the remarks in 2.6. Thus, $A \in \mathfrak{R}(X)$ is clopen precisely when $A = \mathrm{coz}(e)$, for a suitable component e of 1. If this is the case, $\tilde{A} = e^{\perp\perp}$ and $1[\tilde{A}] = e$. Conversely, for any cardinal summand K of $C(X)$, $K \mapsto \mathrm{coz}(1[K])$ describes the inverse of $A \mapsto \tilde{A}$, restricted to summands.

If $\mathfrak{K}$ is a subalgebra of $\mathfrak{R}(X)$ we denote the collection of all the $\tilde{A}$, with $A \in \mathfrak{K}$, by $\tilde{\mathfrak{K}}$. Obviously, $\tilde{\mathfrak{K}}$ is a subalgebra of $\mathcal{P}(C(X))$.

What needs to be discussed next is how $\mathfrak{K}$-covers corrrespond to $\tilde{\mathfrak{K}}$-splitting extensions. To do this in a reasonable fashion we have to consider the passage of polars between $C(X)$ and $C(Y)$, when $g : Y \longrightarrow X$ is an irreducible surjection. First, realize that the surjection g induces an extension $C(X) \leq C(Y) \leq C(X)^e$. To be precise, there is a dense **W**-embedding $C(g) : C(X) \longrightarrow C(Y)$, by $C(g)(f) = f \cdot g$. However, we shall not carry the label "$C(g)$". With such thoughts in mind, here is the technical lemma we need. Its proof is straighforward, albeit tedious. We leave it to the reader.

Lemma 3.3. *Suppose that* $g : Y \longrightarrow X$ *is an irreducible surjection and* $A \in \mathfrak{R}(X)$. *Then*

$$\tilde{A}^{\perp_{C(Y)}\perp_{C(Y)}} = \widetilde{g^{\leftarrow}(A)}.$$

Now here is the expected result.

Proposition 3.4. *Suppose that* $g : Y \longrightarrow X$ *is an irreducible surjection, and* $\mathfrak{K}$ *is a subalgebra of* $\mathfrak{R}(X)$. *Then* Y *is a* $\mathfrak{K}$*-cover of* X *if and only if* $C(Y)$ *is an* $\tilde{\mathfrak{K}}$*-splitting extension of* $C(X)$.

Proof. From Lemma 3.3,

$$\tilde{A}^{\perp_{C(Y)}\perp_{C(Y)}} = \widetilde{g^{\leftarrow}(A)}.$$

Thus, $C(Y)$ is $\tilde{\mathfrak{K}}$-splitting precisely when $\tilde{A}^{\perp_{C(Y)}\perp_{C(Y)}}$ is a summand of $C(Y)$ for each $A \in \mathfrak{K}$, which happens if and only if $g^{\leftarrow}(A)$ is clopen in Y, for each $A \in \mathfrak{K}$. ∎

All this leads to the following theorem.

Theorem 3.5. *Suppose* X *is a space, and* $\mathfrak{K}$ *is a subalgebra of* $\mathfrak{R}(X)$. *Then, in* $\mathrm{Cov}(X)$, *there is a least* $\mathfrak{K}$*-cover, denoted* $X[\mathfrak{K}]$.

Proof. According to Proposition 3.4, $\mathfrak{K}$-covers of X correspond as indicated there to $\tilde{\mathfrak{K}}$-splitting extensions of $C(X)$. Now, consider $G = C(X)[\tilde{\mathfrak{K}}]$, and let $Y = YG$. We have, in $\mathrm{Cov}(X)$, that $X \leq Y \leq Z$, for each $\mathfrak{K}$-cover Z. It suffices then to show that

Y is a $\mathfrak{K}$-cover of X. As per Proposition 3.4, this amounts to showing that $C(Y)$ is a $\tilde{\mathfrak{K}}$-splitting extension of $C(X)$. However, this is clear, as $G \leq C(Y)$ - because the constant 1 is a strong unit - by Corollary 2.3. The cover we want is $X[\mathfrak{K}] = Y$. ∎

We have established the transition from $\mathfrak{K}$-covers to $\tilde{\mathfrak{K}}$-splitting extensions. Theorem 3.5 describes the least $\mathfrak{K}$-cover of a space X in terms of the least $\tilde{\mathfrak{K}}$-splitting extension of $C(X)$. To reverse, we explore how to pass from $\mathcal{X}$-splitting extensions to covers associated with a "derived" algebra of regular closed sets.

4   From Splitting Extensions to Covers

In this section G denotes a fixed **W**-object and $Y = YG$, unless the contrary is specified. We return to the relationship between polars and regular closed sets, now in its most general setting.

Definition & Remarks 4.1. (a) Suppose that P is a polar of G. We associate a regular closed subset $\rho(P)$ of Y as follows:

$$\rho(P) \equiv \mathrm{cl}_Y \left(\cup_{g \in P} \mathrm{coz}(g) \right).$$

ρ defines an isomorphism of boolean algebras from $\mathcal{P}(G)$ onto $\mathfrak{R}(Y)$. Note, by the way, that for $G = C(X)$, X compact and Hausdorff, ρ is the inverse of the map $A \longrightarrow \hat{A}$. In this general context, if $A \in \mathfrak{R}(Y)$,

$$\rho^{-1}(A) = \{ g \in G \, : \, \mathrm{coz}(g) \subseteq A \}.$$

Observe that if $P \in \mathcal{S}(G)$ then $\rho(P)$ is clopen in YG, although the converse may fail, as explained in 2.6.

(b) Now, if H is an extension of G in G^e, then there is an induced irreducible surjection $t : YH \longrightarrow Y$. This happens because of the functorial properties of the Yosida Representation. We explain briefly, as it regards the matter at hand. The reader is referred to [HR77].

Suppose that $\theta : G \longrightarrow H$ is a **W**-morphism. Let $\phi_G : G \longrightarrow G'$ and $\phi_H : H \longrightarrow H'$ denote the Yosida representations, over YG and YH, respectively. Then there is a continuous map $t : YH \longrightarrow YG$ such that for each $p \in YH$ and $g \in G$,

(§) $$\phi_G(g)(t(p)) = \phi_H(\theta(g))(p).$$

Note that θ is an essential embedding if and only if t is an irreducible surjection.

What we need next is the formulation of the analogue of Lemma 3.3; it follows immediately from (§). The subscript on ρ in the display which follows should be self-explanatory:

Suppose that $G \leq H$ is an essential extension; then for each polar P of G,

(§§) $\rho_H(P^{\perp_H \perp_H}) = t^{\leftarrow}(\rho_G(P))$.

With these remarks in hand we now have a general, albeit unbalanced version of Proposition 3.4. The proof is carried over *mutatis mutandis*, using (§§) .

The reader might take note of the obvious notation: if $\mathcal{X} \leq \mathcal{P}(G)$ is a subalgebra, then $\rho_G(\mathcal{X})$ is simply the image subalgebra of $\mathfrak{R}(Y)$.

Proposition 4.2. *Suppose that $H \leq G^e$ is an extension of G, and $\mathcal{X}$ is a subalgebra of $\mathcal{P}(G)$. Then, if H is an $\mathcal{X}$-splitting extension of G, YH is a $\rho_G(\mathcal{X})$-cover of Y. The converse is true provided the designated unit is strong in H.*

There is still some work to be done to establish that, to $G[\mathcal{X}]$, corresponds the least $\rho_G(\mathcal{X})$-cover of Y. Evidently, the imbalance in Proposition 4.2 begs to be fixed. Unfortunately, as the example in 2.6(c) points out, it cannot be. If $G = S(\beta\mathbb{N})$, let H be the extension generated by G and the identity function. Choose $\mathcal{X} = \mathcal{S}(G)$. Then H is not a $\mathcal{X}$-splitting extension of G, while $YH = YG$.

We will actually get around the imbalance in Proposition 4.2.

Definition & Remarks 4.3. G^* stands for the convex ℓ-subgroup of G generated by the designated unit. It is common to refer to G^* as the *bounded* subgroup (or part) of G. Obviously, the designated unit of G is strong if and only if $G = G^*$. The reader should note – see [D95] – that the map $P \mapsto P \cap G^*$ is an ℓ-isomorphism of $\mathcal{P}(G)$ onto $\mathcal{P}(G^*)$. The inverse map carries $K \in \mathcal{P}(G^*)$ to the largest convex ℓ-subgroup K' of G which intersects G^* in K. This fact is a consequence of a more substantial observation, namely [D95], Theorem 12.13, which we now paraphrase below.

Recall, in general, that $G(a)$ stands for the convex ℓ-subgroup of G generated by $a \in G$. $\mathcal{C}^a(G)$ will denote the set of all convex ℓ-subgroups of G that do not contain a.

Suppose that G is any ℓ-group, not necessarily archimedean. Suppose that $a \in G$, then the map $C \mapsto C \cap G(a)$ is an order isomorphism between $\mathcal{C}^a(G)$ and the set of proper convex ℓ-subgroups of $G(a)$. The inverse map assigns to $B \in \mathcal{C}(G(a))$, with $a \notin B$, the convex ℓ-subgroup

$$\overline{B} \equiv \{ x \in G : |x| \wedge g \in B, \forall g \in G(a)^+ \}.$$

$\overline{B}$ is the largest convex ℓ-subgroup of G intersecting $G(a)$ in B.

One should be careful though; even in **W** it happens that B is a summand of G^*, while $\overline{B}$ is not a summand of G. Again, consider $S(\beta\mathbb{N})$, and let G be the subgroup of $C(\mathbb{N})$ generated by $S(\beta\mathbb{N})$ and the identity function. Then $G^* = S(\beta\mathbb{N})$, and the polar $P \in \mathcal{P}(G)$ consisting of all functions which vanish at all even integers is not a summand of G, whereas $P \cap G^*$ is a summand of G^*.

The following lemma ought not surprise the reader. A moment's thought will surely make it transparent.

Lemma 4.4. *Suppose that $P \in \mathcal{P}(G)$. For each $a \in G^*$, $a[P] = a[P \cap G^*]$.*

All of which makes the next proposition fairly clear. Abusing our notation a bit, if $\mathcal{X}$ is a subalgebra of $\mathcal{P}(G)$, let $\mathcal{X}^* \equiv \{ P \cap G^* : P \in \mathcal{X} \}$. Let us also refer to $\mathcal{X}^*$ as the *trace* of $\mathcal{X}$.

The final piece of preparation is contained in the next proposition.

Proposition 4.5. *For each subalgebra $\mathcal{X}$ of $\mathcal{P}(G)$,*

$$(G[\mathcal{X}])^* = G^*[\mathcal{X}^*].$$

Proof. According to Theorem 2.4, each $f \in G[\mathcal{X}]$ can be expressed a finite disjoint sum of components $a[K]$, with $a \in G$ and $K \in \mathcal{X}$. Thus, if $0 \leq f \in (G[\mathcal{X}])^*$ then there is a positive integer n such that $f \leq nu$, where u is the designated unit of G. So if,

$$f = \sum_{i=1}^{m} a_i[K_i],$$

with $K_i \cap K_j = \{0\}$, for $i \neq j$, and each $0 \leq a_i \in G$, we also have

$$f = \sum_{i=1}^{m} (a_i \wedge nu)[K_i],$$

which, in view of Lemma 4.4, means that $f \in G^*[\mathcal{X}^*]$. The converse is clear.   ∎

And so we have arrived at the goal of this section.

Theorem 4.6. *Suppose that $\mathcal{X}$ is a subalgebra of $\mathcal{P}(G)$. Then $YG[\mathcal{X}]$ is the least $\rho(\mathcal{X})$-cover of $Y = YG$.*

Proof. By Proposition 4.2, $YG[\mathcal{X}]$ is a $\rho(\mathcal{X})$-cover of Y. Now suppose that Z is a $\rho(\mathcal{X})$-cover of Y, and consider $C(Z)$. Since its Yosida space is Z and $YG^* = Y$, the same proposition tells us that $C(Z)$ is a $\mathcal{X}^*$-extension of G^*. This implies that $G^*[\mathcal{X}^*] = (G[\mathcal{X}])^* \leq C(Z)$ – by Proposition 4.5 – and, since $G[\mathcal{X}]$ and $(G[\mathcal{X}])^*$ have the same Yosida space, it follows that, in $\mathrm{Cov}(Y)$, $YG[\mathcal{X}] \leq Z$.   ∎

Remark 4.7. Suppose that $\mathcal{X}$ is a subalgebra of $\mathcal{P}(G)$ and $G \leq H$ is an essential extension. Then H^* can be an $\mathcal{X}^*$-splitting extension of G^* while H is not an $\mathcal{X}$-splitting extension of G. Indeed, use the example in 4.3: G is the subgroup of $C(\mathbb{N})$ generated by $S(\beta\mathbb{N})$ and the identity function. Let $\mathcal{X} = S(S(\beta\mathbb{N}))$. Then $G^* =$

$S(\beta\mathbb{N})$ is obviously an $\mathcal{X}^* = \mathcal{X}$-splitting extension of itself, but G is not an $\mathcal{X}$-splitting extension of G^*.

It is true, and easy to see, that if H is an $\mathcal{X}$-splitting extension of G then H^* must be an $\mathcal{X}^*$-splitting extension of G^*. A separate question, which is probably interesting in its own right: when is H^* an $\mathcal{X}^*$-splitting extension of G^*?

Definition 4.8. Assume that G is a **W**-object and $Y = YG$. If the subalgebras $\mathcal{X}$ of $\mathcal{P}(G)$ and $\mathfrak{K}$ of $\mathfrak{R}(Y)$ are related according to $\mathfrak{K} = \rho(\mathcal{X})$, we shall say that $\mathfrak{K}$ is *derived* from $\mathcal{X}$.

We reflect on the examples in 2.7 in the light of the material in this section.

Remarks 4.9. In this discussion α stands for a regular, uncountable cardinal, or else the symbol ∞, to be thought of as larger than every cardinal. As the reader will doubtless appreciate, in the case $\alpha = \infty$ below, the sizes are unrestricted.

(a) Recall that $\mathcal{P}_\alpha^\alpha(G)$ stands for the subalgebra of $\mathcal{P}(G)$ consisting of all α-generated polars with α-generated complements. The reader will also recollect that we have abbreviated $G[\mathcal{P}_\alpha^\alpha] = G[\alpha]$.

Let us denote $\mathfrak{R}_\alpha^\alpha(YG) \equiv \rho(\mathcal{P}_\alpha^\alpha(G))$; this consists of the closures of α-complemented α-cozerosets. (Recall: an open set U is an α-*cozeroset* if it is the union of fewer than α cozerosets. To say it is α-*complemented* means that there is an α-cozeroset V, disjoint from U, such that $U \cup V$ is dense.)

This assertion deserves a little amplification: strictly put, $A \in \mathfrak{R}_\alpha^\alpha(YG)$ if and only if $A = \mathrm{cl}_{YG}(\cup_{i \in I} \mathrm{coz}(g_i))$, for suitable $g_i \in G$, and such that $|I| < \alpha$, and its regular complement is also of this form. On the other hand, the cozerosets of the form $\mathrm{coz}(g)$ ($g \in G$) form a base for the open sets of YG; this means that every cozeroset is a countable union of these. Therefore, if A is the closure of an α-cozeroset, it is in fact the closure of a union of fewer than α sets of the form $\mathrm{coz}(g)$. This makes it clear why the closure of an α-complemented α-cozeroset belongs to $\mathfrak{R}_\alpha^\alpha(YG)$, and conversely.

On the other hand, the $\mathfrak{R}_\alpha^\alpha(YG)$-covers of $Y = YG$ are precisely the covers $g :$ $Z \longrightarrow Y$ for which each $A \in g^{\leftarrow}(\mathfrak{R}_\alpha^\alpha(YG))$ is clopen. Now, a space X having the property that every α-complemented α-cozeroset has clopen closure is called an α-*cloz* space. As we shall see in Example 5.5(a), $YG[\alpha]$ is, in fact, the minimum cover of YG which is α-cloz. This is also the content of Theorem 6.6 of [HM01], there phrased in the context of archimedean f-rings.

Observe that $G[\infty]$ is the strongly projectable hull, and ∞-cloz spaces are simply the extremally disconnected spaces.

(b) Next, recall that $\mathcal{P}_\omega(G)$ denotes the subalgebra of $\mathcal{P}(G)$ generated by the principal polars of G. For $Y = YG$, $\rho(\mathcal{P}_\omega(G))$ is the subalgebra generated by the closures of cozerosets of the form $\mathrm{coz}(g)$, with $g \in G$. The $\rho(\mathcal{P}_\omega(G))$-covers are the irreducible pre-images X in which $\mathrm{cl}_X \mathrm{coz}(g)$ is clopen, for each $g \in G$. According to Theorem 4.6,

$YG[\mathcal{P}_\omega(G)]$ is the minimum $\rho(\mathcal{P}_\omega(G))$-cover of Y. But since $G[\mathcal{P}_\omega(G)]$ is projectable, $YG[\mathcal{P}_\omega(G)]$ must be zero-dimensional.

Finally, in this section, we address the question of when one might expect the $\mathfrak{K}$-cover of a compact space to be zero-dimensional. [HVW89] provides most of what we need. We don't know whether the converse of the following result is true.

Proposition 4.10. *Suppose that $\mathfrak{K}$ is a subalgebra of $\mathfrak{R}(X)$. If $\mathfrak{K}$ is a base for the closed subsets of X, then $X[\mathfrak{K}]$ is zero-dimensional. This implication holds, in particular, if X itself is zero-dimensional.*

Proof. As we have already observed, $X[\mathfrak{K}]$ is a Wallman cover, as specified in [HVW89]. If $\mathfrak{K}$ is a base for the closed sets, then $X[\mathfrak{K}]$ is homeomorphic to the space of ultrafilters on $\mathfrak{K}$, by the comment in [HVW89] just prior to Theorem 2.11. Hence it is zero-dimensional. Since $\mathfrak{K}$ includes all clopen sets of X, the final assertion is obvious. ∎

The examples of subalgebras produced in 2.7 can be identified as canonically associated with an ℓ-group, not with the particulars of the given ℓ-group. More precisely, there is a function $G \mapsto \mathcal{X}(G)$, associating with a **W**-object G a subalgebra $\mathcal{X}(G)$ of $\mathcal{P}(G)$. And what might the significant features be of such an assignment? That is the subject of the next section.

5 Polar Functions

Definition & Remarks 5.1. (a) We consider a function $\mathcal{X}$ which associates to each **W**-object G a subalgebra $\mathcal{X}(G)$ of $\mathcal{P}(G)$; such a function will be called a *polar function*. If, in addition the polar function $\mathcal{X}$ has the feature that, whenever H is an essential extension of G, then $K^{\perp_H \perp_H} \in \mathcal{X}(H)$, for each $K \in \mathcal{X}(G)$, we shall say that such an $\mathcal{X}$ is an *invariant polar function*, and use the abbreviation *ipf*. Let us add a note about notation and terminology: if $\mathcal{X}$ is a polar function and G is a **W**-object, strictly speaking, we ought to write $G[\mathcal{X}(G)]$ for the least $\mathcal{X}(G)$-splitting extension. We think the reader will appreciate it if we abbreviate all this by saying $\mathcal{X}$-*splitting* in place of $\mathcal{X}(G)$-splitting, and by writing $G[\mathcal{X}]$ for the least $\mathcal{X}$-splitting extension of G.

The dual notion for compact spaces is this: consider a function $X \mapsto \mathfrak{K}(X)$ assigning to each space a subalgebra $\mathfrak{K}(X)$ of $\mathfrak{R}(X)$; this is a *covering function*. If $\mathfrak{K}$ also satisfies the provision that, for each irreducible surjection $g : Y \longrightarrow X$ and each $A \in \mathfrak{K}(X)$, $g^\leftarrow(A) \in \mathfrak{K}(Y)$, we say that $\mathfrak{K}$ is an *invariant covering function*, or, briefly, an *icf*. The reader who is familiar with [H89] or [V84a] will recognize the icf criterion as the main preamble in Vermeer's classification of covering types. We will observe the same economy of notation and terminology here that was outlined in the preceding paragraph.

(b) One may iterate a polar function, in the following way. Suppose that $\mathcal{X}$ is a polar function. Define, at the first step of the induction, $\mathcal{X}^1 \equiv \mathcal{X}$. Assume now that λ is an ordinal, and that for each ordinal $\gamma < \lambda$, the polar function $\mathcal{X}^\gamma$ is defined, such that if $\gamma < \delta < \lambda$ and G is a **W**-object, $\mathcal{X}^\gamma(G)$ is a subalgebra of $\mathcal{X}^\delta(G)$.

If λ is a limit ordinal, then let

$$\mathcal{X}^\lambda(G) \equiv \bigcup_{\gamma < \lambda} \mathcal{X}^\gamma(G),$$

for each **W**-object G. On the other hand, if κ precedes λ, put

$$\mathcal{X}^\lambda(G) \equiv \{ G \cap P \ : \ P \in \mathcal{X}(G[\mathcal{X}^\kappa]) \}.$$

Observe that, since $\mathcal{X}(H)$ always contains the summands of H, and the extended polar $K^{\perp_{G[\mathcal{X}^\kappa]}\perp_{G[\mathcal{X}^\kappa]}}$ is a summand of $G[\mathcal{X}^\kappa]$ for each $K \in \mathcal{X}^\kappa(G)$, we do obtain that $\mathcal{X}^\lambda(G)$ is a subalgebra of polars containing $\mathcal{X}^\kappa(G)$.

The above defines a transfinite sequence:

(†) $$\mathcal{X} = \mathcal{X}^1 \leq \cdots \leq \mathcal{X}^\lambda \leq \cdots \leq \mathcal{P},$$

where it should be understood that, for polar functions $\mathcal{X}$ and $\mathcal{Y}$, $\mathcal{X} \leq \mathcal{Y}$ means that $\mathcal{X}(G) \leq \mathcal{Y}(G)$, for each **W**-object G. If $\mathcal{X}$ is an ipf, then each member of the sequence $\mathcal{X}^\lambda$ is also an ipf; this verification we leave to the reader. We state it formally as part of Proposition 5.2, just ahead.

As the reader might expect by now, the sequence (†) must stabilize, owing to cardinality conditions; that is, for each **W**-object G, there is an ordinal λ such that $\mathcal{X}^\mu(G) = \mathcal{X}^\lambda(G)$, for each $\mu > \lambda$. Thus we are able to define the following: for each **W**-object G, let

$$\mathcal{X}^\flat(G) \equiv \mathcal{X}^\tau(G),$$

where τ is the least ordinal such that $\mathcal{X}^{\tau'}(G) = \mathcal{X}^\tau(G)$, for each ordinal $\tau' > \tau$. Then $\mathcal{X}^\flat$ is a polar function, called the *limit* polar function of the sequence $\mathcal{X}^\lambda$. It has to be shown – see the proof of Proposition 5.2 - that $\mathcal{X}^\flat$ is an ipf, when $\mathcal{X}$ is an ipf.

The above transfinite construction is much simplified from its original description. The author thanks Eric Zenk for pointing that it could be done.

Proposition 5.2. *Suppose that $\mathcal{X}$ is a polar function. Then for each ordinal λ, $\mathcal{X}^\lambda$ and $\mathcal{X}^\flat$ are also polar functions. If $\mathcal{X}$ is invariant then so are each $\mathcal{X}^\lambda$ and $\mathcal{X}^\flat$. Moreover, for each ordinal λ, and each **W**-object,*

$$G[\mathcal{X}^\lambda][\mathcal{X}] = G[\mathcal{X}^{\lambda+1}],$$

and if λ is a limit ordinal, then

$$G[\mathcal{X}^\lambda] = \bigcup_{\gamma < \lambda} G[\mathcal{X}^\gamma].$$

Proof. As has already been indicated, the transfinite argument proving the invariance of $\mathcal{X}^\lambda$ is left to the reader. Incidentally, one should convince oneself that $\mathcal{X}^\lambda(G)$ is a subalgebra of polars, for each G. This is clear at the limit step, as the union of a chain of subalgebras of $\mathcal{P}(G)$ is a subalgebra. For ordinals with predecessors $\mathcal{X}^\lambda(G)$ is the inverse image of a subalgebra under a boolean isomorphism, namely that of tracing on an essential ℓ-subgroup, and hence also a subalgebra.

Now suppose that $\mathcal{X}$ is invariant; as previously announced, we let the reader check that each $\mathcal{X}^\lambda$ is invariant. Let us establish that $\mathcal{X}^b$ is invariant. Suppose that $H \leq G^e$ is an extension of G. Let τ be the least ordinal for which (†) in 5.1(b) stabilizes for both G and H. Since $\mathcal{X}^\tau$ is an ipf, it follows that $K^{\perp_H \perp_H} \in \mathcal{X}^b(H) = \mathcal{X}^\tau(H)$, for each $K \in \mathcal{X}^b(G) = \mathcal{X}^\tau(G)$.

As to the two identities involving least splitting extensions of G, the one for limit ordinals follows immediately from Theorem 2.4 and the fact that

$$\mathcal{X}^\lambda(G) = \bigcup_{\gamma < \lambda} \mathcal{X}^\gamma(G);$$

we let the reader check the details. To establish the identity for successors, note that if $K \in \mathcal{X}^{\lambda+1}(G)$ then, by definition, $K^{\perp_{G[\mathcal{X}^\lambda]} \perp_{G[\mathcal{X}^\lambda]}} \in \mathcal{X}(G[\mathcal{X}^\lambda])$, whence

$$K^{\perp_{G[\mathcal{X}^\lambda][\mathcal{X}]} \perp_{G[\mathcal{X}^\lambda][\mathcal{X}]}} \in \mathcal{S}(G[\mathcal{X}^\lambda][\mathcal{X}]),$$

which proves that $G[\mathcal{X}^{\lambda+1}] \leq G[\mathcal{X}^\lambda][\mathcal{X}]$. Conversely, if $K \in \mathcal{X}(G[\mathcal{X}^\lambda])$ we then have $K \cap G \in \mathcal{X}^{\lambda+1}(G)$, from which we conclude that $K^{\perp_{G[\mathcal{X}^{\lambda+1}]} \perp_{G[\mathcal{X}^{\lambda+1}]}}$ is a summand of $G[\mathcal{X}^{\lambda+1}]$, and the reverse inclusion then follows.

This completes the proof. ∎

Definition & Remarks 5.3. As we will illustrate in two of our canonical examples, it is interesting to consider when $\mathcal{X}^b = \mathcal{X}$, for a certain polar function. Evidently this is so precisely when $\mathcal{X}$ is *idempotent*, in the sense that $\mathcal{X}^2 = \mathcal{X}$. We amplify: $\mathcal{X}^2 = \mathcal{X}$ if and only if, for each **W**-object G, $(G[\mathcal{X}])[\mathcal{X}] = G[\mathcal{X}]$. This, in turn, is so precisely when the map $K \mapsto K^{\perp_{G[\mathcal{X}]} \perp_{G[\mathcal{X}]}}$ carries $\mathcal{X}(G)$ onto $\mathcal{X}(G[\mathcal{X}])$. Keep in mind that the latter algebra is then the algebra of all cardinal summands of $G[\mathcal{X}]$.

It is easily verified by transfinite induction that $\mathcal{X}^b$ is the least idempotent polar function exceeding $\mathcal{X}$; we leave the details to the reader. It shall be called the *idempotent closure* of $\mathcal{X}$.

Here is a simple example which ought to highlight the point that, if $\mathcal{X}$ is an ipf and G is a **W**-object, then the associated algebra of regular closed sets $\rho(\mathcal{X}(G))$ very much depends on G! It may seem premature to mention this now, but the reader might as well get used to this.

Example 5.4. Consider $\alpha\mathbb{N}$, the one-point compactification of the discrete natural numbers. This example concerns $G_1 \equiv C(\alpha\mathbb{N}) \geq G_2 \equiv S(\alpha\mathbb{N})$. Observe that in both examples, $\mathcal{P}_0(G)$ is a subalgebra. Now, $\mathcal{P}_0(G_1) = \mathcal{P}(G_1)$ is isomorphic to the algebra of all subsets of $\mathbb{N}$; thus, $\rho(\mathcal{P}_0(G_1)) = \mathfrak{R}(\beta\mathbb{N})$ and, incidentally, $YG_1[\mathcal{P}_0] \cong \beta\mathbb{N}$. On the other hand, $\mathcal{P}_0(G_2) = \mathcal{S}(G_2)$, which is the subalgebra of all finite and cofinite subsets of $\mathbb{N}$. Hence, $\rho(\mathcal{P}_0(G_2)) = \mathfrak{B}(\alpha\mathbb{N})$, and $YG_2[\mathcal{P}_0] = \alpha\mathbb{N}$.

Now some meatier comments, illustrated by the examples of 2.7.

Examples 5.5. In (a) and (b) below, as in 4.9, α denotes a regular, uncountable cardinal or else the symbol ∞.

(a) Recall that $\mathcal{P}_\alpha^\alpha(G)$ stands for the subalgebra of all α-generated polars which have α-generated complements. We argue that $\mathcal{P}_\alpha^\alpha$ is an idempotent ipf. Suppose that $H \le G^e$ is an extension of G. Since the polar of H generated by any α-generated polar of G is α-generated in H, the invariance is clear.

Now, in $G[\alpha]$ a typical element is of the form $a = \sum_{i=1}^n g_i[P_i]$, where each $g_i \in G$, and $P_1, P_2, \ldots, P_n$ are pairwise disjoint elements of $\mathcal{P}_\alpha^\alpha(G)$. A bit of reflection will reveal that

$$
\begin{aligned}
a^{\perp G[\alpha] \perp G[\alpha]} &= \{ |g_1|[P_1], \ldots, |g_n|[P_n] \}^{\perp G[\alpha] \perp G[\alpha]} \\
&= |g_1|[P_1]^{\perp G[\alpha] \perp G[\alpha]} \vee \ldots \vee |g_n|[P_n]^{\perp G[\alpha] \perp G[\alpha]}.
\end{aligned}
$$

The point of this is that we may without loss of generality consider polars of $G[\alpha]$ generated by elements of the form $g[P]$, with $g \in G^+$ and $P \in \mathcal{P}_\alpha^\alpha(G)$. Suppose P is generated in G by the set S of positive elements (with $|S| < \alpha$). We let the reader check that $g[P]^{\perp G[\alpha] \perp G[\alpha]}$ is generated by the elements $g \wedge s$, with $s \in S$. This proves that every principal polar of $G[\alpha]$ is generated by fewer than α elements, and since α, as a cardinal, is regular, it follows that the same is true of every α-generated polar of $G[\alpha]$. Then it is also clear that the map $P \mapsto P^{\perp G[\alpha] \perp G[\alpha]}$ carries $\mathcal{P}_\alpha^\alpha(G)$ onto $\mathcal{P}_\alpha^\alpha(G[\alpha])$. By the remarks in 5.3, we may conclude that $\mathcal{P}_\alpha^\alpha$ is idempotent. Put differently, $\mathcal{P}_\alpha^\alpha = (\mathcal{P}_\alpha^\alpha)^\flat$.

(b) By contrast, let $\mathcal{P}_\alpha$ be the polar function which computes the subalgebra of polars generated by the α-generated polars. It is easy to check that this is an ipf. The reader is referred to [HM99], where it is shown that $\mathcal{P}_\alpha^\flat$ is the polar function which produces, for each G, the α-projectable hull $G[\mathcal{P}_\alpha^\flat]$. (Note: G is α-*projectable* if every α-generated polar is a cardinal summand.) Except for $\alpha = \infty$, the iteration to obtain $\mathcal{P}_\alpha^\flat$ must be carried out. Also note that

$$
(\mathcal{P}_\infty^\infty)^\flat = \mathcal{P}_\infty^\infty = \mathcal{P}_\infty^\flat = \mathcal{P}_\infty = \mathcal{P}.
$$

These produce the strongly projectable hull.

(c) Consider $\mathcal{P}_\omega$, which computes the subalgebra of polars generated by the principal polars. That it is invariant is obvious. What emerges from 2.7(b) is that $\mathcal{P}_\omega^\flat = \mathcal{P}_\omega$, the ipf which produces, for each **W**-object G, the projectable hull.

(d) We consider the polar function $\mathcal{S}$. The discussion of 2.6(c) explains that $\mathcal{S}$ is not invariant. Also $\rho(\mathcal{S}) = \mathfrak{B}$, which is, trivially, invariant. Thus, the converse of Proposition 5.7, ahead, is false.

Some comments on the topological side of "iteration" are in order, without going into the fine detail of the presentation of this section to this point. First, let us set a convention, which extends the notion of duality defined in Definition 4.8.

Definition 5.6. Suppose that $\mathcal{X}$ is a polar function and $\mathfrak{K}$ is a covering function. If for each compact space X we have that

$$(*) \qquad\qquad \rho(\mathcal{X}(C(X))) = \mathfrak{K}(X),$$

we shall say that $\mathfrak{K}$ is *derived from* $\mathcal{X}$. We shall also refer to $\mathfrak{K}$ as the *covering derivative* of $\mathcal{X}$. We write, suggestively, $\mathfrak{K} = \rho(\mathcal{X})$.

Next, we record the following observation.

Proposition 5.7. *Suppose that $\mathcal{X}$ is a polar function and $\mathfrak{K} = \rho(\mathcal{X})$. Then if $\mathcal{X}$ is invariant, so is $\mathfrak{K}$.*

  Proof. Suppose that $g : Y \longrightarrow X$ is an irreducible surjection. We invoke (§§), in 4.1(b): if $A \in \mathfrak{K}(X)$, then $A = \rho_{C(X)}(P)$, for suitable $P \in \mathcal{X}(C(X))$, and we have:

$$g^{\leftarrow}(A) = g^{\leftarrow}(\rho_{C(X)}(P)) = \rho_{C(Y)}(P^{\perp\perp});$$

since $P^{\perp\perp} \in \mathcal{X}(C(Y))$, it follows that $g^{\leftarrow}(A) \in \mathfrak{K}(Y)$. ∎

Definition & Remarks 5.8. Suppose that X is a compact Hausdorff space and that $\mathfrak{K}$ is a covering function. We outline a transfinite iteration process: first, $\mathfrak{K}^1 \equiv \mathfrak{K}$. Next suppose that λ is an ordinal number, and that, for each $\gamma < \lambda$, the tower $\mathfrak{K}^{\gamma}(X)$ is defined, so that for all ordinals $\gamma < \delta < \lambda$, $\mathfrak{K}^{\gamma}(X)$ is a subalgebra of $\mathfrak{K}^{\delta}(X)$.
  If λ is a limit ordinal, we let

$$\mathfrak{K}^{\lambda}(X) \equiv \bigcup_{\gamma < \lambda} \mathfrak{K}^{\gamma}(X).$$

On the other hand, if κ is the predecessor of λ, then set

$$\mathfrak{K}^{\lambda}(X) \equiv \{\, g^{\kappa}(A) \,:\, A \in \mathfrak{K}(X[\mathfrak{K}^{\kappa}]) \,\},$$

where $g^{\kappa} : X[\mathfrak{K}^{\kappa}] \longrightarrow X$ is the covering map associated with $X[\mathfrak{K}^{\kappa}]$.

Now, it is tedious indeed to verify the various details related to the iteration above, but they are straightforward, by transfinite induction. And they are as follows:

Proposition 5.9. *Suppose that $\mathfrak{K}$ is a covering function, and $\mathfrak{K}^\lambda$ (for each ordinal λ) is as defined in 5.8. Then we have the following:*

(i) *$\mathfrak{K}^\lambda(X)$ is a subalgebra of $\mathfrak{R}(X)$ containing all clopen sets of X, and, hence, $\mathfrak{K}^\lambda$ is a covering function.*

(ii) *Each $\mathfrak{K}^\lambda$ is invariant when $\mathfrak{K}$ is an icf.*

(iii) *For each ordinal λ, and each compact Hausdorff space X,*

$$X[\mathfrak{K}^{\lambda+1}] = (X[\mathfrak{K}^\lambda])[\mathfrak{K}],$$

and if λ is a limit ordinal then

$$X[\mathfrak{K}^\lambda] = \bigvee_{\gamma < \lambda} X[\mathfrak{K}^\gamma],$$

with the supremum taking place in $\mathrm{Cov}(X)$.

(iv) *For each space X, there is an ordinal τ such that $\mathfrak{K}^{\tau'}(X) = \mathfrak{K}^\tau(X)$, for each $\tau' > \tau$. Setting $\mathfrak{K}^\flat(X) = \mathfrak{K}^\tau(X)$ defines a covering function. If $\mathfrak{K}$ is invariant then so is $\mathfrak{K}^\flat$.*

We shall let the reader verify the above properties.

Definition 5.10. As with the polar functions, we say that a covering function $\mathfrak{K}$ is *idempotent* if $\mathfrak{K}^2 = \mathfrak{K}$. Note that idempotence is equivalent to the identity $\mathfrak{K}^\flat = \mathfrak{K}$, and, alternatively, to the following: if $g_X : X[\mathfrak{K}] \longrightarrow X$ denotes the covering map associated with the space X, then (for each X) $g_X^\leftarrow : \mathfrak{K}(X) \longrightarrow \mathfrak{K}(X[\mathfrak{K}])$ is surjective. (As we remarked for idempotent ipfs, the reader will note that $g_X^\leftarrow$ surjects onto the algebra of clopen sets of $X[\mathfrak{K}]$.)

As for polar functions, one may verify that $\mathfrak{K}^\flat$ is the least idempotent covering function which exceeds $\mathfrak{K}$. $\mathfrak{K}^\flat$ is the *idempotent closure* of $\mathfrak{K}$.

Let $\mathcal{X}$ be a polar function, $\mathfrak{K} = \rho(\mathcal{X})$. It would be nice if the limit $\mathcal{X}^\flat$ of the transfinite sequence defined above for $\mathcal{X}$ had $\mathfrak{K}^\flat$ as its covering derivative. Unfortunately, this is not so, as the following example demonstrates. The author is grateful to A. W. Hager and his reminder to look in Vermeer's [V84b], where the specific instance mentioned below occurs.

Example 5.11. As we saw in 2.7(b), the polar function $\mathcal{P}_\omega$ is idempotent. On the other hand, its covering derivative is not. Let us recall what this covering derivative is: for a compact space X, $X' \equiv X[\rho(\mathcal{P}_\omega)]$ is the minimum cover of X in which the closure of $\mathrm{coz}(f)$ is clopen, for each $f \in C(X)$. The obstruction to $X' = X'[\rho(\mathcal{P}_\omega)]$ is then that, in spite of this stipulation on the cozerosets of functions from $C(X)$, X'

will fail to be ω_1-disconnected ($\equiv$ basically disconnected). To that end one requires a space X in which the subalgebra $\mathfrak{R}_\omega(X)$ generated by all $\mathrm{cl}_X\mathrm{coz}(f)$ (for $f \in C(X)$) is not σ-complete, as the Yosida space of $C(X)[\mathcal{P}_\omega]$ is precisely $X[\mathfrak{R}_\omega(X)]$, and it is the Stone dual of $\mathfrak{R}_\omega(X)$. (This is shown in [HM∞b].)

Such spaces exist! In [V84b], for example, the author shows that, for $X = \beta\mathbb{N} \setminus \mathbb{N}$, $X' = X[\mathfrak{R}_\omega(X)]$ is not the smallest ω_1-disconnected cover; which is to say, it is not ω_1-disconnected.

Note that to obtain such a space one must avoid the *cozero complemented* spaces; recall that X is cozero complemented if for each cozeroset U there is a cozeroset V such that $U \cap V = \emptyset$ and $U \cup V$ is dense in X. For such a space, $\mathfrak{R}_\omega(X)$ *is* the set of all closures of cozerosets, which is closed under countable suprema and infima.

We are ready to discuss hulls formally, as well as their companion hull classes. Hulls associated with certain completions in archimedean ℓ-groups are amply discussed in [HM99]; in a more abstract setting, completions vs covering classes are taken up in [HM∞a].

Definition & Remarks 5.12. (a) We consider a class $\mathfrak{H}$ of **W**-objects closed under formation of ℓ-isomorphic copies. We call the extension $G \leq hG$ in **W** a $\mathfrak{H}$-*hull* of G if it is an essential extension, $hG \in \mathfrak{H}$, $G \leq H \leq G^e$, and

$$H \in \mathfrak{H} \Rightarrow \exists g : hG \longrightarrow H, \text{ an embedding in } \mathbf{W}, \text{ extending the identity on } G.$$

Proposition 2.4 of [HM99] tells us that each **W**-object G has a $\mathfrak{H}$-hull if and only if $\mathfrak{H}$ is *essentially intersective*; that is, for each essentially closed **W**-object E, and each collection $\mathcal{B}$ of subobjects of E, such that $\mathcal{B} \subseteq \mathfrak{H}$ and $\bigcap \mathcal{B}$ is essential in E, then $\bigcap \mathcal{B} \in \mathfrak{H}$.

A class $\mathfrak{H}$ with these features is called a *hull class*. Observe that any hull class is nontrivial, as it contains all the essentially closed **W**-objects.

(b) Dually, recall that a class $\mathfrak{C}$ of compact spaces is called a *covering class* if for each compact space X, there is a smallest member of $\mathrm{Cov}(X)$ which belongs to $\mathfrak{C}$.

We have come to what may, reasonably, be considered the high point of the article. The pairing of hull class and covering class that the reader might have expected cannot be realized, because of the kind of problem illustrated in Example 5.11. The actual pairing of hull class and covering class is handled in [HM∞a].

Also observe, in the proof of the next theorem, that invariance does appear to be employed in a significant way. In fact, Theorem 5.13 fails for the covering function $\mathfrak{R}_{\mathrm{fin}}^\flat$ introduced in the example following the theorem.

Theorem 5.13.

(a) For any ipf $\mathcal{X}$, the **W**-objects of the form $G[\mathcal{X}^\flat]$ form a hull class, denoted $\mathbb{H}(\mathcal{X})$. Indeed, $G[\mathcal{X}^\flat]$ is the hull of G in $\mathbb{H}(\mathcal{X})$.

(b) *For any icf $\mathfrak{K}$, the compact spaces which satisfy $X = X[\mathfrak{K}^{\flat}]$ form a covering class* $\mathbb{T}(\mathfrak{K})$. *The minimum cover of X in* $\mathbb{T}(\mathfrak{K})$ *is* $X[\mathfrak{K}^{\flat}]$.

Proof. We prove (a) and leave the dual, (b), to the reader.

Suppose that G is a **W**-object. Since $\mathcal{X}^{\flat}$ is idempotent $G[\mathcal{X}^{\flat}]$ has no proper $\mathcal{X}^{\flat}$-splitting extensions. Thus, $G[\mathcal{X}^{\flat}] \in \mathbb{H}(\mathcal{X})$. Suppose now that H is an essential extension of G belonging to $\mathbb{H}(\mathcal{X})$. Since H has no proper $\mathcal{X}^{\flat}$-splitting extensions, each member of $\mathcal{X}^{\flat}(H)$ is a summand, and as the polar function $\mathcal{X}^{\flat}$ is invariant, it follows that H itself is a $\mathcal{X}^{\flat}$-splitting extension of G, whence $G[\mathcal{X}^{\flat}] \leq H$. ∎

Example 5.14. Let X be a compact space. $\mathfrak{R}_{\text{fin}}(X)$ denotes the following subalgebra of regular closed sets. First, say that a closed set A in X is *almost open* if $A \setminus \text{int}_X A$ is finite. (There is an obviously dual notion of an *almost closed* set.) Now define for a regular closed set A,

$$A \in \mathfrak{R}_{\text{fin}}(X) \iff A \text{ is almost open.}$$

It should be clear that the complement of an almost closed set is almost open, and so the regular complement of an $A \in \mathfrak{R}_{\text{fin}}(X)$ is, likewise, in $\mathfrak{R}_{\text{fin}}(X)$. Since it is clear that $\mathfrak{R}_{\text{fin}}(X)$ is closed under (finite) suprema and infima, it follows that $\mathfrak{R}_{\text{fin}}(X)$ is, indeed, a subalgebra.

(a) First, let us settle that $\mathfrak{R}_{\text{fin}}$ is not an icf.

Consider $\alpha\mathbb{N}$, the one-point compactification of the discrete natural numbers. Observe that $\mathfrak{R}(\alpha\mathbb{N}) = \mathfrak{R}_{\text{fin}}(\alpha\mathbb{N})$. Thus, the least $\mathfrak{R}_{\text{fin}}$-cover of $\alpha\mathbb{N}$ is $\beta\mathbb{N}$. Evidently, as $\beta\mathbb{N}$ is extremally disconnected, every regular closed set of it is clopen; in particular, $\mathfrak{R}_{\text{fin}}(\beta\mathbb{N}) = \mathfrak{B}(\beta\mathbb{N})$. However, as we shall see, it is not the least cover of $\alpha\mathbb{N}$ in which $\mathfrak{R}_{\text{fin}}(X) = \mathfrak{B}(X)$. This will show that Theorem 5.13(b) fails without the invariance of the covering function.

Next, construct a cover of $\alpha\mathbb{N}$ as follows. Partition $\mathbb{N}$ into two infinite sets, A and A'. Establish a fixed bijection ψ between them, and let ψ also denote the homeomorphism of βA onto $\beta A'$ which extends this bijection. Note that these are two copies of $\beta\mathbb{N}$. Now let Y be the space obtained by identifying each point $p \in \beta A \setminus A$ with its image $\psi(p)$. Now let $g : Y \longrightarrow \alpha\mathbb{N}$ be the map which extends the identity; g is an irreducible surjection. Now the reader will observe that $g^{\leftarrow}(\text{cl}_{\alpha\mathbb{N}}A) = \text{cl}_Y A$, which is not in $\mathfrak{R}_{\text{fin}}(Y)$, proving that $\mathfrak{R}_{\text{fin}}$ is not an icf.

(b) Here we argue that Y in (a) has the feature $\mathfrak{R}_{\text{fin}}(Y) = \mathfrak{B}(Y)$, thus settling that $\beta\mathbb{N}$ is not minimal in this regard over $\alpha\mathbb{N}$. For this argument the reader ought to refer to Warren McGovern's dissertation ([Mc98]). However, we shall supply some of the background.

McGovern defines a Tychonoff space to be a C^*-*space* if every cofinite subset is C^*-embedded. To get an idea of the basic behavior of C^*-spaces here is his Theorem 5.4.9, conflated with other comments from [Mc98].

Suppose that X is a Tychonoff space. Then the following are equivalent.

(a) *X is a C^*-space.*

(b) *Every dense almost closed subset of X is C^*-embedded.*

(c) *Every open almost closed subset of X is C^*-embedded.*

(d) *For each $p \in X$, $X \setminus \{p\}$ is C^*-embedded.*

(e) *βX is a C^*-space.*

(f) *Every dense almost closed subset is a C^*-space.*

(g) *Every open almost closed subset is a C^*-space.*

It is obvious that every extremally disconnected space is a C^*-space. McGovern uses the example Y under scrutiny here as a typical example of how to manufacture C^*-spaces that are not extremally disconnected.

Furthermore, McGovern shows that every C^*-space X satisfies $\mathfrak{R}_{\text{fin}}(X) = \mathfrak{B}(X)$, and there is a converse proposition, as follows:

Suppose that X is strongly zero-dimensional. Then X is a C^*-space if and only if $\mathfrak{R}_{\text{fin}} \equiv \mathfrak{B}$, and for each finite subset $F \subseteq X$, $X \setminus F$ is also strongly zero-dimensional.

For the unit square I^2, with the usual topology, $\mathfrak{R}_{\text{fin}}(I^2) = \mathfrak{B}(I^2) = \{\emptyset, I^2\}$ ([Mc98]), although I^2 is far from being a C^*-space.

These remarks then show, referring to our space Y, that $\mathfrak{R}_{\text{fin}}(Y) = \mathfrak{B}(Y)$, as promised.

(c) This discussion does not answer the question of whether the condition $\mathfrak{R}_{\text{fin}}(X) = \mathfrak{B}(X)$ defines a covering class; it does not. Neither is the class of compact C^*-spaces a covering class. These facts, among others, are determined in [MMc∞].

Finally, a comment which involves many details if one is going to flesh things out. We prefer not to do that. The reader is referred to [HM99] for the background.

Remark 5.15. Recall, from 5.5(b), that $\mathcal{P}_\alpha$ is the polar function which collects the α-generated polars. We have already remarked that $G[\mathcal{P}_\alpha^\flat]$ is the α-projectable hull of G, and now that we have said what a hull class is, the associated hull class of $\mathcal{P}_\alpha^\flat$ is the class of α-projectable **W**-objects. As explained in [HM99], the associated Yosida spaces are the α-*disconnected spaces*: the spaces in which every α-cozeroset has clopen closure. $\rho(\mathcal{P}_\alpha^\flat)$ is the icf which computes the minimum α-disconnected cover.

References

[BKW77]   A. Bigard, K. Keimel & S. Wolfenstein, *Groupes et Anneaux Réticulés*. Lecture Notes in Math **608**, Springer Verlag (1977); Berlin-Heidelberg-New York.

[Bl74]   R. D. Bleier, *The SP-hull of a lattice-ordered group*. Canad. Jour. Math. **XXVI**, No. 4 (1974), 866-878.

[Ch71]   D. Chambless, *The Representation and Structure of Lattice-Ordered Groups an f-Rings*. Tulane University Dissertation (1971), New Orleans.

[C71]   P. F. Conrad, *The essential closure of an archimedean lattice-ordered group*. Duke Math. Jour. **38** (1971), 151-160.

[C73]   P. F. Conrad, *The hulls of representable ℓ-groups and f-rings*. Jour. Austral. Math. Soc. **26** (1973), 385-415.

[D95]   M. R. Darnel, *The Theory of Lattice-Ordered Groups*. Pure & Appl. Math. **187**, Marcel Dekker (1995); Basel-Hong Kong-New York.

[GJ76]   L. Gillman & M. Jerison, *Rings of Continuous Functions*. Grad. Texts in Math. **43**, Springer Verlag (1976); Berlin-Heidelberg-New York.

[H89]   A. W. Hager, *Minimal covers of topological spaces*. In Papers on General Topology and Related Category Theory and Topological Algebra; Annals of the N. Y. Acad. Sci. **552** March 15, 1989, 44-59.

[HM99]   A. W. Hager & J. Martínez, *Hulls for various kinds of α-completeness in archimedean lattice-ordered groups*. Order **16** (1999), 89-103.

[HM01]   A. W. Hager & J. Martínez, *The ring of α-quotients*. To appear, Algebra Universalis.

[HM∞a]   A. W. Hager & J. Martínez, *Polar functions, II: completion classes of archimedean f-algebras vs. covers of compact spaces*. Preprint.

[HM∞b]   A. W. Hager & J. Martínez, *The projectable and regular hulls of a semiprime ring*. Work in progress.

[HR77]   A. W. Hager & L. C. Robertson, *Representing and ringifying a Riesz space*. Symp. Math. **21** (1977), 411-431.

[HVW89]   M. Henriksen, J. Vermeer & R. G. Woods, *Wallman covers of compact spaces*. Diss. Math. **CCLXXX** (1989), Warsaw.

[MMc∞]   J. Martínez & W. Wm. McGovern, *C*-compactifications*. Ongoing research.

[Mc98]   W. Wm. McGovern, *Algebraic and Topological Properties of C(X) and the 𝔉-topology*. University of Florida Dissertation, 1998; Gainesville, FL.

[PW89]   J. R. Porter & R. G. Woods, *Extensions and Absolutes of Hausdorff Spaces*. Springer Verlag (1989); Berlin-Heidelberg-New York.

[V84a]   J. Vermeer, *On perfect irreducible preimages*. Topology Proc. 9 (1984), 173-189.

[V84b]   J. Vermeer, *The smallest basically disconnected preimage of a space*. Topology and its Appl. **17** (1984), 217-232.

Department of Mathematics, University of Florida, P. O. Box 118105
Gainesville, FL 32611-8105, USA
martinez@math.ufl.edu

A Priestley-type Method
for Generating Free ℓ-Groups [1]

Néstor G. Martínez and Alejandro Petrovich[2]

ABSTRACT.   A Priestley-type topological representation is developed for the class of partially ordered groups which are p. o. subgroups of lattice-ordered groups. The appropriated analog of the spectrum of prime lattice filters is the class of increasing subsets P satisfying $ab \in P$ and $cd \in P$ imply $ad \in P$ or $cb \in P$. In developing this representation, we give new embedding theorems for these groups. In particular, we give a necessary and sufficient condition for a p. o. group in this class to be embedded in an ℓ-group of sets in such a way that the embedding preserves meets and joins that already exist in the group. Our construction gives also an alternative and very natural way to obtain the free ℓ-group generated by a p. o. group in this class.

1   Preliminaries

Recall that $G = (G, \leq, ^{-1}, \cdot, e)$ is a *partially ordered group (p. o. group)* if $(G, ^{-1}, \cdot, e)$ is a group, $(G, \leq)$ is a poset and $x \leq y$ implies $xz \leq yz$ and $zx \leq zy$ for all $z \in G$. A *lattice-ordered group* (ℓ-group for short) is a p. o. group such that the partial order is a lattice order.

A *p. o. subgroup* of an ℓ-group G is a subgroup S such that the positive cone of S is the intersection of the positive cone of G with S. Also recall that P is an *increasing subset* of a poset $(G, \leq)$ if $x \in P$ and $x \leq y$ imply $y \in P$ for all $x, y \in G$. A subset F of a lattice $(G, \wedge, \vee)$ is a *filter* if F is an increasing subset of G and $x, y \in F$ implies $x \wedge y \in F$. F is a *prime* lattice filter if it is a proper filter of G satisfying $x \vee y \in F$ implies $x \in F$ or $y \in F$.

Both Stone duality and Priestley duality for bounded distributive lattices are developed by providing suitable topologies for the family of all prime lattice filters of the lattice. The key property of this family is the Prime Lattice Filter Theorem, the fact that there are enough prime lattice filters to *strongly separate points* of the lattice in the following precise sense:

---

[1]**Keywords.** lattice-ordered groups, isolated groups, Priestley duality, Stone duality, free ℓ-groups.
1991 Mathematics Subject Classification. 06D05, 06E15, 03G25
[2]The authors acknowledge the referees, for their careful reading and for several suggestions to improve the paper.

J. Martínez (ed.), Ordered Algebraic Structures, 301–312.
© 2002 *Kluwer Academic Publishers.*

If $x \nleq y$ then there exists a prime lattice filter P such that $x \in P$ and $y \notin P$.

In the sequel we will need the following general definition.

Definition 1.1. Let $(P, \leq)$ be a partially ordered set. Given a family $\mathcal{F}$ of subsets of P, we say that $\mathcal{F}$ *strongly separates points* of P if whenever $x \nleq y$ then there exists $F \in \mathcal{F}$ such that $x \in F$ and $y \notin F$.

Taking advantage of the fact that the underlying lattice of an ℓ-group is distributive, a Priestley-type duality for ℓ-groups can be developed by adding a rather natural binary function (the product of filters) to the Priestley space [7]. For an account on Priestley spaces, see for example [4]. The topological duality for ℓ-groups was first developed in [7].

In this paper we start the extension of the duality theory to the class of p. o. groups that are p. o. subgroups of lattice-ordered groups. For lack of a better name, we will call them *isolated groups*, since in the abelian case this class coincides with the class of groups whose positive cone is *isolated* (in the sense that $x^n \geq e$ for some positive integer n implies $x \geq e$). Examples of isolated groups are

(1) Polynomials in one variable with real coefficients, with the usual sum and point-wise order. Note that this is a p. o. subgroup of the lattice-ordered group of real continuous functions of one variable which is not an ℓ-group.

(2) The most general example in the abelian case is given in [9]. It is proved there that a p. o. group has an isolated positive cone if and only if it is a subdirect product of totally ordered groups.

P. Conrad, W. C. Holland and others have extensively studied this class of groups. In particular P. Conrad has shown in [3] that the class of isolated groups is precisely the class of p. o. groups for which the positive cone is an intersection of cones of *total right orders*; (see [2] or [1], 4.5, p. 83, for definitions and an exposition of the subject).

In developing a Priestley-type representation the first difficulty we face is the lack of the notion of prime lattice filter (since a p. o. subgroup H of an ℓ-group G is not necessarily a lattice-ordered group, as the first example given above shows). However, the family

$$\{P \cap H : P \text{ is a prime lattice filter of } G\}$$

is a class of increasing subsets that strongly separates points of H. Therefore what we need is to find an intrinsic algebraic condition to capture these subsets.

Definition 1.2. A subset P of a group G is *quasiprime* if and only if it satisfies $ab \in P$ and $cd \in P$ imply $ad \in P$ or $cb \in P$.

If G is a p. o. group, the *spectrum* of G is the family of all increasing and quasiprime subsets of G. We denote it by $S(G)$. Note that $\emptyset$ and G belong to $S(G)$.

We have now, as desired, that if H is a p. o. subgroup of an ℓ-group G and P is a prime lattice filter of G then the subset $P \cap H$ is quasiprime and increasing. This follows at once from the following fact:

Proposition 1.3. *Prime lattice filters of a lattice-ordered group G are quasiprime subsets of G.*

Proof. Let P be a prime lattice filter of G and suppose that $ab, cd \in P$. Then $a(b \vee d) \in P$ and $c(b \vee d) \in P$. Since P is a filter, $a(b \vee d) \wedge c(b \vee d) = (a \wedge c)(b \vee d) = (a \wedge c)b \vee (a \wedge c)d \in P$. Therefore, using the fact that P is prime, $(a \wedge c)b \in P$ or $(a \wedge c)d \in P$. In the first case it follows that $cb \in P$. In the second case it follows that $ad \in P$. ∎

As a consequence, if H is a p. o. subgroup of an ℓ-group G, then the family $S(H)$ strongly separates points of H.

However, not every quasiprime subset of a p. o. subgroup of an ℓ-group arises as the intersection of a prime lattice filter with the given subgroup. The following example shows a proper quasiprime and increasing subset of an ℓ-group that is not a prime lattice filter and that cannot be included in any prime lattice filter.

Example 1.4. Let G be any totally ordered abelian group (for example $\mathbb{Z}$) and take the ℓ-group $G \times G$. Then the subset $P = \{(x, y) : x + y > 0\}$ is an increasing quasiprime subset of $G \times G$.

However P is not a filter because for any $a > 0$ it follows that $(a, 0) \in P$ and $(0, a) \in P$ but $(a, 0) \wedge (0, a) = (0, 0) \notin P$. Moreover, P cannot be extended to a proper lattice filter of $G \times G$. Indeed, since the equality $(a, b) = (a + k, b) \wedge (a, b + k)$ holds for every $k \geq 0$, every element (a, b) of $G \times G$ can be written as the infimum of two elements in P by choosing $k > (-a - b) \vee 0$. Therefore the filter generated by P is not proper.

A closer analog to the notion of prime lattice filter is given by the following.

Definition 1.5. A subset P of a p. o. group G is *prime* if it is a proper and increasing quasiprime subset of G satisfying:

(a) If $x, y \in P$ and there exists $x \wedge y$, then $x \wedge y \in P$.

(b) If there exists $x \vee y$ and $x \vee y \in P$, then either $x \in P$ or $y \in P$.

If G is an ℓ-group it turns out that a subset P of G is a prime lattice filter if and only if it is an increasing and prime subset as defined above.

In section 2 we will prove that an isolated p. o. group can be embedded into an ℓ-group preserving the existing joins and meets if and only if the family of prime and increasing sets strongly separates points.

A simple but crucial fact about quasiprime sets is given by the following proposition. The straightforward proof is left to the reader.

Proposition 1.6. *Let G be a group. For each $a \in G$ and each $P \in S(G)$ the set $P_a = \{x : x^{-1}a \notin P\}$ belongs to $S(G)$.*

Let G be a p. o. group and let $\mathcal{F}$ be a family of quasiprime and increasing subsets of G. For each $a \in G$ let $\sigma_{\mathcal{F}}(a)$ be the set of all members of $\mathcal{F}$ containing a. In particular, we denote by $\sigma(a)$ the family of all quasiprimes and increasing subsets of G containing a. We denote by $\Sigma(\mathcal{F}) = \{\sigma_{\mathcal{F}}(a) : a \in G\}$. In particular, $\Sigma(S(G))$ will denote the family $\{\sigma(a) : a \in G\}$.

Proposition 1.7. *Let G be a p. o. group, and define $g : S(G) \longrightarrow S(G)$ by $g(P) = P_e$. Then:*

(1) *g is an order-reversing bijection from $S(G)$ to $S(G)$ satisfying $g(g(P)) = P$.*

(2) *The usual product of sets*

$$PQ = \{xy : x \in P \text{ and } y \in Q\}$$

defines a binary function on $S(G)$ which is associative and order-preserving in each variable with respect to set theoretical inclusion.

Theorem 1.8. *Let $\mathbf{G} = (G, \leq, \cdot, ^{-1}, e)$ be a p. o. group and let $\mathcal{F}$ be a family of quasiprime and increasing subsets satisfying*

(1) *If P, Q are in $\mathcal{F}$, then $PQ \in \mathcal{F}$.*

(2) *If $P \in \mathcal{F}$, then $P_a \in \mathcal{F}$ for all $a \in G$.*

Then

(i) *$\sigma_{\mathcal{F}}(a^{-1}) = (g^{-1}[\sigma_{\mathcal{F}}(a)])^c$, where ^{c} denotes set-theoretic complementation.*

(ii) *$\sigma_{\mathcal{F}}(ab) = \bigcap_{P \in \sigma_{\mathcal{F}}(a^{-1})} \{Q : PQ \in \sigma_{\mathcal{F}}(b)\}$.*

If, in addition, the family $\mathcal{F}$ strongly separates points (in the sense of Definition 1.1), then the family of sets $\Sigma(\mathcal{F})$ equipped with set-theoretical inclusion, the inverse operation $(\sigma_{\mathcal{F}}(a))^{-1} = (g^{-1}[\sigma_{\mathcal{F}}(a)])^c$, the product

$$\sigma_{\mathcal{F}}(a)\sigma_{\mathcal{F}}(b) = \bigcap_{P \in (g^{-1}[\sigma_{\mathcal{F}}(a)])^c} \{Q : PQ \in \sigma_{\mathcal{F}}(b)\}$$

and the unit $\sigma_{\mathcal{F}}(e)$, is a p. o. group which is order-isomorphic to G.

Proof. (Sketch) For (i) note that $P \in \sigma_{\mathcal{F}}(a^{-1})$ if and only if $a^{-1} \in P$ if and only if $a \notin g(P)$, which is equivalent to say that $P \notin g^{-1}(\sigma_{\mathcal{F}}(a))$.

For (ii) suppose first that $Q \in \sigma_{\mathcal{F}}(ab)$ and take $P \in \sigma_{\mathcal{F}}(a^{-1})$. Then $a^{-1} \in P$ and $ab \in Q$. Therefore $b = a^{-1}ab \in PQ$. For the opposite inclusion suppose that Q belongs to the right member intersection and assume, by the way of contradiction, that $ab \notin Q$. Therefore $a^{-1} \in Q_b$ and hence $Q_b \in \sigma_{\mathcal{F}}(a^{-1})$. By our initial hypothesis $Q_bQ \in \sigma_{\mathcal{F}}(b)$. This means that there exist elements $x \in Q_b$ and $q \in Q$ such that $b = xq$. Therefore $x^{-1}b = q \in Q$ which is a contradiction with the fact that $x \in Q_b$.

Joining conditions (i) and (ii) we obtain that

$$\sigma_{\mathcal{F}}(ab) = \bigcap_{P \in (g^{-1}[\sigma_{\mathcal{F}}(a)])^c} \{Q : PQ \in \sigma_{\mathcal{F}}(b)\}.$$

Then the aplication $a \mapsto \sigma_{\mathcal{F}}(a)$ yields a p.o. group in $\Sigma(\mathcal{F})$ which is order-isomorphic to G provided $\mathcal{F}$ strongly separates points. ∎

In particular, $S(G)$ satisfies conditions (1) and (2) of Theorem 1.8. Therefore if $S(G)$ strongly separates points, then $\Sigma(S(G))$ is a p. o. group which is order-isomorphic to G. In the sequel we will say that a family $\mathcal{F}$ of quasiprime and increasing subsets of a p. o. group G is *complete* if $\mathcal{F}$ satisfies conditions (1) and (2) of Theorem 1.8.

We list now some additional properties of the spectrum $S(G)$ with the goal of providing a complete axiomatization for the dual spaces of isolated groups. The proofs of all these properties are straightforward. As an illustration, we give below the proof of item (3) and of the crucial property stated in (9).

Proposition 1.9. Let G be a p. o. group and let $\mathcal{F}$ be a complete family for G. Then:

(1) $P \subset g(P)$ or $g(P) \subset P$, for all $P \in \mathcal{F}$; (note that if $P \in S(G)$ then $P \neq g(P)$ because $e \in P \setminus g(P)$ or $e \in g(P) \setminus P$.)

(2) $\sigma_{\mathcal{F}}(e) = \{P \in \mathcal{F} : g(P) \subset P\}$.

(3) $g(Q)Q \subset g(g(Q)Q)$, for all $Q \in \mathcal{F}$.

(4) $g(PQ)P \subseteq g(Q)$, for all $P, Q \in \mathcal{F}$.

(5) $Pg(PQ) \subseteq g(Q)$, for all $P, Q \in \mathcal{F}$.

(6) $g(Q) \subset Q$ implies $P \subseteq PQ$ and $P \subseteq QP$ for all $P \in \mathcal{F}$.

(7) If $Q \notin \sigma_{\mathcal{F}}(e)$ and $\sigma_{\mathcal{F}}(a) \in \Sigma(\mathcal{F})$, then there exists $P \in \sigma_{\mathcal{F}}(a)$ such that $PQ \notin \sigma_{\mathcal{F}}(a)$.

(8) Given $Q \in \mathcal{F}$ and $\sigma_{\mathcal{F}}(a) \in \Sigma(\mathcal{F})$, then $Q_a = \{x \in G : x^{-1}a \notin Q\}$ belongs to $\mathcal{F}$ and it is the greatest $P \in \mathcal{F}$ (with respect to inclusion) satisfying $PQ \notin \sigma_{\mathcal{F}}(a)$.

(9) Given $Q \in \mathcal{F}$ and $\sigma_{\mathcal{F}}(a), \sigma_{\mathcal{F}}(b) \in \Sigma(\mathcal{F})$, then $Q_a \subseteq Q_b$ or $Q_b \subseteq Q_a$.

Proof. To prove (3) suppose that $e \notin g(g(Q)Q)$. Then $e \in g(Q)Q$. This means that there exists $x \in g(Q)$ and $y \in Q$ such that $e = xy$. Since $x \in g(Q)$, it follows that $x^{-1} = y \notin Q$, which is a contradiction. Therefore $e \in g(g(Q)Q)$ and using (2) we conclude that $g(Q)Q \subset g(g(Q)Q)$.

In order to prove (9) take $Q \in \mathcal{F}$, elements $a, b \in G$ and suppose that neither $Q_a \subseteq Q_b$ nor $Q_b \subseteq Q_a$. Then there exist elements $x \in Q_a \backslash Q_b$ and $y \in Q_b \backslash Q_a$. This means that $x^{-1}a \notin Q, y^{-1}b \notin Q, x^{-1}b \in Q$ and $y^{-1}a \in Q$. Since Q is a quasiprime set, from the last two conditions we can infer that $x^{-1}a \in Q$ or $y^{-1}b \in Q$, which is a contradiction.   ∎

Theorem 1.10. *Let G be a p. o. group and let $\mathcal{F}$ be a complete family for G. Let $\mathcal{O}$ be a family of increasing subsets of $\mathcal{F}$ (with respect to inclusion) satisfying:*

(a) $\sigma_{\mathcal{F}}(e) = \{P \in \mathcal{F} : g(P) \subset P\}$ *belongs to $\mathcal{O}$.*

(b) *If $U, V \in \mathcal{O}$, then $U^{-1} = (g^{-1}[U])^c \in \mathcal{O}$ and*

$$UV = \bigcap_{P \in (g^{-1}[U])^c} \{Q : PQ \in V\} \in \mathcal{O}.$$

(c) *If $Q \notin \sigma_{\mathcal{F}}(e)$ and $U \in \mathcal{O}$, then there exists $P \in U$ such that $PQ \notin U$.*

Then $\mathcal{O}$ is a p. o. group of sets with the order given by inclusion.

Proof. We prove first that $U(VW) = (UV)W$ for all $U, V, W \in \mathcal{O}$.
For the left inclusion, let

$$Q \in \bigcap_{P \in U^{-1}} \{R : PR \in VW\}$$

and suppose that

$$Q \notin (UV)W = \bigcap_{P \in (UV)^{-1}} \{R : PR \in W\}.$$

Choose $P_1 \in (UV)^{-1}$ such that $P_1Q \notin W$. Since $P_1 \in (UV)^{-1} = (g^{-1}[UV])^c$, it follows that

$$g(P_1) \notin UV = \bigcap_{P \in U^{-1}} \{R : PQ \in V\}$$

and we can choose $P_2 \in U^{-1}$ such that $P_2g(P_1) \notin V$. Therefore, $g(P_2g(P_1)) \in V^{-1}$. Since $P_2 \in U^{-1}$ and

$$Q \in \bigcap_{P \in U^{-1}} \{R : PR \in VW\}$$

we can derive

$$P_2Q \in VW = \bigcap_{P \in V^{-1}} \{R : PR \in W\}.$$

Since $g(P_2g(P_1)) \in V^{-1}$, it follows that $g(P_2g(P_1))(P_2Q) \in W$. Using the fact that the product of quasiprimes is associative, we obtain that $(g(P_2g(P_1))P_2)Q \in W$. Now, using Proposition 1.9(4), we have that $g(P_2g(P_1))P_2 \subseteq g(g(P_1)) = P_1$. Since the product of quasiprimes is order preserving in each variable and the members of $\mathcal{O}$ are increasing sets with respect to inclusion, we obtain that $(g(P_2g(P_1))P_2)Q \subseteq P_1Q$ and therefore $P_1Q \in W$, which is a contradiction. The opposite inclusion can be proved with similar arguments, using Proposition 1.9(5).

We prove now that $\sigma_{\mathcal{F}}(e)V = V$ for all $V \in \mathcal{O}$. From condition (i) of Theorem 1.8, $\sigma_{\mathcal{F}}(e) = \sigma_{\mathcal{F}}(e)^{-1}$ and therefore we will prove that

$$\bigcap_{P \in \sigma_{\mathcal{F}}(e)} \{R : PR \in V\} = V.$$

Suppose first that $Q \in V$ and let $P \in \sigma_{\mathcal{F}}(e)$. Using Proposition 1.9(6), it follows that $Q \subseteq PQ$, and therefore $PQ \in V$. For the opposite inclusion, suppose that

$$Q \in \bigcap_{P \in \sigma_{\mathcal{F}}(e)} \{R : PR \in V\}$$

and note that from Proposition 1.9(3) it follows that $g(Qg(Q)) \in \sigma_{\mathcal{F}}(e)$. Then $g(Qg(Q))Q \in V$. Using Proposition 1.9(4), it follows that $g(Qg(Q))Q \subseteq Q$, and therefore $Q \in V$.

The fact that $U^{-1}U = \sigma_{\mathcal{F}}(e)$ follows at once from Proposition 1.9(6) and from condition (iii) of the hypothesis. It is also immediate that $U \subseteq V$ implies $UW \subseteq VW$ and $WU \subseteq WV$ for all $U, V, W \in \mathcal{O}$.   ∎

Of course, an obvious example of such a family $\mathcal{O}$ is $\Sigma(S(G))$. Our next proposition points to a second interesting example.

Theorem 1.11. *Let G be a p. o. group and let $\mathcal{F}$ be a complete family for G that strongly separates points. Let $\mathcal{O}$ be the distributive lattice generated by $\Sigma(\mathcal{F})$, i.e., the* *family of sets of the form*

$$A = \bigcup_{i=1}^{n} \bigcap_{j=1}^{m(i)} \sigma_{\mathcal{F}}(a_{ij}).$$

Then $\mathcal{O}$ is an ℓ-group of sets, with the lattice operations given by set theoretical union and intersection. The group G is order-isomorphic to a p. o. subgroup of $\mathcal{O}$. Moreover if we take as $\mathcal{F}$ the family $S(G)$ of all quasiprime and increasing sets, then $\mathcal{O}$ is the free ℓ-group generated by G.

Proof. In order to prove that the distributive lattice $\mathcal{O}$ is a p. o. group, it is enough to show that $\mathcal{O}$ satisfies conditions (a), (b) and (c) of Theorem 1.10. Item (a) is straightforward. In order to prove (c) let $Q \in \mathcal{F}$ such that $Q \notin \sigma_{\mathcal{F}}(e)$ and suppose first that $U = \sigma_{\mathcal{F}}(a) \cup \sigma_{\mathcal{F}}(b)$. Then $Q_a \in \sigma_{\mathcal{F}}(a)$ and $Q_b \in \sigma_{\mathcal{F}}(b)$. Since $Q_a \subseteq Q_b$ or $Q_b \subseteq Q_a$, by choosing $Q_U = \min(Q_a, Q_b)$ it follows that $Q_U \in U$ and $Q_U Q \notin U$. In a similar way, if $U = \sigma_{\mathcal{F}}(a) \cap \sigma_{\mathcal{F}}(b)$, then Q_U can be chosen as $Q_U = \max(Q_a, Q_b)$. An easy induction argument based on these two cases shows that if $U = \bigcup_{i=1}^{n} \bigcap_{j=1}^{m_i} \sigma_{\mathcal{F}}(a_{ij})$ and Q_U is chosen as $Q_U = \min_{1 \leq i \leq n}(\max_{1 \leq j \leq m_i} Q_{a_{ij}})$, then $Q_U \in U$ and $Q_U Q \notin U$. It also holds that Q_U is the greatest element P in $\mathcal{F}$ such that $PQ \notin U$. Indeed, let $P \in \mathcal{F}$ such that $PQ \notin U$. Then for all $i \in \{1, \dots, n\}$ there exists $j_i \in \{1, \dots, m_i\}$ such that $PQ \notin \sigma(a_{ij_i})$. Hence for each $1 \leq i \leq n$, we have that $P \subseteq Q_{a_{ij_i}}$. Therefore $P \subseteq \min_{1 \leq i \leq n} Q_{a_{ij_i}}$. Since $Q_{a_{ij_i}} \subseteq \max_{1 \leq j \leq m_i} Q_{a_{ij}}$, it follows that

$$P \subseteq \min_{1 \leq i \leq n} Q_{a_{ij_i}} \subseteq \min_{1 \leq i \leq n} \left(\max_{1 \leq j \leq m_i} Q_{a_{ij}} \right) = Q_U.$$

From this property note that the definition of Q_U does not depend of the $\sigma(a_{ij})$ chosen in the representation.

It is clear from the definition that $U^{-1} = (g^{-1}[U])^c \in \mathcal{O}$ for any $U \in \mathcal{O}$. The last step of the proof is to verify that if $U, V \in \mathcal{O}$ then $UV \in \mathcal{O}$. Note first that for any $U, V, W \in \mathcal{O}$ the following equalities are satisfied:

$$\bigcap_{P \in U \cup V} \{Q : PQ \in W\} = \bigcap_{P \in U} \{Q : PQ \in W\} \cap \bigcap_{P \in V} \{Q : PQ \in W\}.$$

$$\bigcap_{P \in U} \{Q : PQ \in V \cap W\} = \bigcap_{P \in U} \{Q : PQ \in V\} \cap \bigcap_{P \in U} \{Q : PQ \in W\}.$$

$$\bigcap_{P \in U \cap V} \{Q : PQ \in W\} = \bigcap_{P \in U} \{Q : PQ \in W\} \cup \bigcap_{P \in V} \{Q : PQ \in W\}.$$

$$\bigcap_{P \in U} \{Q : PQ \in V \cup W\} = \bigcap_{P \in U} \{Q : PQ \in V\} \cup \bigcap_{P \in U} \{Q : PQ \in W\}.$$

The first two equalities hold simply by set theoretical properties of union and intersection. In order to prove the third equality take

$$R \in \bigcap_{P \in U \cap V} \{Q : PQ \in W\}$$

and suppose that there exist $P_1 \in U$ and $P_2 \in V$ such that $P_1 R \notin W$ and $P_2 R \notin W$. Then, by the property about Q_W proved above, $P_1 \subseteq Q_W$ and $P_2 \subseteq Q_W$. Therefore, $Q_W \in U \cap V$, which is a contradiction. The opposite inclusion is straightforward.

Finally, to prove the fourth equality, take

$$Q \in \bigcap_{P \in U} \{Q : PQ \in V \cup W\}$$

and suppose that there exist $P_1, P_2 \in U$ such that $P_1 Q \notin V$ and $P_2 Q \notin W$. Then $Q_V \in U$ and $Q_W \in U$. Take $R = \min(Q_V, Q_W)$. Then $R \in U$ and $RQ \notin V \cup W$, which is a contradiction. The opposite inclusion is straightforward.

Since each $U \in \mathcal{O}$ is a finite union of finite intersections of sets of the form $\sigma_{\mathcal{F}}(a)$ for some $a \in G$ and since the family of these sets is closed under the product (recall that $\sigma_{\mathcal{F}}(a) \cdot \sigma_{\mathcal{F}}(b) = \sigma_{\mathcal{F}}(ab)$), it follows from the four equalities above that the product, defined in $\Sigma(\mathcal{F})$, can be extended to any pair of members $U, V \in \mathcal{O}$.

Up to this point we have proved that $\mathcal{O}$ is a p. o. group. Since $\mathcal{O}$ is a distributive lattice it is in fact an ℓ-group. It is plain that G is order-isomorphic to $\Sigma(\mathcal{F})$, which is a p. o. subgroup of $\mathcal{O}$. Finally, for the last statement, we suppose now that $\mathcal{F}$ is the family $S(G)$. In order to prove that $\mathcal{O}$, for this family, is the free ℓ-group generated by G, let H be an ℓ-group and let $f : G \longrightarrow H$ be a morphism of p. o. groups. We define $h : \mathcal{O} \longrightarrow H$ as follows

$$h\left(\bigcup_{i=1}^{n} \bigcap_{j=1}^{m_i} \sigma(a_{ij})\right) = \bigvee_{i=1}^{n} \bigwedge_{j=1}^{m_i} f(a_{ij}).$$

In order to prove that h is well defined suppose that

$$\bigcup_{i=1}^{n} \bigcap_{j=1}^{m_i} \sigma(a_{ij}) = \bigcup_{k=1}^{r} \bigcap_{l=1}^{s_k} \sigma(b_{kl}),$$

but $\bigvee_{i=1}^{n} \bigwedge_{j=1}^{m_i} f(a_{ij}) \neq \bigvee_{k=1}^{r} \bigwedge_{l=1}^{s_k} f(b_{kl})$. Then by the Prime Lattice Filter Theorem, there exists a prime filter P of H such that (for example) $\bigvee_{i=1}^{n} \bigwedge_{j=1}^{m_i} f(a_{ij}) \in P$ and $\bigvee_{k=1}^{r} \bigwedge_{l=1}^{s_k} f(b_{kl}) \notin P$. Since P is prime, for some index i we have that $\bigwedge_{j=1}^{m_i} f(a_{ij}) \in P$. It follows that $a_{ij} \in f^{-1}(P)$ for all $1 \leq j \leq m_i$. Note now that $f^{-1}(P)$ is a quasiprime and increasing subset of G. Therefore $f^{-1}(P)$ is a member of $S(G)$ that belongs to $\bigcup_{i=1}^{n} \bigcap_{j=1}^{m_i} \sigma(a_{ij})$. It is easy to check in a similar way that $f^{-1}(P)$ does not belong to

$\bigcup_{k=1}^{r} \bigcap_{l=1}^{s_k} \sigma(b_{kl})$. We have derived a contradiction. Therefore h is well defined. It is straightforward to prove that h is an ℓ-group homomorphism such that $h \circ \sigma = f$. ∎

We can obtain as a corollary the following embedding theorems, as an alternative approach to well known results due to Conrad and Holland [1].

Theorem 1.12. (Embedding theorem for p. o. groups)
Let G be a p. o. group. Then the following conditions are equivalent:

(i) *$S(G)$ strongly separates points of G.*

(ii) *G is order-isomorphic to a p. o. subgroup of an ℓ-group of sets, whose lattice operations are usual union and intersection.*

(iii) *G is order-isomorphic to a p. o. subgroup of an ℓ-group.*

Proof. The fact that (i) implies (ii) follows directly from Theorem 1.11. That condition (iii) implies (i) follows from our comment after Proposition 1.3. ∎

From this we can recover a well known embedding theorem for abelian p. o. groups due to Conrad and Holland and then independently reobtained by others authors. For an exposition of this result see [1], 4.5.6, p. 85. and also Remark 2.2, with further references on pages 88 and 89.

Corollary 1.13. (Embedding theorem for abelian p. o. groups)
Let G be an abelian p. o. group. Then any of the conditions of the previous theorem is equivalent to:

(iv) *G^+ is isolated.*

We refer to [8] for a proof.

2   An Embedding Preserving Joins and Meets

As we pointed out before, our method gives an alternative approach to well known embedding theorems due to Conrad (via right orders) and Holland (via automorphisms of the group). The embeddings that they are able to obtain do not preserve in general the existing joins and meets of the p. o. group [10]. The family of all quasiprime and increasing subsets also fails to provide an embedding preserving joins and meets. Indeed, note that in Example 1.4, taking G as the integers, it follows that

$$\sigma((1,0)) \cap \sigma((0,1)) \neq \sigma((1,0) \wedge (0,1)) = \sigma((0,0))$$

because the quasiprime and increasing subset $P = \{(x,y) : x + y > 0\}$ belongs to the left intersection but $P \notin \sigma((0,0))$. However, if we replace the family of quasiprime sets by the family of *prime* sets as defined in Definition 1.5, a second embedding which does preserve joins and meets can be obtained, provided this family strongly separates points.

Let G be a p. o. group and let $\mathcal{P}$ be the family of all prime and increasing subsets of G. We claim that $\mathcal{P}$ is a complete family for G. In fact, condition (2) of Theorem 1.8 is immediate, and for condition (1) of the same theorem the only non-trivial part is the verification that given P, Q primes and elements $x, y \in PQ$ such that there exists $x \wedge y$, then $x \wedge y \in PQ$. To prove this suppose that $x = p_1 q_1$ and $y = p_2 q_2$. Since $p_1^{-1} x = q_1$ and $p_2^{-1} y = q_2$ it follows that $p_1 \notin Q_x$ and $p_2 \notin Q_y$. Then either $p_1 \notin \max\{Q_x, Q_y\} = Q_{x \wedge y}$ or $p_2 \notin \max\{Q_x, Q_y\} = Q_{x \wedge y}$. In the first case $p_1^{-1}(x \wedge y) \in Q$ and therefore $x \wedge y \in PQ$. In the second case $p_2^{-1}(x \wedge y) \in Q$ and therefore $x \wedge y \in PQ$.

Therefore, by Theorem 1.8, $\Sigma(\mathcal{P})$ is a p. o. group which is order isomorphic to G (provided $\mathcal{P}$ strongly separates points).

Moreover, if $\mathcal{O}$ is the distributive lattice generated by $\Sigma(\mathcal{P})$, it follows from Theorem 1.11 that $\Sigma(\mathcal{P})$ is a p. o. subgroup of the ℓ-group $\mathcal{O}$. It is straightforward to prove that the embedding $a \mapsto \sigma_{\mathcal{P}}(a)$ preserves the joins and meets that exist in G. Also, in this case the group $\mathcal{O}$ is the free ℓ-group generated by G when taking as morphisms the p. o. group homomorphisms preserving joins and meets whenever they exist. We collect these results in the following theorem.

Theorem 2.1. *Let G be a p. o. group and let $\mathcal{P}$ be the family of all prime and increasing subsets of G. Then the following conditions are equivalent*

(1) *$\mathcal{P}$ strongly separates points of G.*

(2) *G can be embedded as a p. o. subgroup of an ℓ-group of sets in such a way that the embedding preserves joins and meets of G.*

Moreover, if this is the case, then the group $\mathcal{O}$ is the free ℓ-group generated by G in the following sense: For any other ℓ-group H and any homomorphism $f : G \longrightarrow H$ preserving joins and meets that exist in G, there exists a homomorphism of ℓ-groups $h : \mathcal{O} \longrightarrow H$ such that $h \circ \sigma = f$.

We conclude with the following comment.

Remark 2.2. In Priestley's duality for distributive lattices, the sets

$$\sigma_L(a) = \{P : P \text{ is a prime lattice filter of } L \text{ and } a \in P\}$$

have a topological characterization as the clopen and increasing subsets of the dual space. In our theory we have not yet found a similar topological characterization of the sets

$$\sigma(a) = \{P \in S(G) : a \in P\}.$$

We claim that such characterization, together with the conditions collected in Proposition 1.9 would be enough to provide a full topological duality for the class of isolated groups.

References

[1] A. Bigard, K. Keimel and S. Wolfenstein, *Groupes et Anneaux Réticulés.* Lecture Notes in Math. **608** (1977), Springer-Verlag, New York.

[2] P. F. Conrad, *Right-ordered groups.* Michigan Math. J. **6** (1959), 267-275.

[3] P. F. Conrad, *Free lattice-ordered groups.* J. Algebra **16** (1970), 191-203.

[4] B. A. Davey and H. A. Priestley, *Introduction to Lattices and Order.* (1990) Cambridge University Press.

[5] A. M. W. Glass and W. C. Holland (eds), *Lattice-Ordered Groups, Advances and Techniques.* (1989) Kluwer Acad. Publ., Dordrecht.

[6] W. C. Holland, *The lattice ordered group of automorphisms of an ordered set.* Michigan Math. J. **10** (1963), 399-408.

[7] N. G. Martínez, *A topological duality for lattice-ordered algebraic structures including ℓ-groups.* Alg. Univ. **31** (1994), 516-541.

[8] N. G. Martínez, *Spectra and embeddings theorems for ordered groups.* Proc. of the IX Latin American Symposium on Mathematical Logic; Notas de Lógica Matemática **39** Part 2 (1994), INMABB-CONICET, UNS, Bahía Blanca, 131-143.

[9] P. Ribenboim, *Théorie des groupes ordonnés.* Monografías de Matemática (1959), Instituto de Matemática, UNS, Bahía Blanca.

[10] C. Tsinakis, *Personal communication.*

Departamento de Matemática, Universidad de Buenos Aires, Ciudad Universitaria, Pabellón 1, (1428) Buenos Aires, Argentina
gmartin@dm.uba.ar
apetrov@dm.uba.ar

On the Flatness of the Epimorphic Hull
of a Ring of Continuous Functions

R. Raphael and R. G. Woods [1]

ABSTRACT.   For commutative semiprime rings R, the classical ring of quotients $Q_{Cl}(R)$ is R-flat, but the epimorphic hull $E(R)$ need not be. An example due to Quentel shows that $E(R)$ can be flat and still not coincide with $Q_{Cl}(R)$. In Proposition 7 below we show that such behaviour is excluded for rings of the form $C(X)$. A related question is addressed, and we characterize, for any cardinal α, the Tychonoff spaces X for which all ideals of $C(X)$ are essentially α-generated.

In this note we will impose the general assumption that all rings are commutative, semiprime with identity, and that a subring has the same identity as its overring. Throughout X will denote a Tychonoff (i.e., completely regular Hausdorff) topological space. Undefined topological notation and terminology can be found in the text by Gillman and Jerison [GJ]. In particular, $C(X)$ denotes the ring of real-valued continuous functions with domain X.

A ring R defines a natural sequence of important ring extensions

$$R \subseteq Q_{Cl}(R) \subseteq E(R) \subseteq Q(R),$$

which are defined as follows:

$Q_{Cl}(R)$ is the classical ring of quotients, obtained by inverting the non zero-divisors of R. The ring $Q(R)$, the complete ring of quotients of R, is constructed as follows. An ideal D of R is called *dense* if its annihilator is the zero ideal. For each dense ideal D, let $\mathrm{Hom}(D, R)$ be the set of all R-module homomorphisms from D to R. Such homomorphisms are called *fractions*. Take the set of all fractions $\mathrm{Hom}(D, R)$ as D ranges over all dense ideals of R. There is a natural definition of addition and multiplication of fractions. The ring $Q(R)$ is the set of all fractions modulo the equivalence relation that identifies two fractions that agree on a dense ideal common to their domains of definition. Since the ring R is itself a dense ideal, the elements of R can be viewed as fractions, and this shows that R is a subring of $Q(R)$. The classical ring of quotients lies between R and $Q(R)$ because it consists of all (equivalence classes of) fractions that

[1] The authors thank the NSERC (Canada) for its support. The first author thanks the conference organizer, Jorge Martinez, for support and manifold conference arrangements.

J. Martínez (ed.), Ordered Algebraic Structures, 313–321.
© 2002 *Kluwer Academic Publishers.*

are defined on the dense ideals that are principal. A full treatment of $Q(R)$ can be found in [[L], §2.3], where it is also shown that $Q(R)$ is self-injective and von Neumann regular.

In the case where R is a domain, $Q(R)$ coincides with $Q_{Cl}(R)$ and the inclusion of R into $Q(R)$ is an epimorphism of rings. However this inclusion is generally not an epi. Since the embedding of R into $Q_{Cl}(R)$ is always an epimorphism of rings, and also a ring of quotients, it was (historically) natural to study rings of quotients that are also epimorphisms. Von Neumann regularity was pertinent for two reasons: the complete ring of quotients of a domain is a field and these are von Neumann regular; secondly, ring epimorphisms emanating from von Neumann regular rings must be surjective [[S], 5.4]. Seminal work on epimorphisms in algebraic categories was undertaken by Isbell in a series of papers called "Epimorphisms and dominions". Other papers followed. There was an interesting series of contributions from France.

Storrer [S] began his studies in this context. He defined the epimorphic hull $E(R)$ of the ring R to be the intersection of all (von Neumann) regular rings between R and $Q(R)$. As such, it is itself regular, and it contains the classical ring of quotients as a subring. The embedding of R into $E(R)$ is an epimorphism of rings. Later it was shown that $E(R)$ is generated as a subring of $Q(R)$ by the ring R and the quasi-inverses of the elements of R [[Ke], 4.2], [[O1], Propositions 5 and 6], [O2], and [[W], Theorem 1, and Corollary to Theorem 6]. It is rare that $E(R)$ coincide with $Q(R)$; for example, it fails for any Boolean algebra that is not complete as a partially ordered set. The rings $Q_{Cl}(R)$ and $E(R)$ coincide when $Q_{Cl}(R)$ is regular but this need not occur [[S], 11.6]. Thus, in general, all of the inclusions between the rings R, $Q_{Cl}(R)$, $E(R)$ and $Q(R)$ are proper.

Recall that an R-module A is called flat if for every inclusion of R-modules $B \subseteq C$, the induced map $A \otimes B \longrightarrow A \otimes C$ is also a monomorphism. Projective modules are flat. Flatness can be formulated in terms of equations [B]. Flatness is a relevant notion because the ring $Q_{Cl}(R)$ is always a flat R-module. Thus it was natural to ask when $E(R)$ is flat as an R-module. Quentel [Q] has given an example of a ring for which $E(R)$ is flat and yet $Q_{Cl}(R)$ is not regular. Such behaviour speaks to the structure of ideals of the form $\text{Ann}(a)$ for $a \in R$. As noticed by Quentel ([Q], Proposition 4) and Cateforis ([C], Theorem 2.1) ideals of the form $\text{Ann}(a)$ must be essentially finitely generated (c.f. Definition 3) if $E(R)$ is flat. Also, having $Q_{Cl}(R)$ regular is equivalent to the $\text{Ann}(a)$ being essentially principal (c.f. Definition 3); see [R], 1.6 and [Q], Proposition 9. Thus in Quentel's example, all ideals of the form $\text{Ann}(a)$ are essentially finitely generated yet at least one of them fails to be essentially principal.

The authors were interested in the application of these ideas to rings of continuous functions, and began by looking for an example like Quentel's that was also a ring of continuous functions. Below we show that this is impossible, because rings of continuous functions satisfy the countable annihilator condition, due to Henriksen and Jerison. This investigation raised two problems in rings of continuous functions – the first, "When are all ideals of $C(X)$ countably generated?", was already solved in the literature (Theorem 11), and the second is solved below in Theorem 17, which connects the topological notion of cellularity with the algebraic notion of essentiality.

This note has been revised in the light of helpful suggestions made in two referees' reports.

Definition 1. (Henriksen and Jerison) A ring R has the *cac (countable annihilator condition)* if given any countable set $\{ x_i : i < \omega \}$ in R, there exists an $x \in R$ such that $\cap_i \mathrm{Ann}(x_i) = \mathrm{Ann}(x)$. This condition is pertinent because any ring of continuous functions has the cac ([HJ], 4.1, p121).

The following is immediate because for each $\mathrm{Ann}(f)$ there is an $a \in R$ such that $\mathrm{Ann}(a) = \mathrm{Ann}(f)$.

Lemma 2. *If R has the cac, so does $Q_{Cl}(R)$, its classical ring of quotients.*

Definition 3. Recall that if $I \subseteq J$ are ideals of R, then I is *essential* in J if for all $j \in J, j \neq 0$, one has $jR \cap I \neq (0)$. We will call an ideal *essentially principal*, *essentially finitely generated*, or *essentially countably generated*, accordingly, as there is a principal ideal, a finitely generated ideal, or a countably generated ideal that is essential in it.

Again we immediately have:

Lemma 4. *Each of the following properties is inherited by $Q_{Cl}(R)$ from R:*

(i) *the ideals of the form $\mathrm{Ann}(a)$ are essentially principal;*

(ii) *the ideals of the form $\mathrm{Ann}(a)$ are essentially finitely generated;*

(iii) *the ideals of the form $\mathrm{Ann}(a)$ are essentially countably generated.*

Proposition 5. *Let R have the cac and suppose that all ideals of the form $\mathrm{Ann}(a)$ are essentially countably generated. Then $Q_{Cl}(R)$ is regular.*

Proof. We will show that the prime ideals in $Q_{Cl}(R)$ are maximal. This is well-known to imply regularity ([B], p. 183). By Lemmas 2 and 4, $Q_{Cl}(R)$ has the cac and annihilators of elements are essentially countably generated. If possible, let P be a prime ideal in $Q_{Cl}(R)$ which is not maximal, and let P lie in the maximal ideal M. Let $f/d \in M \setminus P$, d a non zero-divisor in R. Clearly $\mathrm{Ann}(f/d) \neq (0)$. (All annihilators will be taken in $Q_{Cl}(R)$). Choose a countably generated ideal $\sum_{i<\omega} (g_i/k_i)Q_{Cl}(R)$ that is essential in $\mathrm{Ann}(f/d)$. By the cac in the ring $Q_{Cl}(R)$ we have

$$\bigcap_{i<\omega} \mathrm{Ann}(g_i/k_i) = \mathrm{Ann}(h/k),$$

for some $h/k \in Q_{Cl}(R)$. Clearly $f/d \in \text{Ann}(h/k)$ so $fh = 0$ in R. Since $f \notin P$, $h \in P$, and $f + h \in M$ which means that it is a zero divisor, which we now show to be false. Suppose that $s/b \neq 0$, and $(s/b)(f + h) = 0$. Then $s(f + h) = 0$. Multiply by f to get $sf = sh = 0$. Thus, $s \in \text{Ann}(f/d)$ so there is a $t/t' \in Q_{Cl}(R)$ so that $0 \neq (t/t')s = \sum(r_i/r_i')(g_i/k_i)$, a finite linear combination of some of the $\{g_i/k_i\}$. Since $sh = 0$, $s(h/k) = 0$, so $s \in \cap_i\text{Ann}(g_i/k_i)$, and $s(g_i/k_i) = 0$ for each i. Thus $s(t/t')s = 0$, and $(t/t')s = 0$, which is false.   ∎

Corollary 6. *Let R have the cac and let $E(R)$ be flat over R. Then $Q_{Cl}(R)$ is regular and coincides with $E(R)$.*

Proof. Quentel ([Q], Proposition 4) and Cateforis ([C], Theorem 2.1(d)) have shown that if $E(R)$ is flat over R, then annihilators of elements are essentially finitely generated.

∎

Proposition 7. *Let R have the cac, for example, let R be a ring of continuous functions, or a uniformly closed ϕ-algebra. Then the following are equivalent:*

(i) *for all $f \in R$, $\text{Ann}(f)$ is essentially principal,*

(ii) *for all $f \in R$, $\text{Ann}(f)$ is essentially finitely generated,*

(iii) *for all $f \in R$, $\text{Ann}(f)$ is essentially countably generated,*

(iv) *$E(R)$ is flat as an R-module,*

(v) *$Q_{Cl}(R)$ is regular.*

Proof. Clearly (i) $\Rightarrow$ (ii) $\Rightarrow$ (iii). (iii) $\Rightarrow$ (v) by Proposition 5.

(v) $\Rightarrow$ (iv): In general one has $R \subseteq Q_{Cl}(R) \subseteq E(R)$. If $Q_{Cl}(R)$ is regular, then $Q_{Cl}(R) = E(R)$, and $Q_{Cl}(R)$ is well known to be flat over R.

(iv) $\Rightarrow$ (i): By Corollary 6, $Q_{Cl}(R)$ is regular. Therefore, (i) holds either by ([Q], Proposition 9), or ([R], 1.6).   ∎

Remarks 8. Since Quentel gave a ring in which (iv) holds and (v) fails, his is an example of a ring without the countable annihilator condition. The referees have pointed out that the equivalence of (i) and (v) is implicit in [M].

We now apply Proposition 7 to rings of the form $C(X)$. First recall that if $f \in C(X)$ then the *cozero set of f*, denoted $\text{coz}(f)$, is

$$\text{coz}(f) = \{ x \in X : f(x) \neq 0 \}.$$

Observe that if $f, g \in C(X)$ then $fg = 0$ if and only if $\text{coz}(f) \cap \text{coz}(g) = \emptyset$.

Lemma 9. *Let I and J be ideals of $C(X)$ with $I \subseteq J$, and denote $\cup\{\,\mathrm{coz}(f)\,:\,f \in J\,\}$ by $\mathrm{coz}(J)$. The following are equivalent:*

(i) *I is essential in J.*

(ii) *$\mathrm{coz}(I)$ is dense in $\mathrm{coz}(J)$.*

Proof. Since $I \subseteq J$, clearly $\mathrm{coz}(I) \subseteq \mathrm{coz}(J)$. Suppose $\mathrm{coz}(I)$ were dense in $\mathrm{coz}(J)$, and let $g \in J \setminus \{0\}$. Then $\mathrm{coz}(g)$ is a nonempty open subset of $\mathrm{coz}(J)$, so by hypothesis it intersects $\mathrm{coz}(I)$. Thus, there exists $h \in I$ such that $\mathrm{coz}(g) \cap \mathrm{coz}(h) \neq \emptyset$. As noted above, it follows that $gh \neq 0$ and $gh \in I$ as $h \in I$. Hence I is essential in J. Conversely, if $\mathrm{coz}(I)$ is not dense in $\mathrm{coz}(J)$, since the cozero-sets of X form an open base for X (as X is Tychonoff), there exists $h \in C(X)$ such that $\mathrm{coz}(h) \cap \mathrm{coz}(g) = \emptyset$ for each $g \in I$, while there exists $f \in J$ such that $\mathrm{coz}(h) \cap \mathrm{coz}(f) \neq \emptyset$. Then $hf \in J \setminus \{0\}$, but as $\mathrm{coz}(hf) \subseteq \mathrm{coz}(h) \subseteq X \setminus \mathrm{coz}(I)$, if $k \in C(X)$ and $k(hf) \in I$ then $k(hf) = 0$. Thus, I is not essential in J.   ∎

Recall that the space X is called *cozero-complemented* if for each cozero set U there is a cozero set V so that $U \cap V = \emptyset$ and $U \cup V$ is dense in X.

Corollary 10. *The epimorphic hull $E(C(X))$ is flat over $C(X)$ if and only if X is cozero-complemented.*

Proof. By Proposition 7 we must show that each $\mathrm{Ann}(f)$ in $C(X)$ is essentially principal if and only if X is cozero-complemented. By Lemma 9 the former means that for each f there is a g such that $\mathrm{coz}(g)$ is dense in $\mathrm{coz}(\mathrm{Ann}(f))$. But it is easy to see that $\mathrm{coz}(\mathrm{Ann}(f)) = \mathrm{int}(Z(f))$, so the result follows.   ∎

It seems natural, in the light of Proposition 7 to try to determine both the topological spaces X in which all ideals of $C(X)$ are countably generated, and also those in which all ideals are essentially countably generated. The solution of the first problem is available in the literature and reads as follows:

Theorem 11. (Gillman) *Let X be a Tychonoff space. Every ideal of $C(X)$ is countably generated if and only if X is a finite space.*

Proof. The result is implicit in 5.4 of [G].   ∎

The second question prompted the discussion that follows. As usual, we identify cardinal numbers with the smallest ordinal of that cardinality, and we will use cardinals as index sets.

Definition 12. A *cellular family* of a topological space X is a collection of pairwise disjoint nonempty open subsets of X. The *cellularity* of X, denoted $c(X)$, is defined to be

$$\max\left(\sup\{\alpha\,:\,X \text{ has a cellular family of cardinality } \alpha\},\,\aleph_0\right).$$

If $c(X) = \aleph_0$, X is said to satisfy the *countable chain condition*, or to be a *ccc* space.

Observe that if $c(X)$ is a singular cardinal then X may not have a cellular family of cardinality $c(X)$. The inclusion of $\aleph_0$ in the definition of cellularity is done to ensure that finite spaces have cellularity $\aleph_0$. There is an extensive topological literature on cellularity; see, for example, [H] or [J].

Each Tychonoff space X is dense in its Hewitt realcompactification υX, and the map $f \mapsto f|_X$ is a ring isomorphism from $C(\upsilon X)$ onto $C(X)$; (see chapter 8 of [GJ].) Clearly, $c(X) = c(T)$ if X is dense in T. Since all topological properties of a realcompact space T are determined by the ring-theoretic properties of $C(T)$, it follows that the statement "$c(X) = \alpha$" can be formulated in terms of a ring-theoretic assertion about $C(X)$. We formulate that assertion below.

The reader is alerted that, in the definition that follows, our (algebraic) usage of "α-generated" differs from the way a notion such as this is typically defined in topology.

Definition 13. Let α denote an infinite cardinal. Let I be an ideal of ring R, that is not finitely generated. I is said to be *α-generated* if I contains a subset S of cardinality no greater than α such that each member of I can be written as a (finite) linear combination of members of S. An ideal J of R is said to be *essentially α-generated* if there is an α-generated ideal I that is essential in J.

Lemma 14. *Let J be an ideal of $C(X)$ and let α be an infinite cardinal. If $c(\mathrm{coz}(J)) \leq \alpha$ then J is essentially α-generated.*

Proof. Let $f \in J$ and let F denote the set of all pairwise disjoint families $\mathcal{C}$ of cozero sets, such that

$$V \in \mathcal{C} \Rightarrow V \subseteq \mathrm{coz}(J), \text{ and } \mathrm{coz}(f) \in \mathcal{C}.$$

Then $F \neq \emptyset$ because $\{\mathrm{coz}(f)\} \in F$. Partially order F by inclusion. An easy Zorn's Lemma argument shows that F contains a maximal family $\mathcal{M}$. Then $\bigcup \mathcal{M}$ is a dense subset of $\mathrm{coz}(J)$, for if not, we could choose $k \in C(X)$ and $g \in J$ such that $\mathrm{coz}(k) \cap (\bigcup \mathcal{M}) = \emptyset$ and $\mathrm{coz}(k) \cap \mathrm{coz}(g) \neq \emptyset$. Then $kg \in J \setminus \{0\}$ and $\mathrm{coz}(kg) \cap (\bigcup \mathcal{M}) = \emptyset$, contradicting the maximality of $\mathcal{M}$. As $c(\mathrm{coz}(J)) \leq \alpha$, there exists a subset S of J such that $|S| \leq \alpha$ and $\mathcal{M} = \{\mathrm{coz}(g) : g \in S\}$. If I is the ideal generated by S, then one easily verifies that $\mathrm{coz}(I) = \bigcup \mathcal{M}$ and so by Lemma 9, I is essential in J. Thus J is essentially α-generated. ∎

Corollary 15. *If $c(X) \leq \alpha$ then each ideal of $C(X)$ is essentially α-generated.*

Proof. If $c(X) \leq \alpha$ and W is open in X then clearly $c(W) \leq \alpha$; in particular, $c(\mathrm{coz}(J)) \leq \alpha$ for each ideal J of $C(X)$. Now use Lemma 14. ∎

Lemma 16. *If X has a cellular family of cardinality greater than α, then there exists an ideal of $C(X)$ that is not essentially α-generated.*

Proof. Let $\beta > \alpha$ and let $\{W(i) : i < \beta\}$ be a cellular family of X of cardinality β. Let

$$J = \{g \in C(X) : \text{ there exists } F \subseteq \beta, |F| < \aleph_0, \operatorname{coz}(g) \subseteq \cup\{W(i) : i \in F\}\}.$$

Clearly, J is a proper ideal of $C(X)$. Suppose that I were an ideal of $C(X)$ that was α-generated and contained in J; i.e., suppose $I = \langle g_j : j < \alpha \rangle$, where each $g_j \in J$. If $j < \alpha$ there is a finite subset $L(j)$ of β such that $\operatorname{coz}(g_j) \subseteq \cup\{W(i) : i \in L(j)\}$. If $k \in I$ then there is a finite subset $F(k)$ of α such that $k = \sum\{h_j g_j : j \in F(k)$ and $h_j \in C(X)\}$, and clearly

$$\operatorname{coz}(k) \subseteq \bigcup \Big\{ \operatorname{coz}(g_j) : j \in F(k) \Big\} \subseteq \bigcup \Big\{ W(i) : i \in \cup_j \{L(j) : j \in F(k)\} \Big\}.$$

As $|I| \le \alpha$, if we let $G = \cup \{L(j) : j < \alpha\}$, then $|G| \le \alpha$. Thus $\operatorname{coz}(I) \subseteq \cup_i \{W(i) : i \in G\}$, and as $\alpha < \beta$ and $\{W(i) : i < \beta\}$ is a cellular family, there exists $n \in \beta$ such that $W(n) \cap \operatorname{coz}(I) = \emptyset$. As X is Tychonoff, there exists $f \in C(X)$ such that $\emptyset \ne \operatorname{coz}(f) \subseteq W(n)$. Clearly $f \in J$ and so $\operatorname{coz}(f) \subseteq \operatorname{coz}(J)$. Hence $\operatorname{coz}(I)$ is not dense in $\operatorname{coz}(J)$ and so by Lemma 9, I is not essential in J. Thus J is not essentially α-generated.   ∎

We are indebted to a referee for the following observation which improved the next result. For an infinite cardinal α, call an open set an α-*cozero set* if it is the union of fewer than α cozero sets. It is easy to see that all ideals of $C(X)$ are essentially α-generated if and only if each open set of X densely contains an α^+-cozero set.

Theorem 17. *The following are equivalent for a Tychonoff space X.*

(1) $c(X) \le \alpha$.

(2) *Every ideal of $C(X)$ is essentially α-generated.*

(3) *Each open set of X densely contains an α^+-cozero set.*

Proof. By Corollary 15, (1) implies (2). If (1) fails, then X must have a cellular family of cardinality greater than α, so (2) fails by Lemma 16.   ∎

Corollary 18. *A Tychonoff space X has countable cellularity if and only if every ideal of $C(X)$ is essentially countably generated, if and only if every open set densely contains a cozero set.*

Remark 19. In the presence of the conditions of Theorem 17, (resp. Corollary 18) one obtains the coincidence of $Q(C(X))$ with the so-called rings of α^+-quotients $Q_{\alpha^+}(C(X))$ (resp. $Q_{Cl}(C(X))$). See [HM1] and [HM2] for studies of these phenomena.

References

[B]   N. Bourbaki, *Eléments de mathématique.* Fasc. XXVII: Algèbre commutative; chapitre 1: Modules plats; chapitre 2: Localisation; (1961) Hermann, Paris.

[C]   V. Cateforis, *Flat regular quotient rings.* Trans. AMS **138**(1969), 241-249.

[G]   L. Gillman, *Countable generated ideals in rings of continuous functions.* Proc. AMS **11** (1960), 660-666.

[GJ]   L. Gillman and M. Jerison, *Rings of Continuous Functions.* (1960) Van Nostrand, Princeton.

[HM1]   A. W. Hager and J. Martinez, *Fraction dense algebras and spaces.* Can. J. Math, **45** (5) (1993), 977-996.

[HM2]   A. W. Hager and J. Martinez, *The ring of α-quotients.* To appear; Alg. Universalis.

[HJ]   M. Henriksen and M. Jerison, *The space of minimal prime ideals of a commutative ring.* Trans. AMS **115** (1965), 110-130.

[H]   R. Hodel, *Cardinal Functions, 1.* Handbook of Set-Theoretic Topology (1984) North-Holland, Amsterdam, 1-61.

[J]   I. Juhasz, *Cardinal Functions in Topology.* Math. Centrum Tracts **34** (1971) Amsterdam.

[Ke]   J. F. Kennison, *Structure and costructure for strongly regular rings.* J. Pure and Appl. Algebra **5** (1974), 321-332.

[L]   J. Lambek, *Lectures on Rings and Modules.* (1976) Chelsea Publ. Co., New York.

[M]   J. Martinez, *The maximal ring of quotients of an f-ring.* Alg. Universalis **33** (1995), 355-369.

[O1]   J. P. Olivier, *Anneaux absolument plats universels et epimorphismes d'anneaux.* C. R. Acad. Sci. Paris **266** Serie A (5 fevrier 1968), 317-318.

[O2]   J. P. Olivier, *L'anneau absolument plat universel, les epimorphismes et les parties constructibles.* Bol. de la Soc. Mat. Mexicana **23** (1978) 68-74.

[Q]   Y. Quentel, *Sur la compacité du spectre minimal d'un anneau.* Bull. Soc. Math. France **99** (1971), 265-272.

[R]   R. Raphael, *Injective rings.* Comm. in Algebra **1** (1974), 403-414.

[S]   H. H. Storrer, *Epimorphismen von Kommutativen Ringen*. Comm. Math. Helv. **43** (1968), 378-401.

[W]   R. Wiegand, *Modules over universal regular rings*. Pac. J. Math. **39** (1971), 807-819.

Department of Mathematics, Concordia University, Montreal, Canada H4B 1R6
raphael@alcor.concordia.ca

Department of Mathematics, University of Manitoba, Winnipeg, Canada R3T 2N2
rgwoods@cc.umanitoba.ca

Developments in Mathematics

1. Alladi et al. (eds.): *Analytic and Elementary Number Theory*. 1998
    ISBN 0-7923-8273-0
2. S. Kanemitsu and K. Győry (eds.): *Number Theory and Its Applications*. 1999
    ISBN 0-7923-5952-6
3. A. Blokhuis, J.W.P. Hirschfeld, D. Jungnickel and J.A. Thas (eds.): *Finite Geometries.*
    Proceedings of the Fourth Isle of Thorns Conference. 2001   ISBN 0-7923-6994-7
4. F.G. Garvan and M.E.H. Ismail (eds.): *Symbolic Computation, Number Theory, Special Functions, Physics and Combinatorics*. 2001   ISBN 1-4020-0101-0
5. S.C. Milne: *Infinite Families of Exact Sums of Squares Formulas, Jacobi Elliptic F. Continued Fractions, and Schur Functions*. 2002   ISBN 1-4020-0491-5
6. C. Jia and K. Matsumoto (eds.): *Analytic Number Theory*. 2002
    ISBN 1-4020-0545-8
7. J. Martínez (ed.): *Ordered Algebraic Structures*. 2002   ISBN 1-4020-0752-3

KLUWER ACADEMIC PUBLISHERS – DORDRECHT / BOSTON / LONDON